T0371771

Introduction to Microcontroller Programming for Power Electronics Control Applications

Introduction to Microcontroller Programming for Power Electronics Control Applications

Coding with MATLAB® and Simulink®

Mattia Rossi
Nicola Toscani
Marco Mauri
Francesco Castelli Dezza

CRC Press is an imprint of the
Taylor & Francis Group, an informa business

First edition published 2022
by CRC Press
6000 Broken Sound Parkway NW, Suite 300, Boca Raton, FL 33487-2742

and by CRC Press
2 Park Square, Milton Park, Abingdon, Oxon, OX14 4RN

CRC Press is an imprint of Taylor & Francis Group, LLC

Library of Congress Cataloging-in-Publication Data

ISBN: 978-0-367-70985-3 (hbk)
ISBN: 978-1-032-05303-5 (pbk)
ISBN: 978-1-003-19693-8 (ebk)
eBook + ISBN: 978-1-032-05465-0

DOI: 10.1201/9781003196938

Typeset in LM Roman
by KnowledgeWorks Global Ltd.

eResources are available for this title at www.routledge.com/9780367709853

a Daniela

Contents

Foreword xiii

Preface xvii

Acknowledgments xix

Biographies xxi

1 Advances in Firmware Design for Power Electronics Control Platforms 1
- 1.1 Embedded Control System 1
- 1.2 Selecting a Development Board 3
  - 1.2.1 Key elements of a microcontroller 4
  - 1.2.2 Programming microcontrollers 5
- 1.3 The C2000™ Family of MCU from Texas Instruments™ . . . 7
- 1.4 Scheme of a Power Electronics Control Problem 9

I Embedded Development: Hardware Kits and Coding 13

2 Automatic Code Generation through MATLAB® 15
- 2.1 Model-Based Design and Rapid Prototyping 16
- 2.2 Workflow for Automatic Code Generation 18
- 2.3 Generate Code for C2000™ Microcontrollers 22
- 2.4 TI C2000™ Processors Block-set 24

3 Texas Instruments™ Development Kit 27
- 3.1 TI C2000™ LaunchPad™ : F28069M Piccolo 27
  - 3.1.1 Features 28
  - 3.1.2 Pin muxing 30
  - 3.1.3 Power connectivity 32
  - 3.1.4 Serial connectivity 33
  - 3.1.5 Boot options 33
- 3.2 TI BOOSTXL-DRV8301 BoosterPack 34
  - 3.2.1 BoosterPack PWM signals 35
  - 3.2.2 BoosterPack GPIO signals 36
  - 3.2.3 DC bus and phase voltage sense 37

3.2.4 Low-side shunt-based current sense 37

4 Software Installation **39**

4.1 TI Support Packages:
Code Composer™ Studio and ControlSUITE™ 39

4.2 MATLAB® Support Package:
Embedded Coder for Texas Instruments C2000 Processors . 41

4.3 Installation Procedure 41

II Review of Control Theory: Closing the Loop **47**

Introduction **49**

5 Designing a Closed-Loop Control System **51**

5.1 Dynamical Systems 52

5.1.1 Mathematical laws 52

5.1.2 Dynamical systems in electrical applications 54

5.2 Design a PI Controller in Continuous-Time Domain 54

5.2.1 Serial/parallel form 55

5.2.2 Characterization of the closed-loop dynamics $F(s)$ . . 55

5.3 Derive a PI Controller in Discrete-Time Domain 60

5.3.1 General properties of the discretization process 60

5.3.2 Characterization of the closed-loop dynamics $F(z)$ . . 62

6 Design Example: PI-Based Current Control of an *RL* Load **65**

6.1 Simulink® Simulation 67

6.1.1 Use of standard blocks (continuous/discrete) 71

6.1.2 Use of Simscape™ (specialized power systems) 73

6.1.3 Controller performances 74

6.2 Derive an Anti-Windup PI Controller Scheme 77

6.3 Design Summary 82

7 Manipulate the Variables Format: Data Types **85**

7.1 Fixed Point vs Floating Point Representation 85

7.2 Single vs Double Precision 88

7.3 Use of *Scaling* in Fixed Point Representation 91

7.4 Converting from Decimal Representation to Single Format . 93

7.5 Processing the Data: Implementation Hints 95

III Real-Time Control in Power Electronics: Peripherals Settings **97**

Introduction **99**

8 Basic Settings: Serial Communication COM and Hardware Target **101**
8.1 Virtual Serial Communication through COM port 101

9 Simulink® Configuration **105**
9.1 Simulink® Environments: Firmware vs Testing 107
9.1.1 Overview 107
9.1.2 Execution in Simulink® 108
9.2 MCUs and Real-Time Control with Simulink® 109

10 Serial Communication Interface (SCI) Peripheral **111**
10.1 Hardware Details 112
10.2 Firmware Environment: Send and Receive Data through Serial Communication . . . 113
10.2.1 C2806x SCI receive 113
10.2.2 C2806x SCI transmit 116
10.3 Testing Environment: Send/Receive Data through Serial Communication 117
10.3.1 Serial configuration 117
10.3.2 Serial send 118
10.3.3 Serial receive 119
10.4 Time Variable Settings (Sample Rates) 120
10.5 Examples on Serial Communication 122

11 GPIO Peripheral—Digital Input/Output **131**
11.1 Hardware Details 131
11.2 Firmware Environment: GPIO Peripherals 133
11.2.1 C2806x GPIO digital input (GPIO DI) 133
11.2.2 C2806x digital output (GPIO DO) 134
11.3 Examples with GPIO blocks 135

12 Analog to Digital Converter Peripheral **149**
12.1 Operating Principle 149
12.1.1 Sample & hold 150
12.1.2 Analog to digital converter 150
12.2 Hardware Details 151
12.2.1 Difference between acquisition window and sample time 153
12.3 Firmware Environment: ADC Peripheral 153
12.3.1 C2806x ADC 153
12.4 Example with ADC block 155
12.5 Synchronization between ADC modules 160

13 Pulse Width Modulator Peripheral 163
13.1 Operating Principle 164
13.2 Hardware Details 166
13.2.1 ePWM sub-modules 169
13.3 Generation of PWM signals 171
13.3.1 Counting modes 172
13.3.2 ePWMxA and ePWMxB sub-modules 174
13.3.3 Setting dead bands 175
13.4 Firmware Environment: ePWM Peripheral 178
13.4.1 C2806x ePWM 178
13.5 Example with ePWM block 186
13.6 DAC Peripheral—Filtered PWM 191
13.7 Examples with DAC Peripherals 192
13.8 Synchronization between Multiple ePWM Modules 197
13.9 Synchronization between ADC and ePWM Modules: *Average* Measurements 202
13.10Events Execution within Sample Time 204

14 Encoder Peripheral 207
14.1 Operating Principle of Incremental Encoders 207
14.2 Hardware Details 209
14.3 Optical Rotary Encoder LPD3806 210
14.4 Speed Computation 211
14.5 Firmware Environment: eQEP Peripheral 213
14.5.1 C2806x eQEP 213
14.6 Example with eQEP block 215

IV Real-Time Control in Power Electronics: Applications 219

15 Open Loop Control of a Permanent Magnet DC Motor 221
15.1 Required Hardware 221
15.2 Linear Model of a PMDC Motor 222
15.3 System Simulations 226
15.4 Half-Bridge Configuration 227
15.4.1 Control implementation 231
15.5 Full-Bridge Configuration 234
15.5.1 Modulation strategies 235
15.5.2 Unipolar voltage switching 236
15.5.3 Bipolar voltage switching 241
15.5.4 Control implementation 246

16 Low-Side Shunt Current Sensing 251
16.1 Sensor Characterization: Theoretical Approach 252
16.2 Locked Rotor Test 254
16.3 Sensor Characterization: Experimental Approach 260

17 Current Control of an RL Load **267**
17.1 Required Hardware 267
17.2 Linear Average Model and Controller Design 269
17.3 System Simulations 271
17.3.1 Detailed modeling of the actuation variables 271
17.4 Half-Bridge Configuration 273
17.4.1 Control implementation 280
17.5 Variation of Load Parameters 288
17.5.1 Effects on the transient response 288
17.5.2 Parameters estimation 290

18 Voltage Control of an RLC load **293**
18.1 Required Hardware 293
18.2 Guidelines for the Hardware Design of a *RLC* Load 296
18.3 General State-Space Average Modeling Method 300
18.3.1 Linear average model and controller design 303
18.4 System Simulations 306
18.5 Half-Bridge Configuration 306
18.5.1 Control implementation 314
18.6 Variations of *LC* Filter Parameters 322

19 Cascade Speed Control of a Permanent Magnet DC Motor **325**
19.1 Required Hardware 327
19.2 Linear Model of a PMDC Motor 328
19.3 Cascade Control Architecture and Design 330
19.4 System Simulations 333
19.5 Full-Bridge Configuration 334
19.5.1 Model reference adaptive system (MRAS) observer . . 344
19.6 Single Motor Configuration 347
19.6.1 Parameter identification 348
19.6.2 Control implementation 350
19.7 Back-to-Back (B2B) Configuration 361
19.7.1 Parameter identification 363
19.7.2 Control implementation 364

V Real-Time Control in Power Electronics: Load Emulation **371**

20 Debugging Tools and Firmware Profiling **373**
20.1 Processor-in-the-loop with Simulink® 373
20.1.1 PMDC motor control implementation through PIL . . 375
20.2 External Mode Execution with Simulink® 380
20.2.1 Simulink® setup for external mode execution 381

21 Electric Propulsion Case Studies **385**
21.1 Urban Tramway 385
21.2 Electric Racing Car 390

A Appendix A: Basics of C **401**
A.1 Operations between numbers 401
A.1.1 Sum and differences 401
A.1.2 Shift operation 401
A.1.3 Multiplication 402
A.1.4 Division 402
A.2 Structure of a C program 403

B Appendix B: Custom Expansion Boards and Hardware Kits **405**

Bibliography **423**

Index **427**

Foreword

The new book *Introduction to Microcontroller Programming for Power Electronics Control Applications* contains the fundamental subjects of the interdisciplinary field of power electronic based systems, which draws knowledge from circuit and control theory, (digital) signal processing for embedded implementation, electrical machines/drives, and power semiconductor devices. Written for students and practicing engineers, this book introduces the analysis and design of motor control systems and their implementation on microcontrollers. The requirements and capabilities of the latter influence the structure and design choice of the closed-loop control scheme, a subject particularly relevant for laboratory activities both at university and industry level. This book presents state-of-the-art techniques to implement modulation schemes and control algorithms in a commercial microcontroller (MCU) suitable for rapid prototyping approach, and hint for designing analog circuits, such as low-voltage converter, output filters/load. MATLAB/Simulink® is introduced and used to solve example problems.

The book presents a concise workflow for the reader by using a specific embedded target, which is not a limiting factor for the validity of the suggested approach. The latter can be extended even to different boards. It is valuable to every graduate student, serving as a textbook for classes (looking to create teamwork) and as a starting point for more advanced studies, for industry professionals, researchers, and academics willing to study the broad field of power electronics. The contents of the book are easy to read and presented in an interesting way with good illustrations and solid background on the underlying theory. Solved problems (built-in files) are presented to help the readers.

Introduction to Microcontroller Programming for Power Electronics Control Applications is unique in the synthesis of the main characteristics of electrical drive behavior and in their link to the main implementation aspect into modern MCU, trying to fill the gap between system/circuit design, control techniques, and digital signal processing. The authors are power electronics and drives specialists from the Electrical Machines, Drives and Power Electronics Research Group of Politecnico di Milano, Italy. They collected in the book the several years of teaching and subjects of the laboratory activities for the M.Sc. in Automation and Control Engineering and Electrical Engineering at the same university, with the collaboration and contribution of industrial partner. As industrial partners, we retain this book a great achievement of the last four years collaboration with the authors.

Angelo Strati — Field Application Engineer Italy
Domenico Santoro — Field Application Engineer Italy
Giuseppe Ballarin — Team Leader FAE Italy
Würth Elektronik Group
angelo.strati@we-online.com

Most people go about their day blissfully unaware of the electric motors that are spinning the world around us. We wake up staring upwards at a ceiling fan, silently rotating in a circle. We jump into our car and rely on up to 40 motors—pumps, fans, locks, and lifts—to get us to our destination. We power up our laptop computers and hear the soft whine of fans working to keep the electronics cool. Motors are everywhere because they are one of the main ways that an electronic circuit can interact with the real world, i.e. a power electronic-based system. They are "lectromechanical, turning analog and digital signals into real and visible mechanical motion. It is estimated that electric motors consume 45 percent of the total worldwide electricity—this is a stunning statistic! As we look to reducing energy consumption and enabling a greener future, electric motors present a huge potential for efficiency improvements.

Few engineering students are aware of the impact of electric motors on the world around them, and even less are versed in the design and control of motor systems. This is a problem! We need engineers growing in competency in this field to create better and more efficient motor drive systems.

Motor drive and control is an incredibly multidisciplinary field. Real-time digital processing is implemented in microcontrollers to be the "brain" behind the motor system; controlling speed, power, and efficiency from the digital domain. A wealth of analog components from power management (voltage regulators & gate drivers) and signal chain (amplifiers & sensors) interface the microcontroller to the motor through a power converter while providing sensing, safe operation, and support for the system. Texas Instruments has over 25 years of experience in the field of real-time control and also provides

Texas Instruments Incorporated (Nasdaq: TXN) is a global semiconductor company that designs, manufactures, tests and sells analog and embedded processing chips for markets such as industrial, automotive, personal electronics, communications equipment and enterprise systems. Our passion to create a better world by making electronics more affordable through semiconductors is alive today, as each generation of innovation builds upon the last to make our technology smaller, more efficient, more reliable and more affordable—making it possible for semiconductors to go into electronics everywhere. We think of this as Engineering Progress. It is what we do and have been doing for decades. Learn more at www.ti.com.

a comprehensive analog portfolio covering every block of the motor drive and control system.

This book presents very practical and important lessons to engineers and engineering students alike on the topics of motor drive and control, covering not only general concepts but details on how to create a motor drive system. It provides an excellent resource to encourage the next generation of engineers to grow and develop skills in the area of electric motors and power electronics, introducing them the tools they need to make an impact on the world.

Politecnico di Milano is an outstanding academic partner, and the focus of the *Electrical Machines, Drives and Power Electronics Research Group* on cutting-edge power electronic-based technologies helps shape quality engineering minds. We wish the best of success to this publication and to the continued collaboration between industry and academia.

Olivier Monnier — Marketing Manager, C2000 Real-Time MCU
Matt Hein — Applications Manager, Brushless-DC Motor Drives
Antonio Faggio — Field Application Engineer
Texas Instruments Inc.
a-faggio@ti.com

Can you write the 100 million lines of code that are needed to build an average modern car? The answer is pretty obvious: of course you can, it's just a matter of time. And how would you compare the complexity of this problem to writing the 4501 lines of assembly code needed to build the first version of UNIX in 1971? While both tasks appear to be at a similar level of dauntlessness, the half century separating them has witnessed the emergence of high-level languages that enable programmers to address highly complex problems on their own while reusing the legacy of their peers.

At MathWorks Inc., we relentlessly work on providing the best high-level programming tools to automate the implementation of your ideas into embedded systems. Simulink allows you to design and simulate complex algorithms that you can translate into thousands of lines of embedded code with a click of a button via our code generation technology.

The book *Introduction to Microcontroller Programming for Power Electronics Control Applications* will teach you how to use these modern techniques to create control algorithms for systems involving complex physics. The remarkable work of *Mattia Rossi, Nicola Toscani, Marco Mauri* and *Francesco Castelli Dezza* from Politecnico di Milano, Italy, clearly explains deep concepts to the reader in the field of embedded programming for power electronics applications using Model-Based Design.

While the shift to digital is now largely dominating the industry of motor control, this revolution is just starting for power conversion applications. The material in this book provides state of art techniques to train the many engineers that the world needs tomorrow in a field that is at the core of the indispensable transition to clean energy.

In recent conversations with *Mattia* and *Nicola*, while they politely thanked us for our help, it was clear to us that the quality and the amount of effort in this book deserved much more thanking from our side. With this foreword, we extend all our gratitude to this outstanding contribution to accelerating the pace of engineering and science, our core mission.

Antonin Ancelle — Embedded Targets Development Manager
John Kluza — Embedded Systems Partner Manager
Tom Erkkinen — Code Generation Marketing Manager
Brian McKay — Embedded Systems Partner Manager
MathWorks Inc.
aancelle@mathworks.com

www.we-online.com

www.ti.com

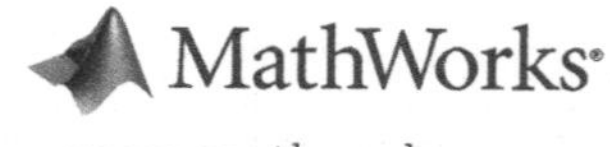

www.mathworks.com

Preface

Power electronics-based systems are the key enabling technology to meet most of the future sustainable challenges from grid to motor applications.

Standard textbooks and courses about power electronics and electrical machines deal with analysis in continuous-time, averaged modeling of switched-mode power converters, and continuous-time control theory. Nevertheless, real control algorithms and management functions around power converters are implemented digitally, thus, extending the field of fundamentals studies to discrete-time modeling and digital control concepts specific to power electronics. The necessary background is achieved by combining specific textbooks and courses from both power electronics and digital control theory. However, students who approach the design of digitally controlled power converters for the first time may not fully understand and successfully practice for a targeted problem due to such fragmented references.

In this book, we attempt to fill this gap by treating the fundamental aspects of digital control implementation for power electronics based systems in a systematic and rigorous manner. Our objectives are to put the reader in the position to understand, analyze, model, design, and implement digital feedback loops around power converters, from system-level transfer function formulations to understand which coding tool may be used when working with microcontroller (MCU or μC) platforms. In particular, the latter belongs to Texas Instruments™ C2000™ family, which is specifically designed for real-time closed loop control such as power supplies, industrial drives, and solar inverters applications. The Simulink® environment is able to automatically generate ANSI/ISO C/C++ code tailored for specific embedded targets through a model-based workflow. Given the settings which enable a background usage of the Code Composer™ Studio IDE, a Simulink® scheme can be directly compiled and executed on C2000™ MCUs. Such automated build and execution procedure speed up the control algorithms implementation, thus, the code generation of software interfaces and MCU peripherals (e.g. ADC, digital I/O, PWM), which can be tested with execution profiling. This makes the reader working in a rapid prototyping manner.

This book is oriented to graduate students of electrical and automation and control engineering pursuing a curriculum in power electronics and drives. Moreover, it aims to be a reference for engineers and researchers who seek to expand on the expertize in design-oriented knowledge for the aforementioned applications. It is assumed that the reader is well acquainted with fundamentals of electrical machines and power converters, along with associated continuous-time modeling and control techniques. Familiarity with sampled-data, discrete-time system analysis and embedded design topics is helpful but not absolutely essential since the scope of the book is to provide a basic knowledge even to whom is approaching these topics for the first time. Key concepts are developed from scratches, including a brief review of control theory and detailed description of the hardware and test bench used. Either custom expansion boards and assembled kits are open to users community. Project files can be shared and/or pre-assembled boards/kit can be directly shipped.

There is no standard notation to cover all of the topics covered in this book. We tried to use the most familiar notation from the literature whenever possible to help the reader's understanding through the chapters.

Acknowledgments

Most of the projects shown in this book have been funded by the Department of Mechanical Engineering of Politecnico di Milano, Italy, with the particular contribution of its Laboratory of Electrical Drives and Power Electronics. Special thanks goes to the university rector Prof. Ferrucio Resta and the department director Prof. Marco Bocciolone for their support to this initiative. We are grateful to our graduate students Marco Gerosa, Matteo Scandella, Andrea Polastri, Matteo Sposito, and Luca Grittini for the precious work in supporting the hardware development, the boards testing and the many suggestions they made. We also thank all the Ph.D. students and researchers who contributed to this book. In particular, Dr. Khaled ElShawarby and Dr. Alberto Bolzoni, who were supporting the project from day zero.

This book is the final outcome of a collaboration between Politecnico di Milano, Italy, Würth Elektronik™ Group Texas Instruments™ Inc. and MathWorks® Inc.. The authors are grateful to them for their support, help and vision. In particular, Tristan De Cande for his encouragement to tight the collaboration between university and company; Antonio Faggio and the US C2000™ product line of Texas Instruments™ Inc. for their commitment to invest in young generation growing at the university; Angelo Strati, Giuseppe Ballarin, Domenico Santoro, Andrea De Gruttola from Würth Elektronik™ Group for helping in components selection and boards design; Antonin Ancelle, John Kluza and their teams from MathWorks® Inc. for their help on embedded development. Moreover, we thank also MathWorks® Inc. for letting us being part of their Book Program.

The authors would like to specially thank Prof. Petros Karamanakos from Tampere University, Finland, and Prof. Ralph Kennel from Technical University Munich, Germany, for their guidance, long discussions and availability to share their high expertize in this field.

We are grateful to Nora Konopka and CRC Press LLC from Taylor and Francis Group for publishing this book. Special thanks goes to Prachi Mishra for her guidance and support.

Finally, we acknowledge the inspiration, patience, and support of our families during the preparation of this book, who allowed us to work during long nights, weekends, and holidays. The book is dedicated to Daniela.

Biographies

Mattia Rossi is a Research Assistant at Politecnico di Milano, Italy. He received the B.Sc. and M.Sc. degrees in Automation and Control Engineering from Politecnico di Milano, Italy, in 2013 and 2015, respectively. In 2015, he was at the ABB MV Drives, Switzerland, working on enhanced motor control design to reduce mechanical vibrations in motor-load couplings. Since 2016, he is a Ph.D. student in Electrical Engineering at Politecnico di Milano, Italy, in collaboration with Tampere University, Finland. In 2019, he was a visiting Ph.D. student at the Technical University of Munich, Germany. His main research activities cover model predictive control (MPC) algorithms for multilevel medium voltage power electronic-based systems and their embedded implementation, aiming to improve power conversion efficiency and system components reliability. He received the Best Student Paper Award at the 2019 IEEE International Symposium on Predictive Control of Electrical Drives and Power Electronics, and the Jorma Luomi Student Forum Award at the 2016 International Conference on Electrical Machines. **Email**: mattia.rossi@polimi.it

Nicola Toscani received the Bachelor (B.Sc.), Master of Science (M.Sc.), and Doctor of Philosophy (Ph.D.) degrees in Electrical Engineering from the Politecnico di Milano, Milan, Italy, in 2013, 2015, and 2019, respectively. He is currently working as a Postdoctoral Research Fellow in the Department of Mechanical Engineering of Politecnico di Milano. His last research activities deal with the development and the control of electrical machines for high-performance vehicles and wireless power transfer. His research interests also include modeling strategies for Electromagnetic Compatibility problems, Power Electronics and Electrical Drives. **Email**: nicola.toscani@polimi.it

Marco Mauri is an Assistant Professor in Electrical Machines and Drives at Politecnico di Milano, Italy. He received the Master of Science (M.Sc.) and Doctor of Philosophy (Ph.D.) degrees in Electrical Engineering from Politecnico di Milano in 1998 and 2002. His research interests mainly include the control of electrical machines and modeling principle of electrical drives and electromagnetic effects. He is a member of the IEEE Power Electronics and Industrial Electronics societies **Email**: marco.mauri@polimi.it

Francesco Castelli Dezza is a Full Professor in Electrical Machines and Drives at Politecnico di Milano, Italy. He received the Master of Science (M.Sc.) and Doctor of Philosophy (Ph.D.) degrees in electrical engineering from the Politecnico di Milano, Milano, Italy, in 1986 and 1990, respectively. He is currently the head of the Electrical Machines, Drives and Power Electronics research group of Politecnico di Milano, Italy, which is mainly located at the Department of Mechanical Engineering of the same university. His research interests include studies on dynamic behavior of electrical machines, electrical drives control and design, and power electronics for energy flow management. He is a member of the IEEE Power Electronics and Industrial Electronics societies. **Email**: francesco.castellidezza@polimi.it

1

Advances in Firmware Design for Power Electronics Control Platforms

The rising complexity and tighter requirements of power electronics-based systems lead to a great complexity of digital control algorithms, which has to be tested in many working conditions, i.e., looking for reliability. Hence, firmware design and testing routines have become increasingly time-consuming. To speed up the proof of concepts and to improve firmware quality assurance, the simulation analysis is combined with several automatic code generation procedures to achieve rapid prototyping. Such approach is aimed to test the control algorithm and the related generated code on real hardware through minimal modifications. This approach also helps to identify coding and platform-specific configuration errors, as well as to get rid of them. In particular, this book refers to microcontrollers (MCU or μC).

In this first, introductory chapter, basics of embedded systems and corresponding coding approaches are presented in detail.

1.1 Embedded Control System

Everyone who encounter these book subjects for the first time may have a very fundamental question: what is an embedded system?
To provide a clear answer, it is necessary to underline the major difference between a *microprocessor* and a *microcontroller* first.

A **microprocessor** is the computational intelligence of complex systems, i.e., the central processing unit (CPU), which is optimized to do logical-mathematical computations. It usually has mainly on-board communications peripherals to interact with the rest of the system. However, in many fields, including power electronics-based systems, there is the need for communicating with rather complex peripherals like Analog-to-Digital Conversion (ADC) channels, Digital-to-Analog Conversion (DAC) modules, Pulse-Width-Modulation (PWM) peripheral, and Inter Integrated Circuit (I2C) communication.

A **microcontroller** (MCU) is designed to include such peripherals. Hence, a MCU is a device built around a microprocessor (or CPU), where the

DOI: 10.1201/9781003196938-1

Standard microprocessor (PCs)	**Microcontroller** (Power electronics)
• High computational power • Several communication peripheral • General usage (e.g., software simulations, document editing)	• "Low" computational power • Communication, actuation and sensing (measurements) peripherals • Target usage: optimizing performances and costs

number of peripherals as well as their type and accuracy/resolution are related to the final market price other than the specific task they have to carry out. Figure 1.1 shows an example of architecture of MCU board. Note that, MCU refers to microprocessor for which the computational power is significantly lower compared to those used inside a personal computer. This is the reason why most of the MCUs for the industrial markets are able to manage a reduced number of computations even if they drive high-speed/high-resolution peripherals.
Therefore, an **embedded system** can be generally defined as a control platform based on a programmable logic (i.e., microprocessor), where the control algorithm and the peripheral interfaces are dedicated to specific tasks/functions or a group of them. All the application requirements, e.g., from ADC resolution to modulator dead time, are given from the beginning. This allows every user to choose the best trade-off between computational power, type of peripherals, and unitary cost, optimizing its design. The algorithm is executed in real-time, generally without using an on-board operating systems (OS). Indeed a custom (light) OS might be needed just for the most demanding cases.

Nowadays, the modern definition of embedded system mainly refers to microcontrollers. Nevertheless, even a single microprocessor which is specialized

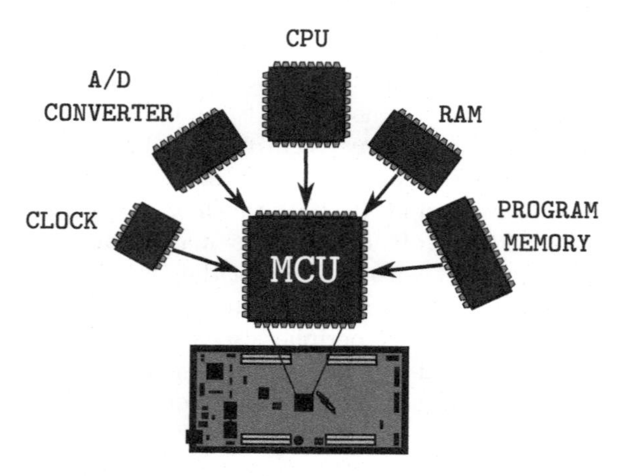

Figure 1.1 Example of the architecture of a MCU board [11].

Figure 1.2 Examples of power electronic-based applications [8].

in certain class of computations, or even custom designed for the application at hand, may be considered as an embedded system. This is the case of digital signal processors (DSPs), which are out of the scope of this book. The wide-ranging insertion of power electronics in many applications with different requirements still creates market for microcontrollers. Some of such applications are reported in Figure 1.2. As an example, the MCU usage in the automotive field is grown about +11% from 2006 to 2015. The popularity of MCUs depends on their ability to work with floating point variables (i.e., to include a floating point unit) as well as their benefit from mass production. Nevertheless, some applications require cost-effective devices,[1] thus, working with integer variables (e.g., 8–16 bit), targeting low cost MCUs. Furthermore, in order to extend the potential market of control platforms, today microcontrollers can be interfaced with different peripheral on the same pins, which can be chosen by setting different values on the corresponding registers. Thus, that allows to have more peripheral channels than pins.

1.2 Selecting a Development Board

Power electronics-based systems like power management and motor drive are going on with their fast evolution toward high-performance and high-efficiency solutions. Likewise, the demand for digital control algorithms managing the electronic component operations and the overall system has been rising, and embedded systems have been increasing in complexity reaching fast execution times. Therefore, it has become a standard engineering practice to test the control algorithm through real-time simulations and to verify its feasibility based on the hardware at disposal. This is the reason why in both research centers and academia development board must be updated according to the new emerging platforms on the market.

The selection of the development board represents the very first task of a project. In particular, development boards are designed to be efficient, portable, and sensored according to a specific applications field.

[1]From the customer's point of view *the "best" microcontroller is the one which matches all the application requirements at the lowest price.*

Since this book focuses on electrical power conversion case studies, from now on the text refers to MCU platforms targeted to power electronics-based applications. Even if this last statement reduces the candidate list, there still are several suitable solutions available on the market which share a common goal of being compact and versatile. Since the definition of a criterion to categorize each board would not be practical (i.e., there might be even deep technology differences), it is recommended to follow a system-level approach like the one presented here:

1. Consider all the components and peripherals that are necessary to run all of the required features. If the board lacks any of them, it is important to identify some supported expansions to include them.
2. Consider the supported programming languages and the level of competence of the final user. Moreover, investigate if any automatic code generation procedure is available as well as the quality of the documentation at disposal for the adopted board. Indeed, community and support are factors of great importance since they are the primary resources when designing a project.
3. Evaluate costs versus adopted components for the considered specific application: is it worth paying for them?

1.2.1 Key elements of a microcontroller

The main features of microcontroller boards that set their performances are reported below:

- **CPU** and **clock speed**: these two values affect the overall performance of the board. Namely, how fast it can perform computations. It should be noted that clock speed comparisons between CPUs coming from different families may not lead to meaningful considerations. Other factors, such as instruction cycles, instruction sets, and pipeline depth, also affect MCU performances.
- **RAM**: the size of this memory affects the number of tasks that can be run simultaneously. It also impacts how fast data can be processed, as swapping it from RAM to nonvolatile storage incurs large performance overheads.
- **Graphical processing unit (GPU)**: it allows development board to run video output (e.g., VGA/HDMI). High-performance GPUs are needed while processing video/images with the development board.
- **Data memory**: it affects the size of programs, operating systems, and generated/downloaded data that can be stored on development boards.
- **General purpose input/output pins**: these pins are used to connect external components to the development board in use. Hence, more pins typically means more possible simultaneous connections. These pins are usually assigned specific functionalities by the manufacturer being compliant with some standards, such as integrated circuit (I2C), serial peripheral

interface (SPI), and universal asynchronous receiver/transmitter (UART). These standards facilitate interoperability with third-party devices such as displays or sensors.

- **Analog-in pins**: these pins are necessary for any data acquisition from sensors and they are classified through the resolution and sampling rate of the peripheral to which they are connected. In particular, resolution refers to the number of discrete levels to which the input signal is quantized, while sampling rate is the number of data points that can be obtained in a reference time interval. High resolution is required for accurate measurements, while high sampling rates are needed for the acquisition of fast changing signals.
- **Pulse width modulator**: pulse width modulation is a kind of digital signal that is ideal for mimicking analog signals. Such signals are generated through modulators which vary the duty cycle of square waves depending on the desired output.
- **Power supply and consumption**: both of them can play a major role in design choices. For portable products, the runtime requirements and power demand of the board/associated components must be considered when selecting an energy source. For example, a development board serving as a data logger in a remote location must run uninterrupted for months, while another one working as a mobile personal computer only needs to run for several hours before recharging. Power consumption can be difficult to quantify as many boards have varying modes of operation, allowing some of them to work in very low power consumption modes. When considering power consumption, the designer must take the following factors into account: required computational power, maximum run time without recharging, and the cost/size of the energy source. The size of power sources is easier to determine since it depends on the rated values of the adopted board

1.2.2 Programming microcontrollers

The development environment of boards affects how the designer interacts with the hardware. The support for multiple operating systems, languages, and integrated development environments creates a rich programming framework and more appealing development board as well.

Depending on the specific application, all the onboard peripherals as well as the main algorithm features have to be correctly set while programming the MCU for executing required tasks.

Across the years, different coding approaches have been developed:

1. **Low-level**, based on machine code (e.g., Assembler language);
2. **Intermediate-level**, based on a combination of C-code (to define algorithm features and peripheral settings) and the usage of an Integrated Development Environment (IDE, e.g., Code Composer™

Studio from Texas Instruments™) to *compile*, *link*, *debug*, and *download* the script on target;

3. **High-level**, based on a combination of objective-oriented languages (e.g., Simulink®, to define algorithm features and peripherals settings) and the usage of specific target toolbox to *generate C-code*, *compile*, *link*, *debug* and *download* the algorithm into platform.

C code is considered as the universal language for microcontroller programming, with C++ following closely behind. In addition, more and more boards have custom IDEs associated with them, providing pre-configured device support and libraries, aimed at offering an all-around user-friendly workspace. More versatile IDEs (i.e., those not designed for a specific development board) allow users to explore alternative programming languages.

Each of the previously mentioned approaches has advantages and drawbacks according to the end-user type. Some of the main differences between *low* and *high* level coding are listed here in the following:

- **Platform Dependencies**

  Low-level programming languages are platform-dependent, which means that programs are written in such way they can run on the one hardware configuration only. Instead, high-level programming languages are platform-independent, that means programs written in such way can run on different hardware configurations.
  Remark: platform-independent does not mean operating system-independent. Indeed, hardware configuration may change, but OS (if present) should be the same for every new setup.

- **Speed**

  Low-level language programs are faster than those written in high-level language since they do not need to convert any algorithm in code executable on the hardware. In addition, low-level languages have less syntax, functions, keywords, and class libraries compared to other coding approaches.

- **Easy Programming and Flexibility**

  Low-level languages are not as easy and user-friendly to manage as high-level ones. Indeed, there are only two low-level programming languages, which are Binary and Assembly. Binary language foresees codes made by zeros and ones only, whereas Assembly requires some symbols knows as mnemonics which are difficult to type. On the other hand, high-level language programs are easy to write, read, modify, and understand. Moreover, high-level languages have huge libraries with a rich set of functions, allowing the development of algorithms for several applications with low effort.

- **Performance**

  Since low-level language programs are faster than any other, their performance are for sure better than those of high-level ones.

- **Translation**

  Programs written in Binary code does not need any translation as this language is a machine code already. Namely, the hardware is capable of understanding them without any translation. Instead, Assembly codes need an Assembler to translate programs to their equivalent counterpart in Machine Code. High-level languages are always translated by compilers or interpreters. Some of them required both compilers and interpreters to get the Object/Binary file.

- **Support**

  Low-level languages have less support than high-level ones. There may be lower number of communities for low-level languages than for high-level ones.

This book focuses on high-level programming languages which are user-friendly both for students and for everyone who is approaching microcontroller programming for the first time. In particular, coding/programming through Simulink® is proposed in the following chapters.

1.3 The C2000™ Family of MCU from Texas Instruments™

Texas Instruments™ (TI) offers free software development tools for LaunchPad™ boards, such as Code Composer™ Studio (IDE) and Energia[2]. In addition, TI allows free use (limited size) of cloud-based development tools. All these tools accept C/C++ code for programming.

Texas Instruments™ has a wide range of embedded processors from very low-cost, limited performance up to high-cost, very high performance. Its main product families are:

- MSP43x™ ultra-low-power microcontroller family;
- C2000™ microcontroller family;
- ARM-based microcontroller family;
- C5000™ low-power signal-processing DSPs;
- C6000™ high-perfomance signal-processing DSPs.

Among them, the C2000™ microcontroller family (also known as the TMS320C2000™ family) is a product line aimed at high-performance control

[2]Energia is a rapid prototyping platform based on Arduino IDE.

applications in the fields of motor control, digital power supplies, lighting, renewable energy, and smart grids. This family is made up of several subfamilies, from which it is worth mentioning:

- C24xx: a 16-bit microcontroller that evolved from the TMS320C2x family of digital signal processors.
- C28xx: a 32-bit microcontroller, fixed or floating point, with a robust set of peripherals and I/Os which match the classical needs typical of power electronics applications.

In particular, the C28xx chips are from low to high performance MCU. Piccolo™ (which main features and applications are summarized in Figure 1.3) and Delfino™ are the families for low and high performance microcontrollers, respectively. Their main characteristics are reported here in the following:

1. Piccolo™
   - MCU with floating-point unit;
   - CPU frequency: from 40 to 120 MHz;
   - Core: 1xC28x;
   - Memory:
     from 60 kbit up to 512 kbit flash;
     from 12 kbit up to 100 kbit SRAM;
   - Main peripherals: ADC, PWM, QEP, DMA, SPI, UART, I2C, CAN, USB.
2. Delfino™
   - MCU with floating-point unit;
   - CPU frequency: from 100 to 200 MHz;
   - Core: from 1xC28x up to 2xC28x + 2xCLA + ARM Cortex-M4;
   - Memory:
     from 512 kB up to 1.5 MB flash;
     from 68 kB up to 338 kB SRAM;
   - Main peripherals: ADC, PWM, QEP, DMA, SPI, UART, I2C, CAN, EMIF.

In particular, the following families of C28xx MCUs (which are also available as LaunchPad™ development kits) are supported with a dedicated library available in Matlab® Simulink®: *F2802x Piccolo™, F2803x Piccolo™, F2805x Piccolo™, F2807x Piccolo™, F2806x Piccolo™, F2837xS Delfino™, F2837xD Delfino™, F28004x, F2823x Delfino™, F28M3x.*
From Chapter 2 on, this book refers to a specific model of development board, that is Texas Instruments™ LaunchXL F28069M Piccolo™, which is shown in

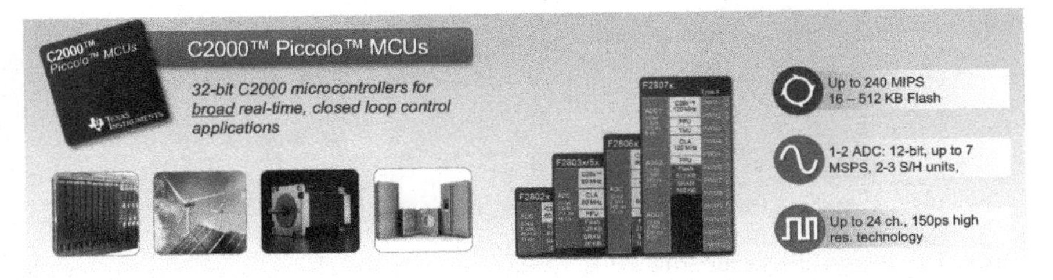

Figure 1.3 LaunchPad™ Piccolo MCU framework [10].

Figure 1.4. The reason behind this choice is its low-cost, low-power, and simple development environment of the board. Moreover, it has both a 256 kbit flash memory and a 96 kbit RAM. Finally, the community and project support for this device is relatively sparse compared to other boards. TI offers several plug-in expansion boards to expand the capabilities of the F28069M LaunchPad™ Piccolo™.

1.4 Scheme of a Power Electronics Control Problem

Most of the MCU-based closed loop control schemes for power electronics applications can be summarized in a structure like the one reported in Figure 1.5, which refers specifically to C2000™ processors. More in details, the key elements/features of this scheme (i.e., the ones necessary to close a loop) are:

- **Reference**: the controller must act to let the system follow a reference signal. The latter can be both internally generated on the MCU or provided as an external signal.
- **Processing**: this relates to the control logic implemented on the MCU. The control structure is the main part of the scheme, since it is aimed at generating the desired output based on the error between the reference and the feedback signals. This stage may also comprise scaling routines.

Figure 1.4 F28069M LaunchPad™ Piccolo™ board.

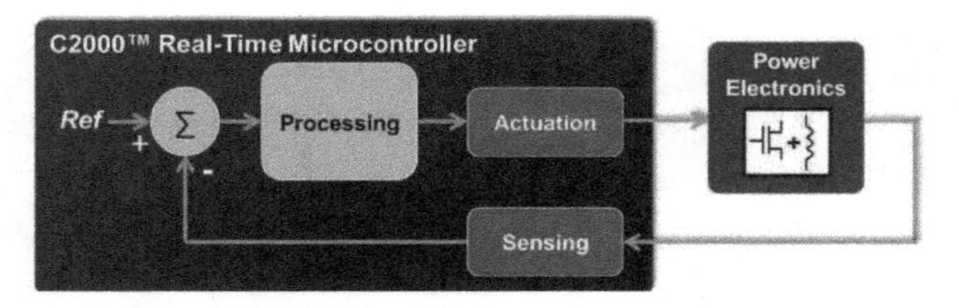

Figure 1.5 Classical structure of a closed-loop control scheme for implementations on C2000™-based boards and its main elements [13].

- **Actuation**: the controller output must be translated into a signal which must be suitable and consistent with the actuation logic used to operate the power converter. The modulator stage is the key element of the actuation block, which also comprise the safety routines.
- **Power electronics**: this refer to the switching devices used to realize a energy conversion. This system is used to drive different loads based on the reference signal. Generally, this stage is not embedded into MCU boards. Typically, they are built up on expansion boards.
- **Sensing**: sensors may be embedded on the MCU board or physically installed in close proximity of the switching device. In both case, analog signals must be acquired and digitalized through analog to digital converters (included in the architecture of the MCU), being available to be processed for closing the loop.

Note that, the C2000™ MCU family includes several parts of this structure representing a complex ecosystem. Its main features are summarized in Figure 1.6.

Sensing	Processing	Actuation
• Accurately sample signals with 12-bit and 16-bit analog-to-digital (ADC) converters • Run systems at high frequencies with ADC conversion rates up to 12.5 MSPS • Protect systems with responsive analog (30 ns) that can directly shut down PWMs • Accurately measure current with sigma delta filter modules — great for motor drives and resolver position decoding • Interface with high-performance external sensors using C2000™ high resolution captures	• Get more performance per MHz with 32-bit C28x™ DSP core optimized for complex single cycle operations common to control theory • Meet the demands of a wide range of applications with optimized processing options from 40 MIPS to 800 MIPS of performance • Add parallel loop control with the Control Law Accelerator (CLA) processing engine — great for controlling multiple motors, power stages and more • Accelerate complex control theory and signal processing, such as trigonometric math, FFTs and complex math, with built-in hardware accelerators	• Achieve higher system performance with micro edge positioning PWM outputs, including support of PWM phase, duty cycle and period • Control a variety of applications and power stage topologies with ultra-configurable PWM generation • Minimize power losses with fully configurable, high resolution PWM dead band • Protect your system with responsive and asynchronous PWM shutdown logic

Figure 1.6 C2000™ MCU feature highlights [13].

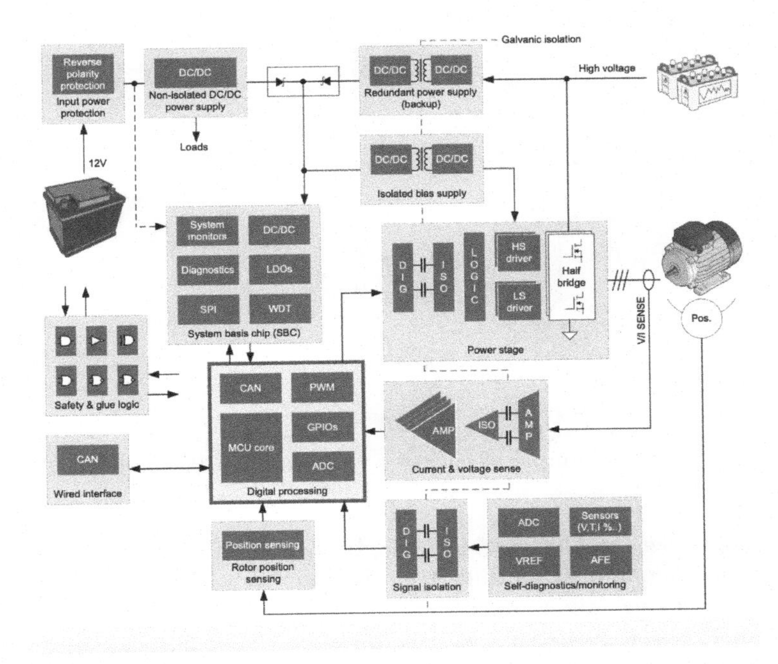

Figure 1.7 Traction converter and motor control for an efficient conversion DC to AC to drive an electric motor [9].

Examples of fields of application for MCU boards

The wide-ranging integration of power electronics in many applications with different requirements creates many control challenges. Several constraints means large embedded platform capabilities (in addition to ad-hoc power electronics design) which is reflected in the demand of specific number/type of peripherals, resolution/accuracy, computational power, safety target, power absorbition, etc . Some examples of power electronic-based applications that can be controlled by an MCU-based control scheme are:

- **Industrial drives:** they are aimed to control a wide variety of motors, e.g., AC and servo drives, equipped with sensing technologies (e.g., encoder or sensor-less solutions) and communications networks (like real-time Ethernet communications and functional safety topologies), which may be included in a complex production system (such as robotic lines or rolling mills) or targeted to a specific industrial task.
- **Electric vehicles (auxiliaries and traction systems):** advanced power control technology is needed for the electrification of vehicles. In particular, a car environment comprises motor control technologies, e.g., traction or propulsion motors, auxiliary motors, power steering, and digital power

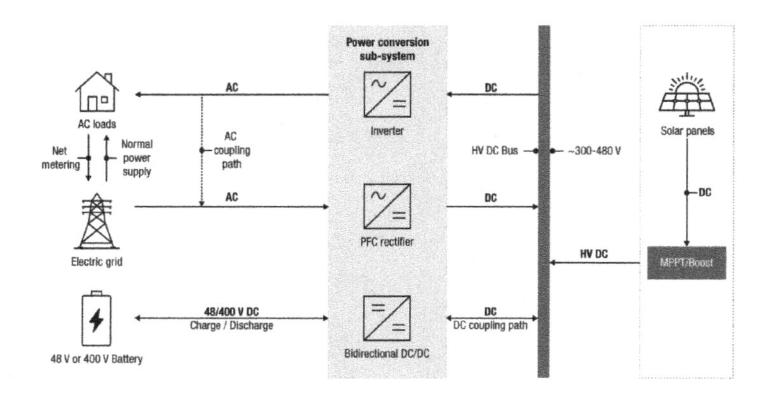

Figure 1.8 Example of energy collection from PV panels [8].

technology, like charging stations (AC/DC), DC/DC power conversion, on-board charging, and AC output stage. In order to achieve high performance while maximizing driving range and minimizing energy loss, more attention is needed on the choice of the right motor and the most efficient control techniques. As more electric vehicles enter the market, it is challenging to provide on-board charging solutions that can support multiple regions with different power grid infrastructures [27].

Several aspects and subsystems may be defined and analyzed, thus, electric vehicles represent a complex environment, as shown in Figure 1.7.

- **Renewable and digital power:** this kind of applications addresses the growing need for high efficiency, smaller form factor, and adaptability in digitally controlled power electronic converters driving energy productions, such as photovoltaic panels and wind turbines.

  Digital power management and control capabilities provide real-time intelligence allowing for the development of adaptable high frequency switching power supplies that automatically tune to their environment, improving efficiency and performance. This automatic adjustment allows for changes in input voltage, output load current and system temperature, delivering energy savings with dynamic voltage scaling, and advanced control techniques for optimal system performance [39].

  Several aspects and subsystems (like batteries and power conversion substations) may be defined and analyzed, as reported in the renewable energy-based distribution system shown in Figure 1.8.

Part I

Embedded Development: Hardware Kits and Coding

2

Automatic Code Generation through MATLAB®

Any time engineers design digital controls for power electronic-based applications, there are many good reasons to perform modeling and simulation:

- Test system behavior an possible variations, e.g., in the topology, power supply, load.
- Test different passive (e.g. resistors, capacitors, inductors) and active (e.g. semiconductor technology) elements to find suitable components.
- Test if the feedback control algorithms are able to meet the currents/voltages/speeds regulation requirements.

Then, move to the coding stage aimed to embedded implementation for a specific target. MathWorks® provides tools which bring the simulation stage together with the implementation one, creating a powerful ecosystem which allow to speed up the workflow from idea to practice.

In this book, the concepts of rapid prototyping and digital control techniques for power electronics-based systems are explained by programming a TI C2000™ based MCU platform through the MathWorks® MATLAB® and Simulink® frameworks. Both MATLAB® and Simulink® are commonly used for the analysis, design, simulation and optimization of models, including power electronic circuits. For the latter, the Simscape™ Power Systems[1] toolbox allows to model all the parts of power networks and to take into account the realistic behavior of each component. Moreover, MathWorks® along with various MCU manufacturers, such as Texas Instruments®, developed several Simulink® toolboxes aimed at automatically generating C/C++ targeted code for a specific CPU. Such toolboxes work together with the IDE of the microcontoller supplier, i.e., TI Code Composer™ Studio in this case.

[1]Simscape™ toolbox enables a quick creation of models of physical systems within the Simulink® environment. With Simscape™, component models based on physical connections are built directly integrated with block diagrams and other modeling paradigms. Thus, it is possible to model systems such as electric motors, bridge rectifiers other than hydraulic actuators. Their interactions can be included in the same environment.

DOI: 10.1201/9781003196938-2

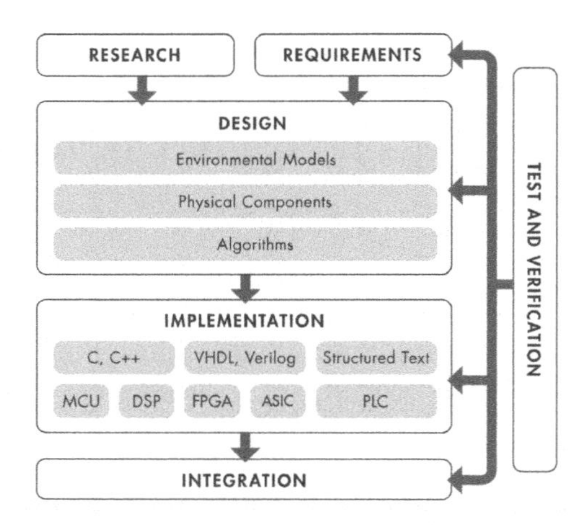

Figure 2.1 Model-Based Design workflow [34].

2.1 Model-Based Design and Rapid Prototyping

Rather than relying on physical prototypes and textual specifications, Model-Based Design approach proposed by MathWorks® in [34] uses a system model as an executable specification throughout development. It supports system- and component-level design and simulation, automatic code generation, and continuous test and verification.

Anyway, the best way to understand what a Model-Based Design is, it is through an example (the reader is referred to [34]):

> *A team of engineers sets out to build a motor control unit for an AC electrical drive. Because they are using the Model-Based Design approach, they begin by settling an architecture model from the system requirements; in this case, the system comprises the AC electrical machine, the power converter, the sensing/actuation stage and the control unit. A simulation/design model is then derived. This high-level model includes portions of the controls software that will be running in the control unit, and the load–in this case, the motor and the power converter. The team performs initial system and integration tests by simulating this high-level model under various scenarios to verify that the system is represented correctly and that it properly responds to input signals. They add detail to the model, continuously testing and verifying the system-level behavior against specifications. If the system is large and complex, the engineers can develop and test individual components independently but still test them frequently in a full system simulation. Ultimately, they build a detailed model of the*

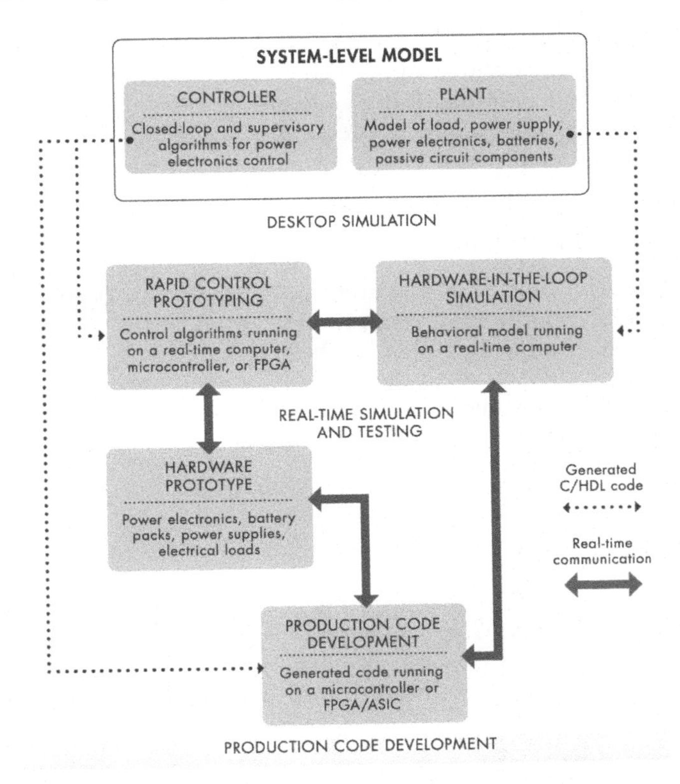

Figure 2.2 Model-Based Design workflow specified to power electronic-based applications [33].

system and the specific environment in which it operates. This model captures the accumulated knowledge about the control unit (e.g. considering the peripheral of the targeted MCU). The engineers generate code automatically from the model of the control algorithms for firmware testing and verification. Then, they download the generated code onto production hardware (e.g., MCU) for testing in an real hardware.

As this example shows, Model-Based Design uses the same elements as traditional development workflows, but with two key differences:

- A system model is at the heart of development, from requirements capture through design, implementation, and testing;
- It requires to follow this modeling approach to enable the automated routines, e.g., automatic code generation.

The workflow reported in Figure 2.1 can be specified for power electronic-based applications, as shown in Figure 2.2.

Simulations allow to analyze system performances in conditions otherwise too expensive, risky, or time-consuming to consider. This aspect combined with the possibility to automatically generate C, C++, HDL, or Structured Text from the model (the generated code can be optimized and combined with hand-written code) defines the *rapid prototyping* approach.

Goals of Real-Time Rapid Prototyping

Assuming that the functional requirements, the system hardware and the targeted control platform are given, rapid prototyping (in terms of real-time simulation, testing and coding) can be used to:

- Refine and verify the functional operations of control system design with the exploited hardware by rapidly iterating between algorithm design and prototyping;
- Continuously explore and test new ideas using a flexible, scalable platform; validate whether a component can adequately control the physical system in real time;
- Evaluate system performance, investigate scenarios and hardware interactions that are complex, expensive, or dangerous to perform with production hardware (i.e. before laying out hardware, coding production software);
- Test hardware cutting the development time from idea to practice, to avoid costly design flaws by detecting errors early when they are still cost-effective to correct.

For more information on this topic, the reader is referenced to [34].

2.2 Workflow for Automatic Code Generation

Making changes to the controller code during hardware testing can be time-consuming. Indeed, modifying the code, recompiling and deploying it to the microcontroller frequently may take long time depending on the extent of the required improvements. As an alternative, Simulink® can generate C/C++ code for a specific MCU starting from the scheme which models the controller, achieving rapid prototyping. By simulating the complete power electronic system (controller included) over all its possible operating and fault conditions, the ability to handle those same scenarios in the real system is increased.

Assuming to consider a generic power electronics-based systems, e.g., a two-level converter driving a combined resistive-inductive (RL) load, it is quite easy (and fast) to simulate the system behavior by representing it as a transfer function or by means of physical components from the Simscape™

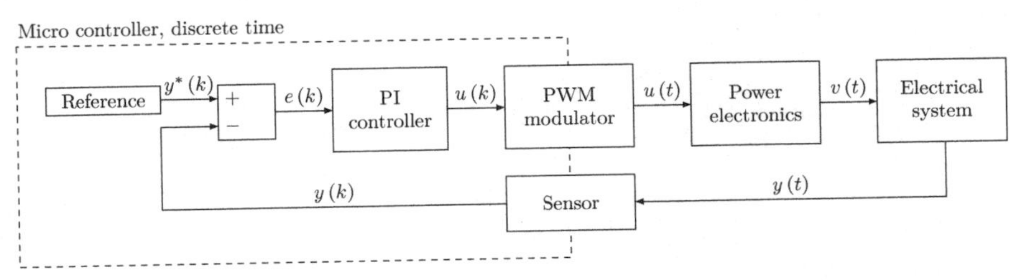

Figure 2.3 Block diagram of a microcontroller-based closed-loop system for power electronics applications.

Power Systems library. In both cases, the system receives an actuating variable/signal $u(t)$ and it returns an output variable/signal $y(t)$. Supposing to design a closed-loop control, e.g., a PI-based current control, the regulator processes the error $y^*(t) - y(t)$ and it returns the control input $u(t)$. The whole control scheme is reported in Figure 2.3.

Due to the switching nature of power converters (i.e., discrete on-off behavior), a modulation stage such as pulse width modulation (PWM), translates the control input $u^*(t)$ into the actuating signal $u(t)$ by a suitable switching pattern of the power converter. In this framework, the controller has to be implemented into the processor of the MCU which cannot handle continuous signals. Thus, the control algorithm must be discretized. Namely, its input/output signals are sampled at discrete time instances, i.e., $u(k)$ and $y(k)$, while the discrete controller form is derived from the continuous-time one through a discretization method. These latter aspects are deeply discussed in Part II of this book.

The overall control implementation process can be summarized in three main steps:

1. The Simulink® file is used to test and to optimize the controller through simulation before its deployment on the selected hardware[2];
2. Once the control design is ready, the control input and output are substituted by the related MCU peripherals which are given in Simulink® as block-set. According to the peripheral requirements, it may be necessary to edit the data type of the signals;
3. Finally, the overall scheme (controller + I/O block-sets) can be deployed on the MCU, i.e., translated into binary code and uploaded into the MCU. The whole procedure is summarized in Figure 2.4.

This model translation is not simple since it involves several hidden steps. Simulink® converts the model to a C programming code through MathWorks®

[2]Given that the converter behavior is assumed to be faster than the load dynamics, it is common practice to design the controller without taking the converter dynamics into account

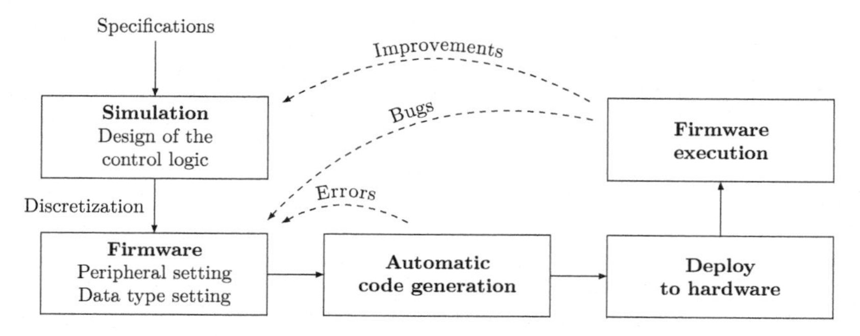

Figure 2.4 Workflow for firmware design.

Embedded Coder®. Then, the executable C code is fed to the IDE Code Composer Studio™ in which it is:

1. Sequentially compiled to assembly language exploiting the Texas Instruments libraries;
2. Assembled (e.g., ASM source code);
3. Lik-edited;
4. Downloaded on the TI C2000™ MCU flash memory.

Figure 2.5 shows such steps with a flowchart.

Benefits

The key advantage of such rapid prototyping technique is a seamless integration capability over multiple processors. This can be achieved by just replacing

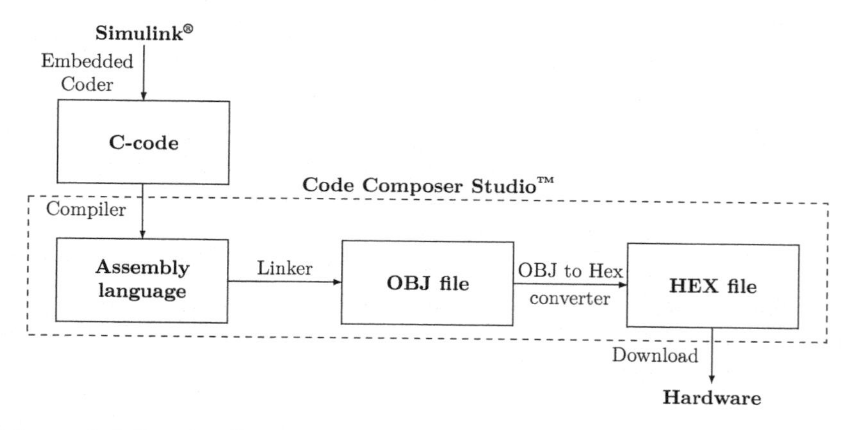

Figure 2.5 Simulink® workflow for firmware deployment on MCU boards.

the processor specific block-set, making the necessary changes in their configuration rather than rewriting or rebuilding the whole model. This is valid not only for hardware made by the same manufacturer, but by different producers as well. Hence, designers do not have to worry about the compatibility of the code. To validate changes made on the controller, it is enough to run the simulation model first and, then, to verify that no errors are generated. Hence, this approach is naturally oriented to research and development activities (i.e., for academic and industries).

Drawbacks

It is important to underline how such procedure may imply performance bottlenecks. Skipping the effort of low-level coding may limit the computational efficiency of the generated code. The resulting C/C++ code is *numerically equivalent* to the previously validated algorithms in Simulink®, but these latter has to be prepared for code generation, e.g., introducing implementation considerations needed for low-level C code and using functions for code generation support.

To clarify such concept, the code generation of a simple MATLAB® function which multiplies two inputs is investigated here in the following:

```
function c = Prod(a,b)
% multiply two inputs
c = a * b;
```

(a) MATLAB® source code

```
#include "Prod.h"
double Prod(double a, double b)
{
  return a * b;
}
```

(b) generated C code

Given two *scalar* inputs, the automatically generated C code maps clearly back to the MATLAB® environment, as shown above.

Nevertheless, as any MATLAB® algorithms intended for code generation, implementation constraints due to the differences between the two programming languages must be considered. These mainly include:

- **Memory allocation:** in MATLAB®, memory allocation is automatic while in C code is manual. Namely, it may be allocated either statically (using `static`), dynamically (using `malloc`), or on the stack (using local variables).
- **Array-based language:** MATLAB® provides a rich set of array operations that allow concise coding of numerical algorithms; C code requires explicit `for`-loops to express the same algorithms.
- **Data type:** MATLAB® automatically determines the data types and sizes as the code runs; C language requires explicit type declarations on all the exploited variables and functions (e.g., `int`, `uint`, `single`, `double`).

- **Polymorphism:** MATLAB® functions can support many different input types and they adapt the use of proper operators, while C code requires clear data type declarations.

Given the purpose of this book, the last two points require particular attention. The polymorphism can give a single line of MATLAB® code different meanings depending on the inputs. For example, the function shown previously could mean scalar multiplication, dot product, or matrix multiplication. In fact, the inputs could be of different data types (logical, integer, floating-point, fixed-point) or either real or complex numbers. If two matrices are multiplied, the automatic procedure produce many lines of C code, even with 3 `for`-loops, as shown here in the following:

```
void Prod(const double a[12],
const double b[20],
double c[15])
{
  int i0; int i1; int i2;
  for (i0 = 0; i0 <3; i0++)
  {
    for (i1 = 0; i1 <5; i1++)
        {
          c[i0 + 3 * i1] = 0.0;
          for (i2 = 0; i2 <4; i2++)
          {
            c[i0 + 3 * i1] =...
            ...a[i0 + 3 *i2] * b[i2 + (i1 << 2)];
          }
        }
  }
}
```

Thus, this piece of code looks quite different from that one reported before. For further information on this topic, the reader is referenced to [5].

2.3 Generate Code for C2000™ Microcontrollers

MATLAB® Coder™, Simulink® Coder™, and Embedded Coder® generate ANSI/ISO C/C++ code that can be compiled and executed on Texas Instruments™ (TI) C2000™ microcontrollers (MCUs) using Code Composer™ Studio IDE. Embedded Coder® allows an easy configuration of the code generation from MATLAB® and Simulink® algorithms to control software interfaces, optimize execution performance, and minimize memory consumption.

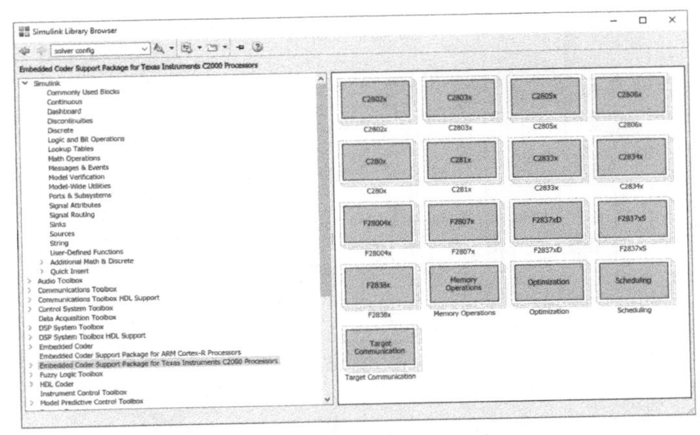

Figure 2.6 The Embedded Coder® Support Package for Texas Instruments™ C2000™ Processors Simulink® library.

In addition, Embedded Coder® Support Package for Texas Instruments™ C2000™ Processors provides additional support for the Piccolo™ and the Delfino™ F28x 32-bit microcontrollers that includes:

- Automated build and execution;
- Block libraries for on-chip and on-board peripherals such as ADC, digital I/O, ePWM, SPI, I2C, and more;
- Real-time parameter tuning and logging using the external mode execution;
- Processor optimized code including DMC and IQMath libraries;
- Ability to perform processor-in-the-loop (PIL) tests with execution profiling;
- Examples for PMSM FOC motor control and DC/DC buck converter.

Embedded Coder® provides additional support packages for the TI processors, IDEs, and real time operating system (RTOS) listed previously. The features vary within each support package and they may include automated build and execution, processor-optimized code, the ability to perform processor-in-the-loop (PIL) tests with execution profiling, block libraries for on-chip and on-board peripherals, and deployment support using bare-board or RTOS. To learn more about the specific processors and evaluation boards supported for each MathWorks® release, it is necessary to examine the settings in the Hardware Implementation section of the Code Generation Configuration Parameters provided with Embedded Coder®.

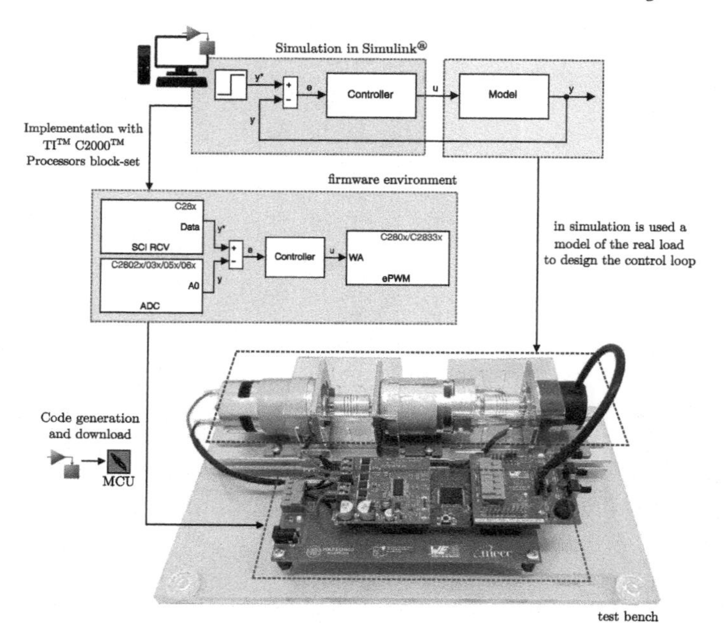

Figure 2.7 Example of a rapid-prototyping on a motor control test bench based on the LaunchPad™ F28069M Piccolo™ combined with TI BOOSTXL-DRV8301 and custom extPot3 board (B2B-PMDC hardware kit, see Appendix B). This hardware is driven through Simulink®.

2.4 TI C2000™ Processors Block-set

The installation of the Embedded Coder® Support Package for Texas Instruments™ C2000™ Processors results in the availability of a new Simulink® block-set library in the Library Browser, as shown in Figure 2.6. To assess the Library Browser, click on the icon 

in the simulation bar of the Simulink® window. Since this manuscript focuses on implementations with the LaunchPad™ F28069M, the block of interest for this book lay in the C2806x subfamily. This set includes all the functionalities of the main peripherals of the MCU in a user-friendly manner, i.e., through a graphical user interface (GUI). As an example, the C/C++ implementation of a PWM peripheral is already included in the built-in ePWM block. Figure 2.7 reports an example of two Simulink schemes that can be used to control a motor test bench and set the communication with the host pc. Nevertheless, their use requires an essential know-how on the block configurations, otherwise the coding simplification is useless. Part III shows in detail how to set the most exploited blocks in power electronics control problems. They can be used together with the standard Simulink® blocks, generating complete control algorithms and real-time interfaces. An example of the given workflow is denoted in Figure 2.7.

3

Texas Instruments™ Development Kit

3.1 TI C2000™ LaunchPad™ : F28069M Piccolo

The C2000™ LaunchPad™ F28069M Piccolo™ is a complete low-cost development board produced by Texas Instruments™. This board is shown in Figure 3.1 and it is equipped with a TMS320F28069M 32-bit MCU featuring a 90 MHz C28x CPU (as reported in Figure 3.2), 256 KB Flash, 12-bit ADC, 2 encoder interfaces (eQEP) and 16 PWM pins. It offers an on-board JTAG emulation[1] tool which provides direct interface to a PC for easy programming, debugging and evaluation of the firmware. In addition, the USB interface provides a UART[2] serial connection from MCU to the host PC.

The LaunchPad™ includes also two further synchronous serial communication protocols called I^{2}C and SPI. Those protocols are not used for PC-device communication but rather for communication between single modules and sensors. In particular, SPI is a synchronous full- duplex serial communication protocol used for the communications between all the modules mounted on the LaunchPad™ board. Its registers contains useful information about the working status of the board as long as many protection parameters (e.g., the current protection modes). However, these registers cannot be easily accessed from the Simulink® environment. For example, the access to *nFault GPIO* (which notifies whether the LaunchPad ™ experienced a fault or not) in Simulink® environment is enabled. Instead, the information on the nature/origin of the fault is stored in different internal SPI registers, which are not easily accessible through Simulink® blocks.

[1]JTAG emulators are the "umbilical cord" between PC software tools and DSP boards. It is a common hardware interface that provides the computer with a way to communicate directly with the chips on a board. It was originally developed by a consortium, the Joint (European) Test Access Group, in the mid-80s to address the increasing difficulty of testing printed circuit boards (PCBs).

[2]UART stands for Universal Asynchronous Receiver/Transmitter. It is not a communication protocol like SPI and I^{2}C, but it is a physical circuit in a microcontroller, or a stand-alone IC. The main purpose of a UART is to transmit and receive serial data.

DOI: 10.1201/9781003196938-3

Feature	LAUNCHXL-F28069M
MCU	TMS320F28069MPZT
Speed	90 MHz
Flash	256 kB
RAM	96 kB
EEPROM	N/A
Timers	3× 32-bit
Serial com	2 SPI, I²C, 2 UARTs, CAN
ADC channels	12-bit, 16 channels
BoosterPack	2× 40
Energia	No
Extra features	High-resolution PWM

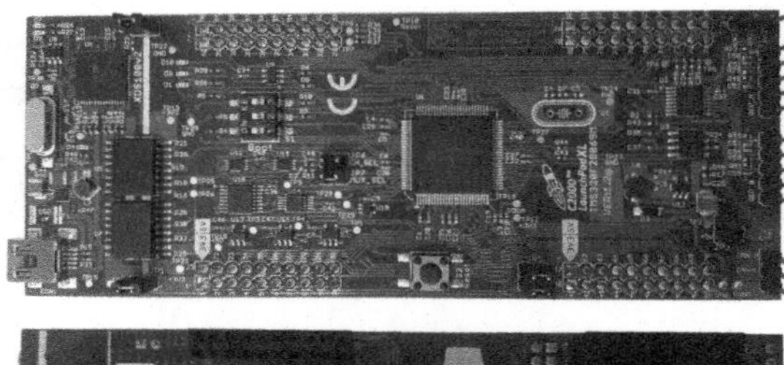

Figure 3.1 LaunchPad™ F28069 Piccolo™ board: main characteristics [12], top and bottom view.

3.1.1 Features

The LaunchPad™ F28069M Piccolo™ allows applications to easily migrate to lower cost devices. As a matter of fact, this board shows the features reported here in the following and in Figure 3.3:

- *XDS100v2 On Board Emulator:* it is the native JTAG emulator. It enables JTAG debugging/programming as well as it provides serial communication back to the PC. The XDS100 can also supply the target MCU through the USB supply and in Figure 3.3;
- *Power Domain:* the LaunchPad™ F28069M has different power domains to enable JTAG isolation. Jumpers JP1, JP2, JP3, JP4, JP5 can be arranged to reconfigure the power flow on the board as reported in Table 3.1;

Table 3.1 Jumper functions for the LAUNCHXL F28069M [16].

Jumper	Power Domain
JP1	Enable 3.3 V from USB (disable isolation)
JP2	Enable GND from USB (disable isolation)
JP3	Enable 5 V switcher (powered off 3.3 V supply of target device)
JP4	Connects target MCU 3.3 V to second set of BoosterPack headers
JP5	Connects target MCU 5 V to second set of BoosterPack headers

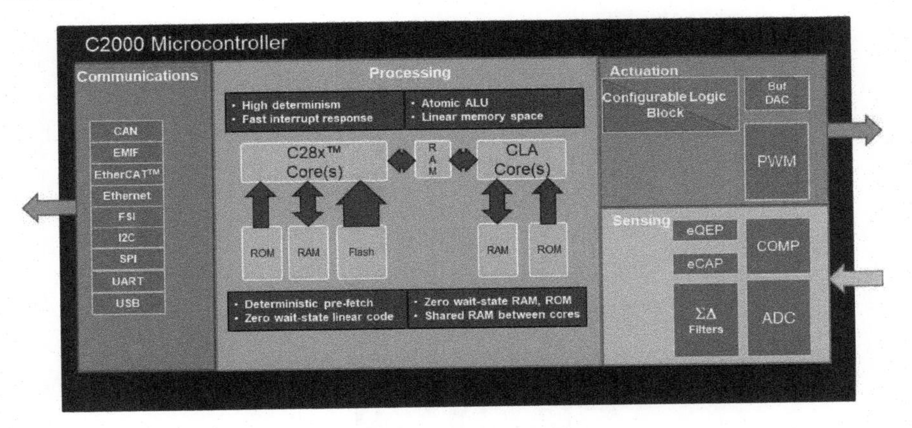

Figure 3.2 The C2000 MCU uses the C28x core as the main processing unit. This is a 32-bit floating point (single precision) core with dedicated instructions tailored to real-time control applications. Complementing the C28x core is a Control Law Accelerator (CLA) a 32-bit floating point co-processor capable of independent code execution increasing the system bandwidth versus a C28x core alone. There are both dual and single core implementations across the C2000 MCU family of devices [15].

- *Electrically Isolated PC Interface*: when the LAUNCHXL F28069M is externally supplied with power through the headers of other expansion boards, jumpers JP1 and JP2 may be removed to enable electrical isolation of the LaunchPad™ from the PC. Hence, the board will not be supplied through the USB cable any more but only by the DC supply connected to the expansion board, provided that $V_s > 5.5\,\text{V}$;
- *Connecting a BoosterPack:* all the pins mounted on the the LAUNCHXL F28069M are aligned in a $0.1\,\text{in} = 2.54\,\text{mm}$ grid to allow easy and inexpensive plugging of add-on expansion boards called BoosterPacks. These satellite boards can access all the GPIOs and analog channels;
- Two user LEDs: D9/D10 are two LEDs (blue and red) driven by GPIO34 and GPIO39 channels. Both of them work with negative logic, i.e., GPIO34 HIGH →LED D9 OFF, GPIO34 LOW →LED D9 ON;
- Device reset push button;
- InstaSPIN library in ROM allowing implementation of InstaSPIN-MOTION and InstaSPIN-FOC solutions;
- Dual 5 V enhanced quadrature encoder (eQEP) interfaces: the eQEP module is used for direct interface with a linear or rotary incremental encoder to get position, direction, and speed information from a rotating machine for use in a high-performance motion and position-control system;

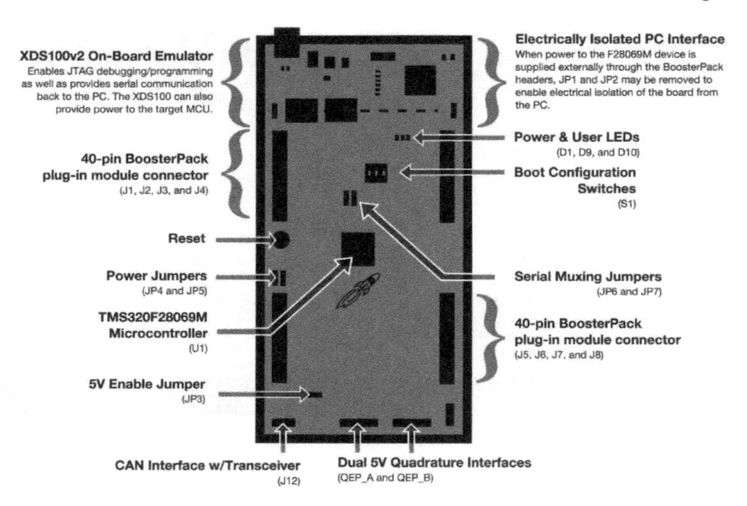

Figure 3.3 Overview of the main components and parts of the LAUNCHXL F28069M [11].

- CAN Interface with integrated transceiver;
- Boot selection switches, that is, switches whose combination let the user select the Boot Mode of the board.

3.1.2 Pin muxing

The C2000™ Piccolo™ LAUNCHXL-F28069M has more functions than pins. Hence, an hardware muxing through jumper selection is used to exploit all pin functionalities. The pinout of the LAUNCHXL F28069M is reported in Figure 3.4, whereas the GPIO-PIN matching is best explained by the schematics reported in Figure 3.5. More information can be found by looking at the board schematics (Figure 3.5 shows the connector pinout) and from the function summary reported in Figure 3.6.

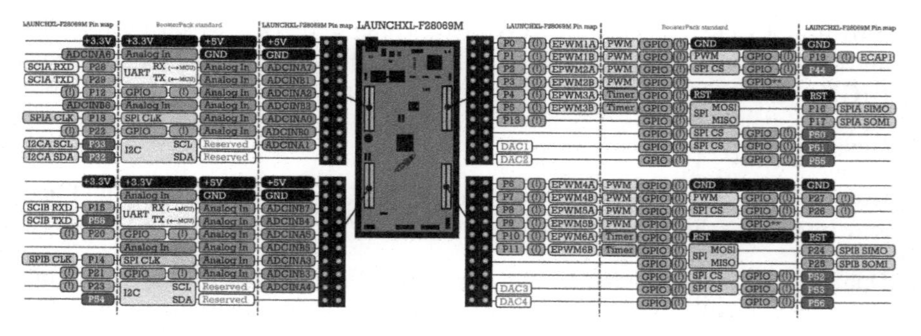

Figure 3.4 Pinout of LaunchPad™ F28069M [11].

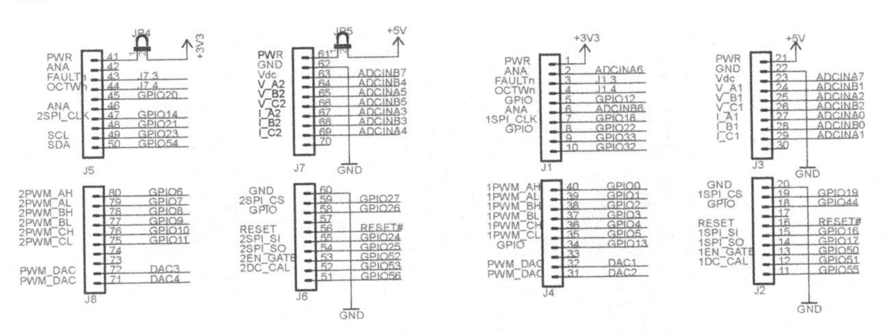

Figure 3.5 Correspondence between pin and GPIO channels for the LaunchPad™ F28069M [16].

Mux Value 3	Mux Value 2	Mux Value 1	Mux Value 0	J1 Pin	J3 Pin	Mux Value 0	Mux Value 1	Mux Value 2	Mux Value 3
			+3.3V	1	21	+5V			
			ADCINA6	2	22	GND			
			J1.3	3	23	ADCINA7			
			J1.4	4	24	ADCINB1			
SPISIMOB	SCITXDA	TZ1	GPIO12	5	25	ADCINA2			
			ADCINB6	6	26	ADCINB2			
XCLKOUT	SCITXDB	SPICLKA	GPIO18	7	27	ADCINA0			
SCITXDB	MCLKXA	EQEP1S	GPIO22	8	28	ADCINB0			
ADCSOCBO	EPWMSYNCO	SCLA	GPIO33	9	29	ADCINA1			
ADCSOCAO	EWPMSYNCI	SDAA	GPIO32	10	30	NC			

Mux Value 3	Mux Value 2	Mux Value 1	Mux Value 0	J4 Pin	J2 Pin	Mux Value 0	Mux Value 1	Mux Value 2	Mux Value 3
Rsvd	Rsvd	EPWM1A	GPIO0	40	20	GND			
COMP1OUT	Rsvd	EPWM1B	GPIO1	39	19	GPIO19	SPISTEA	SCIRXDB	ECAP1
Rsvd	Rsvd	EPWM2A	GPIO2	38	18	GPIO44	MFSRA	SCIRXDB	EPWM7B
COMP2OUT	SPISOMIA	EPWM2B	GPIO3	37	17	NC			
Rsvd	Rsvd	EPWM3A	GPIO4	36	16	RESET#			
ECAP1	SPISIMOA	EPWM3B	GPIO5	35	15	GPIO16	SPISIMOA	Rsvd	TZ2
SPISOMIB	Rsvd	TZ2	GPIO13	34	14	GPIO17	SPISOMIA	Rsvd	TZ3
			NC	33	13	GPIO50	EQEP1A	MDXA	TZ1
			DAC1	32	12	GPIO51	EQEP1B	MDRA	TZ2
			DAC2	31	11	GPIO55	SPISOMIA	EQEP2A	HRCAP1

Mux Value 3	Mux Value 2	Mux Value 1	Mux Value 0	J5 Pin	J7 Pin	Mux Value 0	Mux Value 1	Mux Value 2	Mux Value 3
			+3.3V	41	61	+5V			
			NC	42	62	GND			
			J7.3	43	63	ADCINB7			
			J7.4	44	64	ADCINB4			
COMP1OUT	MDXA	EQEP1A	GPIO20	45	65	ADCINA5			
			NC	46	66	ADCINB5			
SPICLKB	SCITXDB	TZ3	GPIO14	47	67	ADCINA3			
COMP2OUT	MDRA	EQEP1B	GPIO21	48	68	ADCINB3			
SCIRXDB	MFSXA	EQEP1I	GPIO23	49	69	ADCINA4			
HRCAP1	EQEP2A	SPISIMOA	GPIO54	50	70	NC			

Mux Value 3	Mux Value 2	Mux Value 1	Mux Value 0	J8 Pin	J6 Pin	Mux Value 0	Mux Value 1	Mux Value 2	Mux Value 3
Rsvd	Rsvd	EPWM4A	GPIO6	80	60	GND			
COMP1OUT	Rsvd	EPWM4B	GPIO7	79	59	GPIO27	HRCAP2	EQEP2S	SPISTEB
Rsvd	Rsvd	EPWM5A	GPIO8	78	58	GPIO26	ECAP3	EQEP2I	SPICLKB
COMP2OUT	Rsvd	EPWM5B	GPIO9	77	57	NC			
Rsvd	Rsvd	EPWM6A	GPIO10	76	56	RESET#			
ECAP1	Rsvd	EPWM6B	GPIO11	75	55	GPIO24	ECAP1	EQEP2A	SPISIMOB
			NC	74	54	GPIO25	ECAP2	EQEP2B	SPISOMIB
			NC	73	53	GPIO52	EQEP1S	MCLKXA	TZ3
			DAC3	72	52	GPIO53	EQEP1I	MFSXA	Rsvd
			DAC4	71	51	GPIO56	SPICLKA	EQEP2I	HRCAP3

Figure 3.6 Pin multiplexing on LaunchPad™ F28069M [11].

Figure 3.7 Connection between the LAUNCHXL F28069M and a host PC.

3.1.3 Power connectivity

The F28069M LaunchPad™ has several different power domains to enable JTAG isolation, see Figure 3.7. Jumpers JP1, JP2, JP3, JP4, and JP5 are adopted to configure the power flow through the board. It is important to note that the board is powered through the BoosterPack only if JP1 and JP2 are removed (see Table 3.1). This solution makes the operation with the board safe. However, safety comes with a couple of severe drawbacks on the Piccolo™ F2806M:

- Resetting the board entails not only the wipe-out of all data collected, but also of the whole firmware. Further deployment of the firmware is necessary;
- If the power on thr BoosterPack is cut off, the firmware is erased as well.

Thus, this means that the firmware can operate in stand alone operations without JP1 and JP2 (Section 9 explains how to set the board for such operation) only if it is constantly supplied; any fault/power interruption may erase the whole firmware.

Table 3.2 Configurations and functions of jumpers JP6/JP7 [16].

Jumper	Function
JP6 ON / JP7 OFF	USB/UART: GPIO15-58; FAULT/OCTW: GPIO28-29; J7.3-J7.4: Hi-Z
JP6 OFF / JP7 OFF	USB/UART: GPIO15-58; FAULT/OCTW: GPIO28-29; J7.3-J7.4: Hi-Z
JP6 OFF / JP7 ON	USB/UART: GPIO28-29; J1.3-J1.4 – Hi-Z; J7.3-J7.4 – GPIO15-58
JP6 ON / JP7 ON	USB/UART Disabled; J1.3-J1.4 – GPIO28-29; J7.3-J7.4 – GPIO15-58

Table 3.3 Boot modes and corresponding switch arrangements [16].

Boot Mode	S1-Switch 1 (GPIO34) H=Pulled to 1 L=Pulled to 0	S1-Switch 2 (GPIO37/TDO) H=Pulled to 1 L=Pulled to 0	S1-Switch 3 (TRSTn) H=XDS100v2 L=Tied to 0
Emulation Boot	L	H	H
Parallel IO	L	L	L
SCI	H	L	L
Wait	L	H	L
GetMode	H	H	L

3.1.4 Serial connectivity

The F28069M device on this LaunchPad™ contains two serial communication interface (SCI) peripherals. Because of this, a serial connectivity mux has been added to the board to make configuration of the SCI routing easy. Routing is configured via two jumpers (JP6 and JP7). These jumpers should be configured as shown in Table 3.2 to achieve the desired serial connectivity.

- JP6 ON/JP7 OFF means SCI channel B ON, SCI channel A OFF
- JP6 OFF/JP7 OFF means SCI channel B OFF, SCI channel A OFF
- JP6 OFF/JP7 ON means SCI channel B OFF, SCI channel A ON
- JP6 ON/JP7 ON means SCI channel B ON, SCI channel A ON

3.1.5 Boot options

The LaunchPad™ F28069M includes a boot ROM that performs some basic start-up checks and it allows the device to boot in many different ways. Most

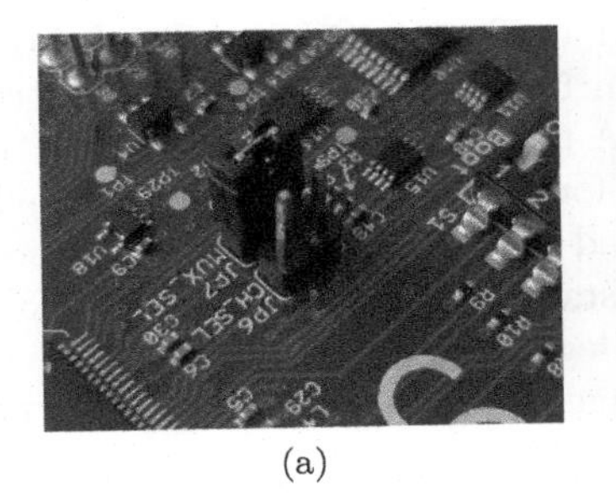

(a)

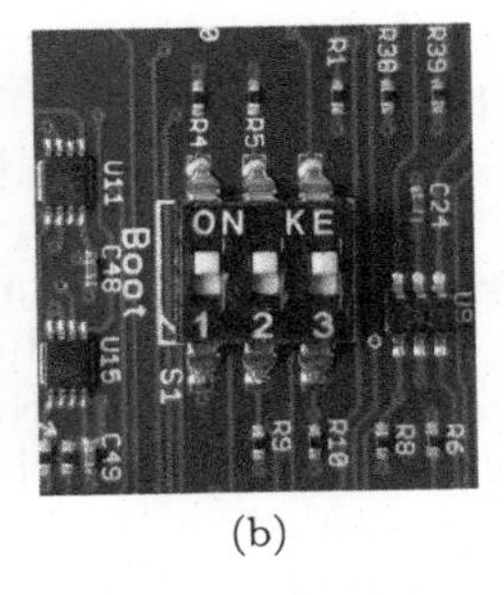

(b)

Figure 3.8 Details of jumper configuration for communication on channel SCIA (a), e.g., JP7 is ON (SCI A is ON) and JP6 is OFF (SCI B is OFF), and boot switches in high state (b).

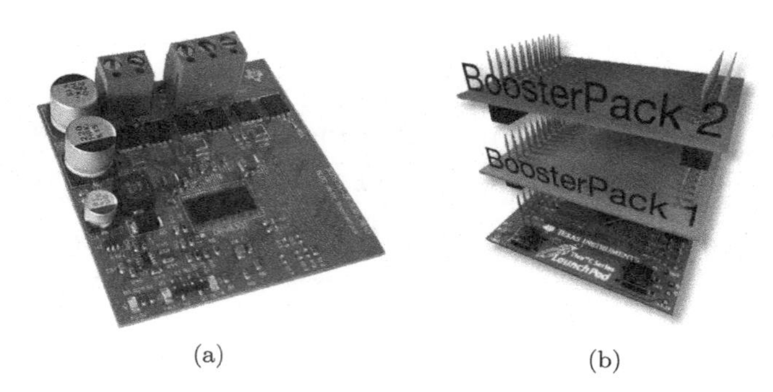

(a) (b)

Figure 3.9 The Motor Drive BoosterPack BOOSTXL-DRV8301 (a) and other BoosterPack boards stacked one over the other [12].

users may either want to perform an emulation boot or a boot to flash if they are running the application standalone. The three switches reported in S1 (see Figures 3.8 (b) and Table 3.3) which allow users to easily configure the pins that the boot ROM checks before the start-up of the board.
Remark: in digital circuits, a **high impedance** (or **Hi-Z**) output is not driven to any defined logic level by the output circuit. The signal is neither driven to a logical high nor low level; this third condition leads to the description "tri-stated". Such a signal can be seen as an open circuit or "floating".

3.2 TI BOOSTXL-DRV8301 BoosterPack

The Motor Drive BoosterPack BOOSTXL-DRV8301 reported in Figure 3.9 (a) is a complete drive stage for 3-phase applications based on the DRV8301 three-phase gate driver and CSD18533Q5A N-channel NexFET™ power MOSFETs, which allows independent control of each switch through dedicated PWM signal and 100% duty cycle operation. The BOOSTXL-DRV8301 can be fed by a power supply in a range from 6 V to 60 V and the board can bear up to 10 A continuous current and 14 A peak current. This BoosterPack includes 3 low-side current sense amplifiers for current measurement scaled for ±14 A peak operation. Thus, the integrated amplifiers support bidirectional current sensing and they provide an adjustable output offset up to 3.3 V. Current is sensed on the low-side of each leg using a 0.01 Ω current sense resistor. Individual phase and DC bus voltage dividers are available to measure these voltages and are scaled for 6 – 60 V operation.

This expansion board requires a compatible LaunchPad™ to receive the appropriate control signals and a power supply suitable to load needs. The BOOSTXL-DRV8301 Motor Drive BoosterPack supplies 3.3 V power to the LaunchPad™ through the onboard 1.5 A step-down buck converter. TI recommends removing the jumper JP1 on the LaunchPad™ to separate the

Figure 3.10 TI BOOSTXL-DRV8301 converter mounted on the left-hand side of the TI LaunchXL F28069M Piccolo™. On the right-hand side the extPot3 board is installed. The reader is referenced to Appendix B for further information on this custom board.

3.3 V power supply from the USB connection (see Table 3.1). Figures 3.9 (b) and 3.10 show how to plug the converter BoosterPack on the LaunchXL-F28069M board, i.e., it is connected to pins in rows from J1 to J4 (the reader shuld refer to Figures 3.9 (b) and 3.5 for the pinout of the adopted LaunchPad™).

3.2.1 BoosterPack PWM signals

The BOOSTXL-DRV8301 can be configured for two different PWM control modes outlined further in the DRV8301 device datasheet. A possible motor control example is reported in the block scheme shown in Figure 3.11. These control modes give PWM control through 3 or 6 independent inputs. Six-input PWM mode gives independent control of the low-side and high-side gates for each converter leg. Three-input PWM mode gives control only of the 3 high side gates of the legs, while low-side switches are driven by an internally generated complementary signal with minimum internal dead time.

PWM modules on the three phase legs (which will be denoted from now on as Leg A, B, C) can be accessed by using the ePWM block in Simulink® and by selecting module 1, 2, or 3. Further details on the management of PWM signals will be discussed in Chapter 13.

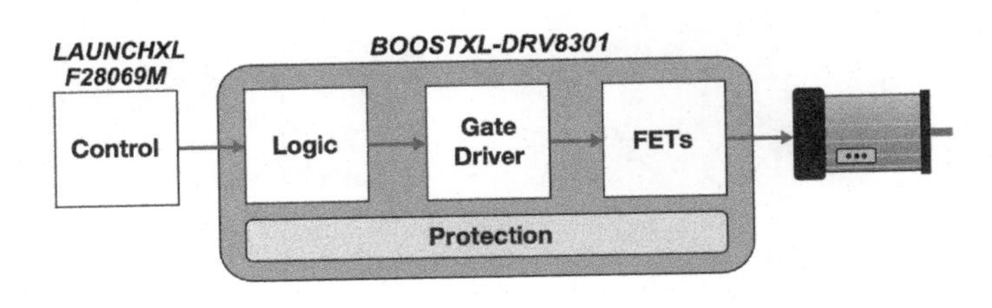

Figure 3.11 Block scheme of the interactions between a LaunchPad™ and the BOOSTXL-DRV8301 aimed to a motor control application [14].

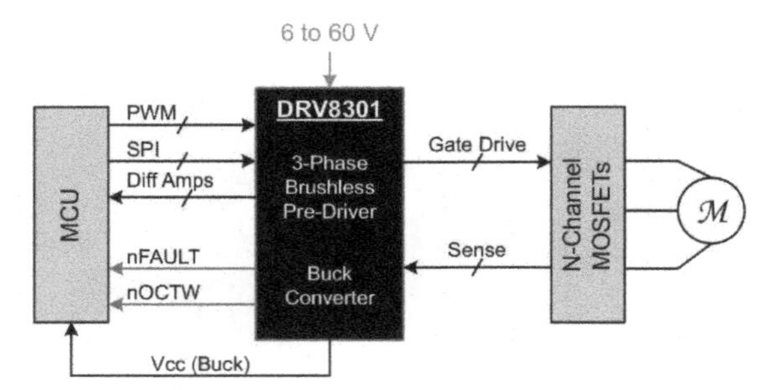

Figure 3.12 Scheme showing the possible interactions between a LaunchPad™ and the BOOSTXL-DRV8301 [14].

3.2.2 BoosterPack GPIO signals

This board exploits the interface with the input/output peripherals of the LaunchPad™ to output diagnostic signals or to set some enables, see Figure 3.12 and 3.13. Namely, these signals are:

1. **nFAULT** – Fault report indicator. Specific fault status can be obtained through the status registers
2. **nOCTW** – Over-current, over-temperature, or both warning indicator. Specific OCTW status can obtained through the SPI status registers
3. **EN_GATE** – Enables gate driver and current shunt amplifiers of the converter board. The corresponding pin on the LaunchPad™ must be set high to use the expansion board
4. **DC_CAL** – When high, the device shorts inputs of shunt

Figure 3.13 Top and bottom view of the BOOSTXL-DRV8301 BoosterPack [22].

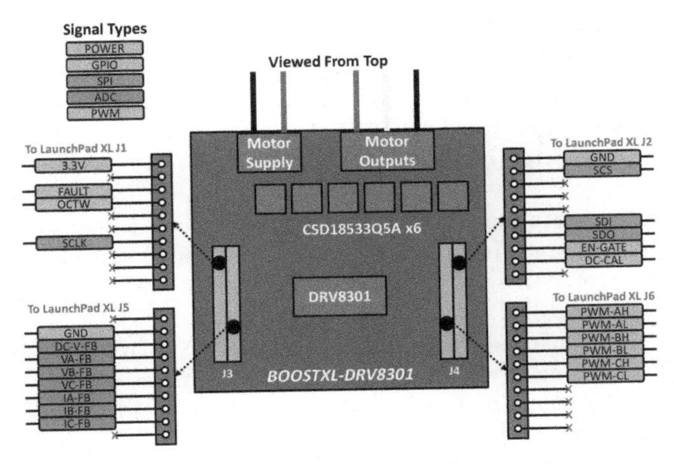

Figure 3.14 Pinout of the BOOSTXL-DRV8301 converter board [22].

amplifiers and disconnects possible loads so that the LaunchPad™ can calibrate DC offsets

In addition, the BOOSTXL-DRV8301 nFAULT and nOCTW signals share the GPIO channels with the UART-RX and UART-TX signals on the LaunchPad™. To see the proper nFAULT and nOCTW signals reporting on the onboard LEDs and MCU peripheral, disconnect the UART lines from the emulation side of the LaunchPad™ to the microcontroller. To this aim, refer to Table 3.2 to set the LaunchPad™ F28069M accordingly. **Software Enable**: it must be noted that the pin of the enable gate (EN-GATE, see Figure 3.14) of this expansion board must be set high to perform any operation with the converter. This is carried out by setting the voltage on this pin to the logic state 1 through a suitable peripheral.

3.2.3 DC bus and phase voltage sense

The BOOSTXL-DRV8301 is designed with voltage sense circuits on the DC bus (PVDD) and each half-bridge outputs (phases A, B, and C). These circuits consist of a voltage divider with a filtering capacitor to reduce high-frequency noise on the ADC pins. These circuits are scaled for 26.314 V, with a filter pole location at 364.692 Hz. The high-side resistors for the phase outputs are located near the output header J11, whereas the low side resistors and filtering capacitors are located near the ADC inputs J3 (for improved noise reduction).

3.2.4 Low-side shunt-based current sense

Low side current shunt resistors are installed on each leg of the converter (i.e., on phase A, B, and C). The current sense setup takes advantage of the dual shunt current amplifiers (phases A and B) of DRV8301 driver and an external TI OPA2374 CMOS operation amplifier (phase C). Each amplifier senses voltage across a $0.01\,\Omega$ sense resistor through a differential connection. The current measurement is scaled on $\pm 16.5\,\mathrm{A}$. Further details on low-side current shunt sense are reported in Chapter 16.

Remark: all the examples proposed in this book refer to the hardware presented in this chapter. It must be said that Texas Instruments™ started the production of a new LaunchPad™ F280049C and new BoosterPacks BOOSTXL-DRV8320RS, BOOSTXL-DRV8323RS, which will be available along with the LaunchPad™ F28069M and BOOSTXL-DRV8301 on the market. Even if a new hardware will imply differences in the settings of the programming environment as well as some modifications in the pinout of the adopted boards, the proposed design approach/workflow of closed-loop control strategies for power electronic applications is still valid. The reader is referenced to [7] for checking the compatibility between LaunchPads and BoosterPacks and [31] to verify the Embedded Coder support of the adopted MCU board from Simulink® blockset.

4

Software Installation

To the aim of programming the microcontroller mounted on the LAUNCHXL F28069M, different software packages need to be installed. The latter are available on the MATLAB® Add-Ons menu and on the website of Texas Instruments™ [10]. Namely, they are:

1. Texas Instruments™ Code Composer™ Studio.
2. Texas Instruments™ ControlSUITE™.
3. Embedded Coder® Support Package for Texas Instruments™ C2000™ Processors.

It is important to verify the installation of MinGW-64 in MATLAB®.

4.1 TI Support Packages: Code Composer™ Studio and ControlSUITE™

An embedded engineer must answer many critical questions when selecting a new MCU platform, such as those reported in Figure 4.1. Texas

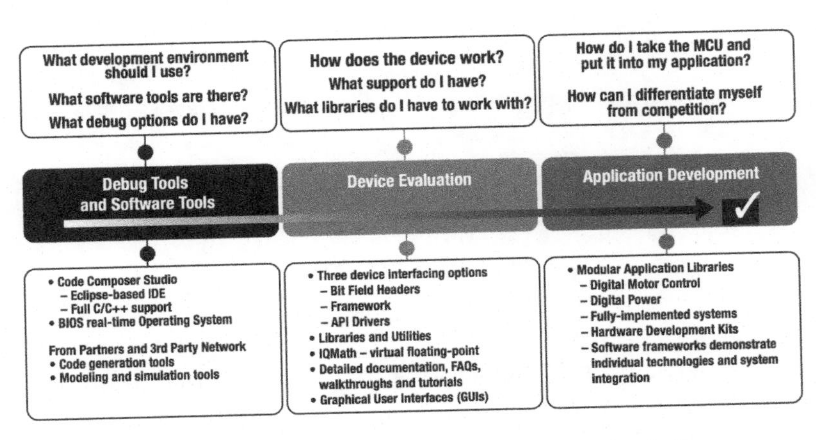

Figure 4.1 MCU platform selection criteria [22].

DOI: 10.1201/9781003196938-4

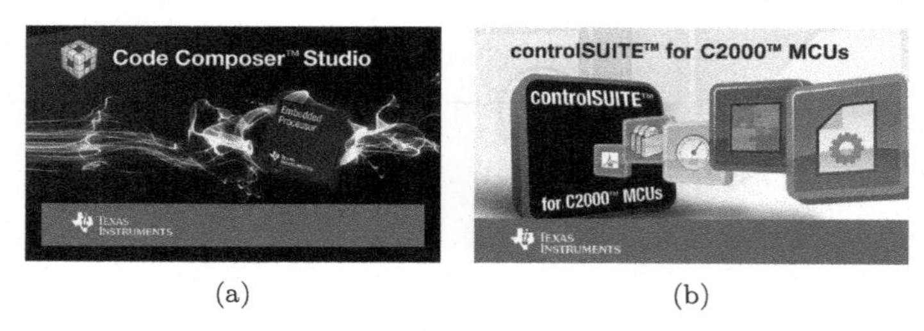

(a) (b)

Figure 4.2 Texas Instruments™ Code Composer™ Studio (a) and ControlSUITE™ (b).

Instruments™ provides two different software aimed at MCU programming: Code Composer™ Studio and ControlSUITE™ (see Figure 4.2).

Code Composer™ Studio is an IDE that supports Microcontroller and Embedded Processors portfolio of TI. Code Composer™ Studio comprises a suite of tools used to develop and debug embedded applications. It includes an optimizing C/C++ compiler, source code editor, project build environment, debugger, profiler, and many other features. The intuitive IDE provides a single user interface bringing the users through each step of the application development flow. Familiar tools and interfaces allow users to get started faster than ever before. Code Composer™ Studio combines the advantages of the Eclipse software framework with advanced embedded debug capabilities, resulting in a compelling feature-rich development environment for embedded developers.

On the other hand, ControlSUITE™ for C2000™ microcontrollers is a cohesive set of software infrastructure, tools, and documentations designed to minimize system development time, as shown in Figure 4.3. From device-specific drivers and support software to complete examples in sophisticated system applications, ControlSUITE™ provides the key resources to make a feasible automatic code generation procedure. Indeed, it contains all the needed

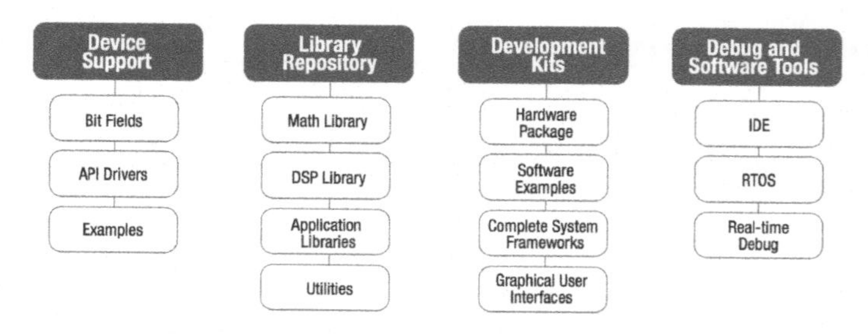

Figure 4.3 ControlSUITE™ overview: device drivers, APIs, utilities and libraries are used to build technology examples and system frameworks [22].

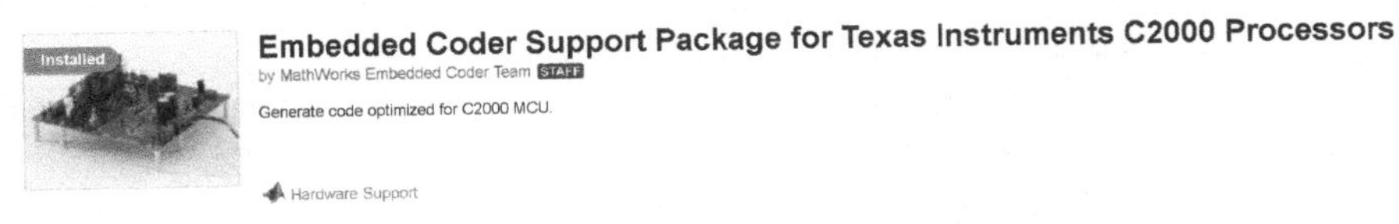

Figure 4.4 Embedded Coder Support Package for Texas Instruments C2000 Processors available in the Add-Ons menu of Matlab®.

settings for the MATLAB® environment (e.g. Simulink® blocks) to characterize the peripherals of the MCU.

4.2 MATLAB® Support Package: Embedded Coder for Texas Instruments C2000 Processors

The **Embedded Coder Support Package for Texas Instruments C2000 Processors** enables the generation of a real-time executable code and its download to MCU boards (from 8 bit to multi-cores) through the Simulink® environment. Such package automatically generates C code (background usage of Code Composer™ Studio) and it makes available a Simulink® library with built-in blocks. Different blocks allows to set different I/O drivers of the device (background usage of ControlSUITE™).

4.3 Installation Procedure

All the previously mentioned softwares can be installed starting from the MATLAB® Add-Ons window following the guided procedure shown below (reference to MATLAB® release 2019b):

- Click on the lower part of the **Add-Ons** icon 

 in the home bar of the Matlab® window and select **Get Hardware Support Package.**
- Search for **Embedded Coder Support Package for Texas Instruments C2000 Processors** (see Figure 4.4) and click on it.

- Check whether the required packages are already installed Requires Simulink Embedded Coder MATLAB Coder Simulink Coder or not. In case any of them is missing, install the required software by clicking on its corresponding name. A guided installation procedure starts after clicking on the **Install** icon Install.
- Once all the required packages are properly installed, click on the **Install** icon Install ▾ back in the window of the **Embedded Coder Support Package for Texas Instruments C2000 Processors** Add-On and select install.
- Accept the license agreement. After that, click the Next button Next to proceed with the installation of the following Third-Party Softwares:
  - TI C2000 Code Generation Tools;
  - Ostermiller Circular Buffer.

  Their download and installation should start immediately.
- After that, click on the Setup Now button Setup Now to go on with the guided procedure.
- The window **Select Processor Family** opens. The user is asked to select the processor family that is going to be used from a list (see Figure 4.5 (a)). Only the **TI Piccolo F2806x** is needed for the solution of the exercises proposed in this book. After this selection, click on the **Next** button.
- The window **Install Third-Party Software** opens as shown in Figure 4.5 (b). Depending on the previous selection, the user may be required to install different softwares. In particular, the TI Piccolo F2806x family requires **TI controlSUITE** and **TI Code Composer Studio** only. TI C2000Ware is needed in case other boards are selected. Click on the corresponding cell in the column Status to install their latest version.
  - **Texas Instruments™ ControlSUITE™**: the latest available version of this software is automatically suggested by the guided installation procedure. In particular, it is required to specify the ControlSUITE™ installation folder. The default path is: **C:\ti\controlSUITE**. Moreover, all the needed libraries are installed by default.
  - **Texas Instruments™ Code Composer™ Studio**: the latest available version of this software is automatically suggested by the guided procedure. In particular, it is required to:
    * Specify the Code Composer™ Studio installation folder. Default path: **C:\ti\ccsvx**, where **x** refers to the release of the software;
    * Specify the product families to be installed:

· Select **C2000™32bit Real-time MCUs**, compiler tools

1. TI C28x Compiler;
2. TI ARM Compiler;
3. GCC ARM Compiler;

* leave the default debug probe: **TI™ XDS Debug Probe Support**;
* Allow access for *Tclsh Application* and *ccstudio* from the settings of Windows Defender Firewall.

It is important to note that warning messages could appear depending on the installed Windows 10 release. Ignore them. Install any other additional driver/software if requested during the installation process.

- The window **Validate Control Suite Installation** opens. In this phase, the installation of the previously mentioned softwares is checked inserting their installation path:
  - ControlSUITE™ installation folder. The default path is: **C:\ti\controlSUITE**
  - Code Composer™ Studio installation folder. The default path is: **C:\ti\ccsv*x***, where ***x*** refers to the installed release.
    * Click the Validate button to check if the version of Code Composer™ Studio is compatible with the installed MATLAB® release.
  - Specify the C2000™ Compiler installation folder. The default path is: C:\ProgramData\MATLAB\SupportPackages\R2019b\3P.instrset\tic2000codegentools.instrset\ti-cgt-c2000_***xx.x.x***.LTS, where all the ***x*** depend on the installed release.

    This compiler is downloaded during the installation of the support package.
- At the end of the validation procedure, the **Summary of third-party installation** opens. A report table showing the installed softwares appears.
- Finally, the **Hardware Setup Complete** window opens. The guided procedure is finished.
- In case any modification of this installation is needed, it is possible to click on the manage button Manage on the window of this add-on (it replaces the Installation button). A window showing all the installed Add-Ons opens and the guided procedure of any of them can be repeated by clicking on the setup icon

A comprehensive summary of all these installation steps is available in [28].

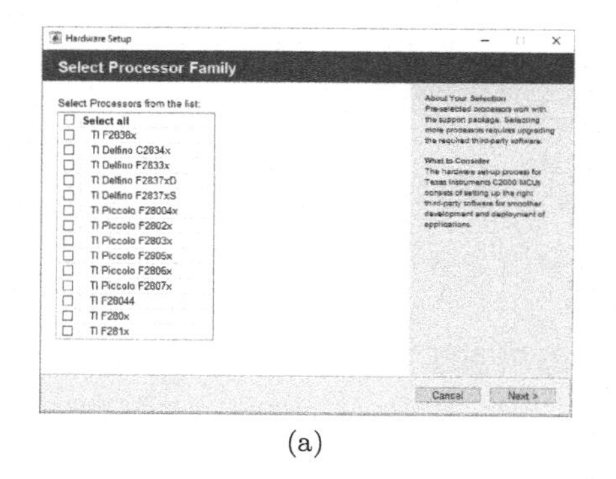

(a)

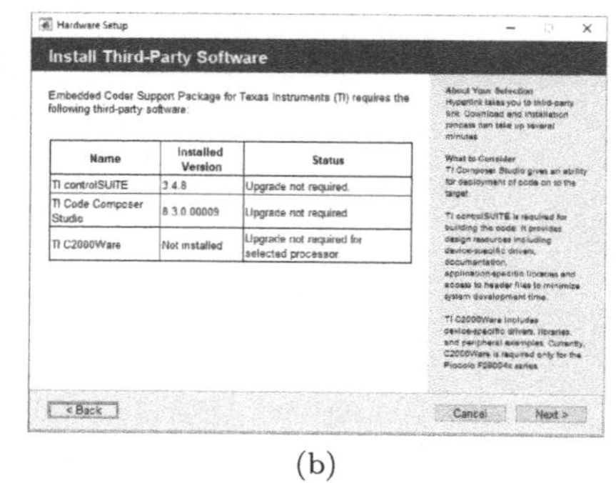

(b)

Figure 4.5 Example of MATLAB® windows that appear during the installation: (a) shows the processor family selection windows (only the required TI Piccolo F2806x is selected), while (b) reports the needed third party softwares. In this latter, all of them have been installed already.

Remark

Check if the **Instrument Control Toolbox**™ is installed on the adopted MATLAB® release. This toolbox is needed to set Serial Communication between the LaunchPad™ and the host PC during the execution of the firmware. More information on serial communication are reported in Chapter 8.

Potential Errors

For MATLAB® releases earlier than the 2019b (like the 2015b one) in which Code Composer™ Studio v5 is used, it may be required to run `checkEnvSetup()` tool to complete/adjust the installation:
To this aim, the following path should be checked by typing `checkEnvSetup ('ccsv5','f28069','setup')` in the MATLAB® prompt:

- ```
  Checking CCSv5 (Code Composer Studio) version
  Required version: 5.0 or later
  Required for : Code Generation
  Your Version : 5.5.0
  ```
  ### Setting environment variable "TI_DIR" to **"C:\ti\ccsv5"**

  - ```
    Checking CGT (Texas Instruments C2000 Code Generation Tools) version
    Required version: 5.2.1 to 6.0.2
    Required for : Code generation
    Your Version : 6.2.0
    ```
    ### Setting environment variable "C2000_CGT_INSTALLDIR" to **"C:\ti\ccsv5\tools\compiler\c2000_6.2.0"**

  - ```
    Checking DSP/BIOS (Real Time Operating System) version
    Required version: 5.33.05 to 5.41.11.38
    Required for : Code Generation
    Your Version : 5.42.01.09
    ```
    ### Setting environment variable "CCSV5_DSPBIOS_INSTALLDIR" to **"C:\ti\bios_5_42_01_09"**
    ```

- Checking XDC Tools (eXpress DSP Components) version
  Required version: 3.16.02.32 or later
  Required for : Code Generation
  Your Version : NOT FOUND $\surd$
  The required version of XDC Tools is not found in $(XDCROOT) $\surd$

- Checking 2806x C/C++ Header Files version
  Required version: 1.36
  Required for : Code Generation
  Your Version : 1.36
  ### Setting environment variable "DSP2806x_INSTALLDIR" to
  "C:\ti\controlSUITE\device_support\f2806x\v136"

- Checking Flash Tools (TMS320F2806x Piccolo Flash API) version
  Required version: 1.00
  Required for : Flash Programming
  Your Version : 1.00
  ### Setting environment variable "FLASH_2806X_API_INSTALLDIR" to
  "C:\ti\controlSUITE\libs\utilities\flash_api\2806x\v100"

It is important to note that this procedure could be also used to check the correct installation of Code Composer Studio™ for other Matlab® releases up to the 2018b one. For newer versions, this function has been removed.

Part II

Review of Control Theory: Closing the Loop

Introduction

The process of designing a control system usually makes many demands of the engineer or engineering team. In general, they can be summarized in the following step by step procedure:

1. Study of the system to be controlled and definition of the required control objectives.
2. Modeling of the system and simplification of the model.
3. Analysis/determination of the system properties (open-loop characteristics) and, if necessary, scaling of the variables.
4. Selection of the variables to be controlled (e.g., control input of a modulator).
5. Identification of the required measurements and manipulated variables: what sensors and actuators should be used and where they should be placed in the real system.
6. Choice of the control configuration (e.g., power electronics systems are typically based on nested control loops).
7. Selection of the controller type to be used (e.g., PI, PID controllers).
8. Setting of the performance specifications based on the overall control objectives (closed-loop characteristics).
9. Design of the controller parameters (e.g., pole/zero cancellation in case a PI controller and a first-order system are considered).
10. Analysis of the resulting closed-loop controlled system to see if the specifications are satisfied. If not, modifications in the specifications or in the controller type are needed.
11. Simulation of the closed-loop controlled system (e.g., MATLAB®- or Simulink®-based simulations).
12. Choice of the adopted software and implementation of the regulator on an embedded system (e.g., a microcontroller).
13. Test and validation of the resulting real-time control driving a pilot/prototype of the system.

Control courses and textbooks usually focus on steps 9 and 10 in the above procedure, that is, on controller design and control system analysis methods.

DOI: 10.1201/9781003196938-4

Interestingly, many real control systems are designed without any consideration of these two points. For example, even for complex systems with many inputs and outputs, it may be possible to design workable control systems using only heuristic tuning (i.e., involving steps 1, 4 5, 6, 7, 12, and 13). However, in this case a suitable control structure may not be known at the outset, and there is a need for systematic tools and insights to assist the designer at least with steps 4, 5, and 6, based on step 2 and 3. This book aims to provide guidelines for steps 12 and 13, for which the control structure design has to deal with microcontroller implementations.

Simply stated, "*even the best control system cannot make a Ferrari out of a Volkswagen.*" Therefore, the process of control system design should in some cases also include a step 0, which involves the design of the system equipment itself. Regarding the purpose of the book, this means to deal with the power electronics design in term of sizing passive components, consider magnetic and saturation effects, evaluate power losses and heat exchange and many more aspects.

Both control and power electronics engineers have to remember that the idea of looking at hardware/system equipment and control system design as an integrated whole is not a novel approach, as it is clear from the following quote taken from [44]:

> In the application of automatic controllers, it is important to realize that controller and system to be controlled form a unit; credit or discredit for results obtained are attributable to one as much as the other. A poor controller is often able to perform acceptably on a system which is easily controlled. The finest controller made, when applied to a miserably designed system, may not deliver the desired performance. True, on badly designed systems, advanced controllers are able to eke out better results than older models, but on these systems, there is a definite end point which can be approached by instrumentation and it falls short of perfection.

The aim of this chapter is to show how to design a development-oriented PI controller for microcontroller applications. Starting from a continuous-time model based on the physics of the system, the most important design guidelines of a PI-based closed-loop control system are revised. In particular, much attention is paid to how to guarantee certain steady-state and dynamic performances, how discretize the controller, and how to cope with real-time computation bottlenecks on physical device. Due to their digital nature, all physical devices are designed in order to operate exclusively with discrete-time variables. Thus, it is necessary to identify a procedure which transform the continuous-time control problem into an equivalent discrete-time one. The book uses simple examples to clarify the transition from "theory" to "practice", without claiming to be exhaustive with regard to the whole detailed aspects of control theory.

5

Designing a Closed-Loop Control System

The objective of a closed-loop control system is to make the output $y(t)$ behave in a desired way by manipulating the input $u(t)$. This "desired way" is defined by a proper formulation of the control problem, which could refer to:

- **Command-tracking**: $u(t)$ is manipulated to keep the output close to a given reference input (set point) $y^*(t)$.
- **Disturbance rejection**: $u(t)$ is manipulated to counteract the effect of disturbances;

The evolution in time of $u(t)$ is given by the controller type and its design, which relates to the control target. Note that the controller should be selected according to the system to be controlled. Starting from the physics of the circuits models, i.e., based on equivalent electrical ports, the following sections denote that both electrical machines and power conversion systems may be described as linear or linearized models. Hence, linear controllers represent the easiest control candidate. Furthermore, the mathematical description of such electrical dynamics refers to multi-input multi-output (MIMO) models. Nevertheless, this manuscript shows how the control configuration/architecture may lead to translate a MIMO model into a combination of single-input single-output (SISO) models. This is possible by taking into accounts feedforward terms. For SISO linear models (like those treated in this book), a proportional–integral–derivative controller (PID) results in control loop feedback mechanisms widely used in industrial power electronics-based systems and a variety of other applications requiring continuously modulated control. A PID controller continuously calculates an error value $e(t)$ as the difference between the set point $y^*(t)$ and the measured variable $y(t)$. Based on this evaluation, the controller applies a correction based on proportional, integral, and derivative actions. Note that, PID controllers are present in the industrial fields since the early 1920s (automatic steering systems for ships).

In general, any closed-loop control scheme for SISO models can be represented by the block diagram shown in Figure 5.1.

DOI: 10.1201/9781003196938-5

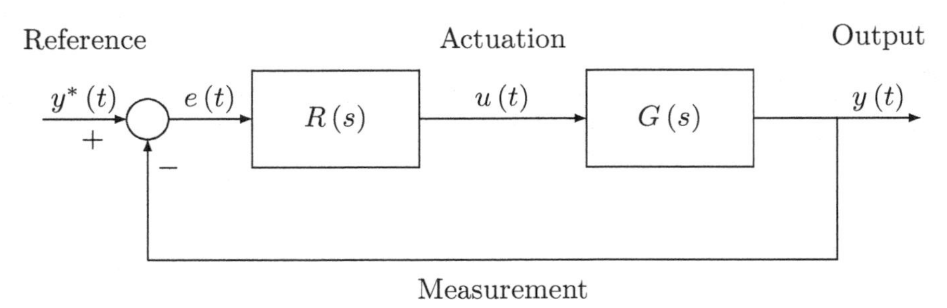

Figure 5.1 Example of closed-loop control system.

5.1 Dynamical Systems

From control theory, the dynamical model of a system is a set of mathematical laws explaining in a compact form (i.e., it must be simple enough) and in a quantitative way (i.e., it must be descriptive enough) how the system evolves over time. The latter is fully described by the evolution of the state $x(t)$. It is worth noting that the main questions about dynamical models are:

- How to built the model?
  ("How $x(t), u(t)$ and $y(t)$ influence each other?")
- How to use such model to design a control system?
  ("How to make the system behave the way I want?")
- How to make simulations?
  ("Which tool is more suitable for a given application?")

Note that simulating a dynamical model has zero-cost compared to real experiment and it allows to get rid of many practical complexities of the real system. In addition, many simulations allows to reduce the order of the system or to focus just on some parts of the whole, allowing more agile and easy runs. To this purpose, it is important to take in mind the well-known aphorism:

> *"Make everything as simple as possible, but not simpler."*
> Albert Einstein

5.1.1 Mathematical laws

Considering a continuous-time domain $t \in \mathbb{R}$, the behavior of the state $x(t)$ is described by ordinary differential equations (ODEs). This book refers to applications whose input-output electrical responses are governed by linear ODEs with constant coefficients and, thus, linear dynamics. The continuous-time state-space model can be formalized as:

$$\frac{\mathrm{d}x(t)}{\mathrm{d}t} = Ax(t) + Bu(t) \tag{5.1}$$

$$y(t) = Cx(t) + Du(t) \tag{5.2}$$

where A, B, C, and D are (time-invariant) system matrices, which sizes results after choosing the system variables. Every time the output is entirely matching the state $x(t)$ via C, the term $Du(t)$ is removed from (5.2). Given the initial conditions $x(0) = x_0 \in \mathbb{R}^n$, the evolution of the state described in (5.1) is equivalent to

$$x(t) = \underbrace{e^{At}x_0}_{\substack{\text{natural response} \\ \text{(effect of } x(0)\text{)}}} + \underbrace{\int_0^t e^{A(t-\tau)}Bu(\tau)\,d\tau}_{\substack{\text{forced response} \\ \text{(effect of } u(t)\text{)}}}. \tag{5.3}$$

Moreover, the system is asymptotically stable on condition that $A < 0$; this is a key aspect.

This manuscript makes extensive use of transfer functions and of the frequency domain, which are very useful because from simple frequency-dependent plots it is possible to derive useful insights such as bandwidth and peaks of closed-loop transfer functions. In particular, Laplace transforms convert integral and differential equations into algebraic equations through the Laplace operator s. A general time-dependent function $f(t)$ is converted as:

$$F(s) = \mathcal{L}\left[f(t)\right] \quad \leftarrow \quad s = \frac{\mathrm{d}}{\mathrm{d}t} = j\omega \tag{5.4}$$

In addition, it can be demonstrated that

$$y(s) = \underbrace{C\left(sI - A\right)^{-1}x_0}_{\substack{\text{Laplace transform} \\ \text{of the natural response}}} + \underbrace{\left[C\left(sI - A\right)^{-1}B + D\right]u(s)}_{\substack{\text{Laplace transform} \\ \text{of the forced response}}} \tag{5.5}$$

where I denotes the identity matrix. Assuming $x_0 = 0$ and $D = 0$, the transfer function $G(s)$ of a continuous-time linear system defined as in (5.1) and (5.2) is:

$$G(s) = \frac{y(t)}{u(t)} = \frac{y(s)}{u(s)} = C(sI - A)^{-1}B. \tag{5.6}$$

Namely, $G(s)$ is defined through the ratio between the Laplace transform of the output $y(s)$ and the Laplace transform of the input $u(s)$. Note that, the mapping $t \leftarrow s$ holds. Thus, $y(t)/u(t) = y(s)/u(s)$. This equality will be considered in the definition of other transfer functions in the following. Equation (5.6) directly links to Figure 5.1.

5.1.2 Dynamical systems in electrical applications

In general, power conversion systems and electrical machines behave as:

- RL (ohmic-inductive) or RC (ohmic-capacitive) load.
- RLC or RLE (including the motor back-electromotive force or higher-order filter) load.

Thus, $G(s)$ results in a 1st or a 2nd order transfer functions in the vast majority of the cases.

A 1st order $G(s)$ is now considered. It can be defined as:

$$G(s) = \frac{1}{a + sb} = \frac{1/a}{1 + s\,(b/a)} = \frac{1/a}{1 + s\tau_G} = \frac{1/\,(a\,\tau_G)}{s - (-1/\tau_G)} \tag{5.7}$$

where a and b are constant (time-invariant) coefficients. The denominator of $G(s)$ can be factorized such that it has the form $(s - p)$, where p defines a pole, e.g., $p = -1/\tau_G$. When $s = p$, the value of $G(s)$ becomes unbounded, i.e., if $1 + s\tau_G = 0$ is set, it follows $s = p = -1/\tau_G$, which characterizes the natural response of the system. Moreover, if $\mathrm{Re}(p) < 0$ the system is stable.

The timing required by the system output to react against transient is a key factor either for uncontrolled, e.g., $G(s)$, and controlled systems, e.g., $L(s)$, $F(s)$ (defined later). Since different ways to define time quantities are present in literature, it is important to specify that this book adopts the following definition:

> given a step-wise input with steady-state value Y_{ss}, the settling time T_a is defined as the time interval required to enter in a specific tolerance bandwidth without further violation. Namely, the steady-state value Y_{ss} is reached with a certain degree of accuracy given by $\pm\epsilon$, i.e., entering in a region $[(1 - 0.01\epsilon)Y_{ss}\ (1 + 0.01\epsilon)Y_{ss}]$. In practice, $T_a = 4 \div 5\tau$ holds (counting from the instance in which a transient happens) with $\epsilon \approx 1\%$, without loss of generality.

Note that, the time constant τ is specified by the transfer function parameters.[1] As an example, referring to $G(s)$, the *natural* settling time is defined as $T_{a,G}$ and it follows that $T_{a,G} = 5\tau_G$, where the time constant is $\tau_G = b/a$. Note that subscripts refer to the settling time of a specific transfer function.

5.2 Design a PI Controller in Continuous-Time Domain

In power electronics-based systems, PI controllers are more common than PIDs since the derivative action is sensitive to measurement noise (e.g., high

[1] Be careful, τ must not be confused with the rise time T_r which is defined as the time interval between the time instances related to 10% Y_{ss} and 90% Y_{ss}.

frequency noise entering in the system is highly amplified). However, if sudden variations in the reference input occur, the derivative part may help to achieve fast transients, e.g., current reference for high speed motors. Nevertheless, the integral action could lead to wind-up effects both for PI and PID solutions, which have to be taken into account.

5.2.1 Serial/parallel form

Using the Laplace operator, the PI controller transfer function $R(s)$ (which is equivalent to $u(t)/e(t)$) could be written both in series and parallel form:

$$R(s) = \frac{u(s)}{e(s)} = \frac{k_p\,(1 + sT_i)}{sT_i} \quad \text{(series)} \qquad \text{or} \qquad R(s) = k_p + \frac{k_i}{s} \quad \text{(parallel)} \tag{5.8}$$

where k_p, k_i are the proportional and integral gains, respectively, while T_i is the integration time defined as $T_i = k_p/k_i$.

5.2.2 Characterization of the closed-loop dynamics $F(s)$

In this subsection, an easy analytical procedure to identify the controller parameters starting from the knowledge of the natural response of $G(s)$ and the desired requirements for the final closed loop-system is recalled. Those requirements are usually given in terms of:

1. Zero steady-state error (static performance)
2. Time response to transients (dynamic performances), i.e., control bandwidth ω_c
3. Robustness level against disturbances (dynamic performances), i.e., phase margin ϕ_m

Once $R(s)$ and $G(s)$ are known, the open- $L(s)$ and closed-loop $F(s)$ transfer functions are defined as:

$$L(s) = \frac{y(s)}{e(s)} = R(s)G(s) \qquad F(s) = \frac{y(s)}{y^*(s)} = \frac{L(s)}{1 + L(s)} \tag{5.9}$$

Note that the definition $e(s) = y^*(s) - y(s)$ holds. If $G(s)$ shows a positive gain and it has Re(p) < 0, it is called minimum phase system (mps). Then, it is enough that $|L(s)|$ cuts the 0 dB axis with a slope of -20 dB/decade to guarantee $\phi_m \cong 90°$ and asymptotic stability. This means that the cut-off frequency is imposed at ω_c (i.e., at $s = j\omega_c$):

$$\left[|L(s)|_{s=j\omega_c}\right]_{\text{dB}} \triangleq 0\,\text{dB} \quad \rightarrow \quad |L(s)|_{s=j\omega_c} \triangleq 1. \tag{5.10}$$

Note that the phase margin is computed as $\phi_m = \pi - \arg L(j\omega_c) = \pi - \phi_c$.

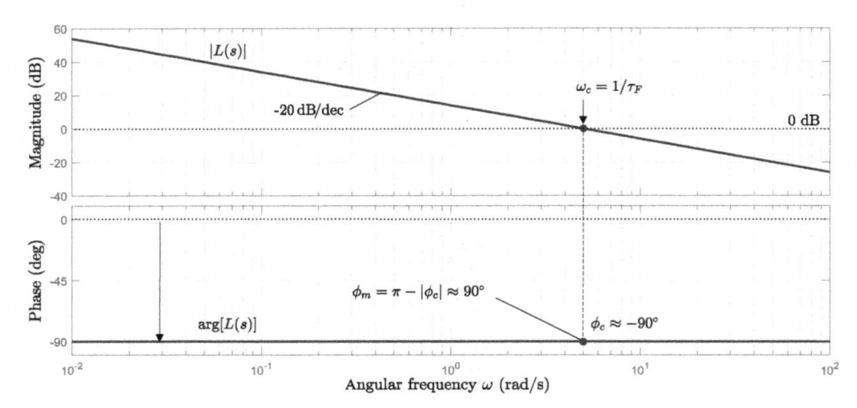

Figure 5.2 Example of $L(s)$ obtained through pole/zero cancellation and its properties detailed through a Bode diagram.

Pole/Zero Cancellation

By considering (5.7), the open-loop transfer function can be computed as:

$$L(s) = \frac{k_p(1 + sT_i)}{sT_i} \frac{1/a}{1 + s(b/a)} \tag{5.11}$$

If $T_i = \tau_G = b/a$ is imposed, the so called pole/zero cancellation is performed, which is rather typical for first-order systems:

$$L(s) = \frac{k_p/a}{sT_i} \tag{5.12}$$

This transfer function behaves as a pure integrator as shown in Figure 5.2, thus $\phi_m \cong 90°$, and it has a cut-off frequency which is defined by the controller parameters

$$\left|\frac{k_p/a}{sT_i}\right|_{s=j\omega_c} \triangleq 1 \quad \rightarrow \quad \omega_c = \frac{k_p}{aT_i} = \begin{cases} k_p/b & (\text{if } T_i = b/a) \\ k_i/a & (\text{if } T_i = k_p/k_i) \end{cases} \tag{5.13}$$

Note that $|x + jy| = \sqrt{x^2 + y^2}$ holds.

In practice, $L(s)$ and $F(s)$ are subjected to the same time constant, which can be denoted as τ_F. Having an explicit form, k_p and k_i can be computed as functions of ω_c. For instance, by considering $\omega_c = k_p/b \rightarrow k_p = \omega_c b$ from (5.13), k_p may be replaced in the expression of $T_i = k_p/k_i = \omega_c b/k_i$, which is equivalent to $T_i = b/a = \omega_c b/k_i$, which leads to $k_i = \omega_c a$. Finally:

$$k_p = \omega_c b \qquad k_i = \omega_c a \tag{5.14}$$

How to use the explicit forms: assume that $y(t)$ is required to follow a step-wise reference $y^*(t)$ within a specific settling time $T_{a,F}$. For instance, the requirement might be $T_{a,F} \ll T_{a,G}$, i.e., to speed up the open loop dynamics by adopting a (closed) control loop. Since the final (closed-loop) settling time is defined as $T_{a,F} = 5\tau_F$, $\tau_F = T_{a,F}/5$ can be computed, which represents the dominant time constant (i.e., associated to the dominant pole) of $F(s)$. Therefore, by choosing the cut off frequency ω_c as $\omega_c = 1/\tau_F$ in rad/s, and considering the other time-invariant parameters, all the ingredients are available to derive k_p, k_i by substitution.

Generalized approach

The pole/zero cancellation is particularly suitable for 1st order linear systems. If the system is higher-order, a different procedure should be adopted. In general, the following formulas can be exploited:

$$k_p = \frac{\cos(\phi_m - \pi - \arg(G(j\omega_c)))}{|G(j\omega_c)|} \tag{5.15}$$

$$k_i = -\omega_c \frac{\sin(\phi_m - \pi - \arg(G(j\omega_c)))}{|G(j\omega_c)|} \tag{5.16}$$

where each parameter is function of the cut-off frequency ω_c and the phase margin ϕ_m. Of course, by using the same hypothesis on T_i for the zero pole/-cancellation, the same results shown in (5.14) can be derived. Note that much flexibility on the performance of the system is achieved compared to the pole/zero tuning

Mathematical Proof

A generalized open loop transfer function cuts the 0 dB axis at ω_c with phase margin ϕ_m. In this point, its frequency response can be evaluated as:

$$L(j\omega_c) = R(j\omega_c)G(j\omega_c) = 1e^{j(\phi_m - \pi)} \tag{5.17}$$

Thus, the PI controller can be evaluated in the same frequency point as well:

$$R(j\omega_c) = k_p + \frac{k_i}{j\omega_c} = k_p - j\frac{k_i}{\omega_c} \tag{5.18}$$

Instead, a generalized $G(s)$ is considered:

$$G(j\omega_c) = |G(j\omega_c)|\, e^{\arg(G(j\omega_c))} \tag{5.19}$$

Then, (5.17) is exploited to derive the following expression:

$$k_p - j\frac{k_i}{\omega_c} = \frac{1}{|G(j\omega_c)|} e^{j(\phi_m - \pi - \arg(G(j\omega_c)))} \tag{5.20}$$

Recalling the Euler formula $e^{jx} = \cos x + j \sin x$, it follows:

$$k_p - j\frac{k_i}{\omega_c} = \frac{1}{|G(j\omega_c)|}[\cos(\phi_{\rm m} - \pi - \arg(\rm G(j\omega_c))) + ...$$

$$... + j\sin(\phi_m - \pi - \arg(G(j\omega_c)))] \qquad (5.21)$$

By computing the real and imaginary parts of the previous equation, k_p and k_i are obtained as reported earlier. All the angles must be considered in radians.

$$k_p = \text{Re}\left(k_p - j\frac{k_i}{\omega_c}\right) = \frac{\cos(\phi_m - \pi - \arg(G(j\omega_c)))}{|G(j\omega_c)|} \qquad (5.22)$$

$$k_i = -\omega_c \,\text{Im}\left(k_p - j\frac{k_i}{\omega_c}\right) = -\omega_c\frac{\sin(\phi_m - \pi - \arg(G(j\omega_c)))}{|G(j\omega_c)|} \qquad (5.23)$$

■

MATLAB® interface: `pidTuner()` and `pidtune()`

A third alternative for PI tuning is to use control design tools from Simulink®, e.g., `pidTuner()` and `pidtune()` from *Control System Toolbox*™. In particular, `pidTuner()` allows to work with a user friendly tool (with a graphic user interface as well) for simple step-response-based identification of a process model, fast PID controller tuning and effective quality evaluation of the control in the MATLAB® framework.
Regarding the control design, either time-domain or frequency-domain approaches can be chosen. As an example, the MATLAB® commands `pidTuner(sys)` or `pidTuner(sys,PI)` launch the user interface which allows to design a desired controller for the system `sys` under investigation, which in our case relates to $G(s)$. The properties of the tuned controller can be simply and visually judged by the PID Tuner window using a simulation of the closed-loop step response. Indeed, this tool displays the plot of the simulated control response and the time behavior of the manipulated variable. It is easy to compare several step responses generated with different values of k_p and k_i, set points, disturbances and constraints on the manipulated variable.

Hint: By defining the Laplace operator as `s=tf('s')` and the parameters comprised in $G(s)$ (e.g., `R=2.5e-3; L=1e-3`), the system transfer function can be defined simply as `G=1/(R+s*L)` in MATLAB® through the command window and standard syntax. Then, it is enough to call `pidTuner(G)` to open the interface reported in Figure 5.3.

The key commands available in this window allow to:

1. Define the system to be controlled $G(s)$;
2. Select the type and form of the controller (e.g., PI Parallel);

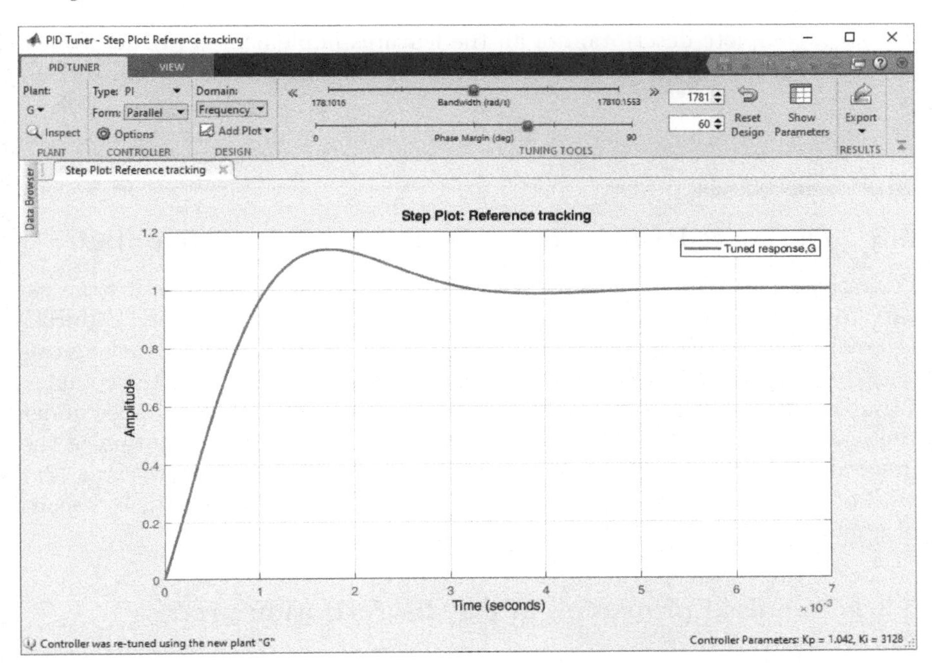

Figure 5.3 `pidTuner(G)` interface window.

3. Select the design domain, i.e., frequency or time;
4. Adjust the tuning parameters (e.g., bandwidth and phase margin for the frequency domain) and check the resulting response to a unitary reference step;
5. Copy or export the resulting k_p and k_i coefficient.

Alternatively, the user can automatically obtain the required parameters using the command `pidtune` in a script or in the command window, which runs the PID Tune software in background. The complete syntax is

```
pidtune(sys,'PI',bandwidth,opt);
```

with

```
opt=pidtuneOptions('PhaseMargin',phase_margin);
```

The result is a structure called *pid* which contains the k_p and k_i coefficients as field of this new variable type.

Note that booth approaches use the generalized formulas presented in the previous section. In particular, the `pidTuner()` interface use `pidtune()` as sub-function to compute the controller parameters.

For a complete description of all the features implemented in PID Tuner, the reader is referred to its documentation on the MathWorks® Website [29].

5.3 Derive a PI Controller in Discrete-Time Domain

Once the design of the controller is performed in continuous time, it is necessary to define an equivalent expression that shows a similar behavior during the real-time execution of the control algorithm. To do so, a functional equivalence between the Laplace operator s for the continuous time and the complex variable z for discrete time series must be defined. This operation goes under the name of "root mapping" and it corresponds to the approximation of the integral in continuous-time domain (expressed by the Laplace operator as $1/s$) with a suitable numerical integration method in discrete time, i.e., by means of a finite set of points/samples.

5.3.1 General properties of the discretization process

The numerical integration can be carried out directly on the transfer function. Since the starting point is in the s-plane, a continuous-time transfer function $G(s)$ can be mapped to the z-domain by substituting the Laplace operator as follows:

$$G(z) = G\left(s = \frac{z-1}{T_s\,(\alpha z + 1 - \alpha)}\right) \tag{5.24}$$

where T_s is the sampling time and $0 \leq \alpha \leq 1$ defines the type of discretization. Indeed, equation (5.24) refers to a generalization of a bilinear transformation which could result in:

- Forward Euler integration $\quad \alpha = 0 \quad \rightarrow \quad s = \frac{z-1}{T_s}$
- Backward Euler integration $\quad \alpha = 1 \quad \rightarrow \quad s = \frac{z-1}{zT_s}$
- Trapezoid integration $\quad \alpha = \frac{1}{2} \quad \rightarrow \quad s = \frac{2}{T_s}\frac{(z-1)}{(z+1)}$

The trapezoid rule is also called Tustin's method in digital control community. The continuous-time transfer function $G(s)$ shows a stable behavior when the poles have negative real part. In discrete-time, this translates in a $G(z)$ with poles inside the unitary circle. The coefficient α influences such stability because a different integration method results in a different mapping, with different boundaries for the pole/zero locations. The three mappings are summarized in Figure 5.4. However, it is possible to find a large number of other

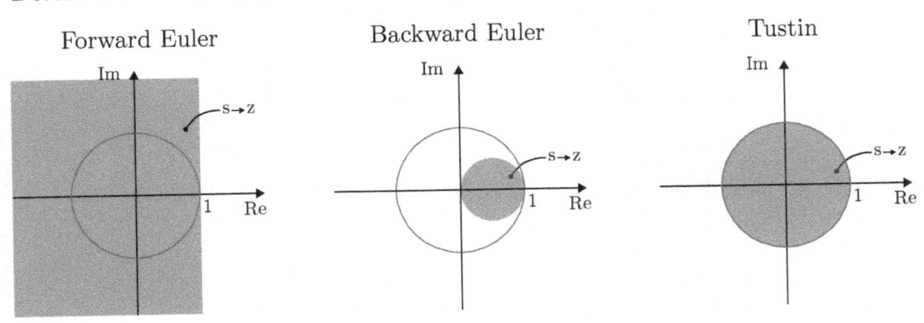

Figure 5.4 Stability regions for different discretization methods.

integration methods with different properties in terms of convergence and accuracy in literature. The three methods reported in this section represent the most common choice and they refer to *single step methods* in which the output depends only on the input at the same time-step and on input/output at the previous time instant.

Hence, they require a limited computational effort, targeting most of the microcontroller-based platforms.[2] However, keep in mind that:

- Forward Euler - if the discrete form is stable, also the continuous one is stable too, but not vice versa;
- Backward Euler - if the continuous form is stable, also the discrete one is stable too, but not vice versa;
- Trapezoid if the continuous form is stable, also the discrete one is stable too, and vice-versa.

Note that the convergence region depends both on the integration method itself and on the considered sampling time width as well. High T_s enlarges the mapping areas. This might create a great amount of mapping distortion, even leading to instability problem if the unitary circle is violated.

Another important property of the z-domain lies in the fact that the quantity z^{-1} is equivalent to a unitary delay in the discrete time domain. This correspondence allows to calculate the output of a generic regulator as function of the inputs/output at the previous time steps.

In particular, Figure 5.5 shows how to interpret the described Backward Euler integration method, where k denotes the discrete-time steps. This method is adopted in the following sections.

For an exhaustive explanation of the z-transform theory, the reader is referred to [2].

[2]In *multi-step methods*, several points are used to evaluate the output at each time step. These ones are characterized by a higher accuracy as the number of steps increases, but they imply a big number of variables to be managed by the microcontroller, i.e., high computational burden.

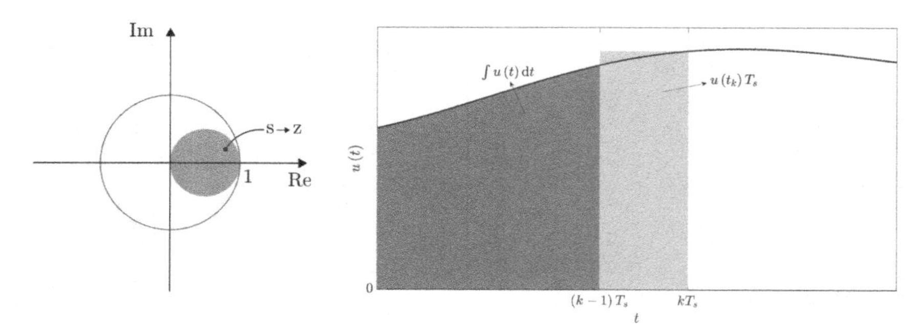

Figure 5.5 Time representation of the Backward Euler integration method.

5.3.2 Characterization of the closed-loop dynamics *F(z)*

The transformation reported in equation (5.24) is used to map $R(s)$ and $G(s)$ in the z variable, i.e., $s \to z$. By considering a Backward Euler integration method $s = \frac{z-1}{zT_s}$ for the transfer function defined in the previous subsection, their resulting discrete counterparts are:

$$R(z) = \frac{u(z)}{e(z)} = k_p \frac{zT_s + (z-1)T_i}{(z-1)T_i} \qquad G(z) = \frac{y(z)}{u(z)} = \frac{zT_s/a}{zT_s + (z-1)\,(b/a)} \tag{5.25}$$

The considered mapping into z-domain produces stable trajectories (i.e., inside the stability region) even if the corresponding continuous-time poles show a positive real part. Then, the resulting open- and closed-loop transfer functions are computed as for the continuous-time case:

$$L(z) = \frac{y(z)}{e(z)} = \frac{zT_s k_p/a}{(z-1)\,T_i} \qquad F(z) = \frac{y(z)}{y^*(z)} = \frac{L(z)}{1+L(z)} \tag{5.26}$$

Note that now the mapping $k \leftarrow z$ holds, where $k \in \mathbb{Z}$ is the discrete-time step index. For instance, this imply $y(k)/e(k) = y(z)/e(z)$. Moreover, the same values for k_p and k_i parameters computed in section 5.2 are still valid. Hence, one may take advantage of the vast body of knowledge available on continuous-time control theory since the discretization method does not change the design approach, which can be summarized as:

- Perform a design based on continuous-time transfer functions. Select a criterion (e.g., pole/zero cancellation) to compute k_p, k_i;
- Select a discretization method (without changing k_p, k_i) and an appropriate sampling time T_s;
- Represent the closed loop system as shown in Figure 5.6 (a) or (b).

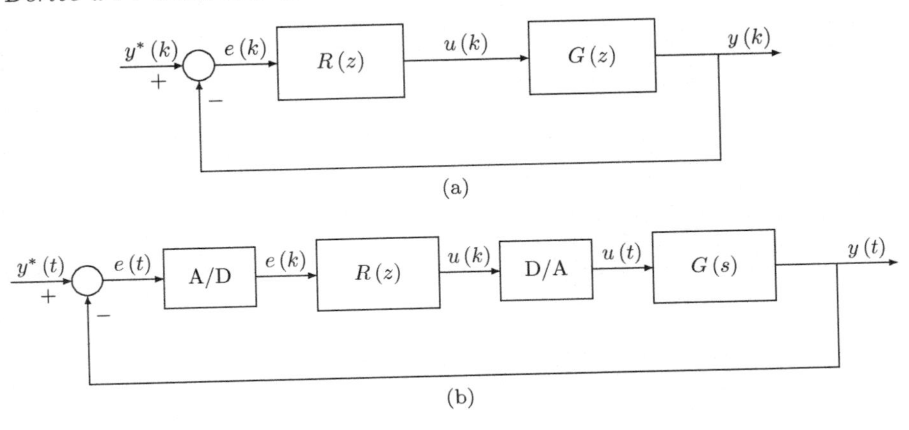

Figure 5.6 Example of discrete closed-loop control system (a) and discrete controller (b). The scheme reported in (b) includes analog to digital (A/D) and digital to analog (D/A) converters as interfaces between the hardware and the real system.

Even if Figure 5.6 (a) and (b) includes the same controller $R(z)$, they differ on how the rest of the scheme is defined. In particular, Figure 5.6 (a) considers a discrete transfer function $G(z)$, which is then used to compute $L(z)$ and $F(z)$ as reported in equations (5.25) and (5.26). Differently, in Figure 5.6 (b) the discretized controller is interfacing the system model in continuous-time domain, i.e., $G(s)$, through analog to digital (A/D) and digital to analog (D/A) converters operating at the sampling time T_s used for the discretization. In Simulink®, the operations of both converters can be mimicked with Rate Transition blocks.

In any case, both methods starts from the results of the design in continuous-time domain, i.e., based on $G(s)$ via s.

As a reminder, note that $G(z)$ may be directly derived by discretizing (5.1) and (5.2), thus, deriving discrete-time differential equations such as

$$x(k+1) = A^{'}x(k) + B^{'}u(k) \tag{5.27}$$

$$y(k) = C^{'}x(k) \tag{5.28}$$

The new matrices $A^{'}, B^{'}$ and $C^{'}$ have to be computed according to one of the previously described discretization method.

6

Design Example: PI-Based Current Control of an RL Load

In order to apply the design rules presented in the previous chapter and to show how to tailor them to a power electronics-based control problem (as a complete exercise), a simple *RL* (ohmic-inductive) load circuit fed by a controllable voltage source is considered as a case study (see Figure 6.1).

The system parameters and the desired characteristics of the controller are reported here in the following:

- **Data:**
  $R = 25\,\mathrm{m\Omega}$ and $L = 100\,\mathrm{mH} \rightarrow \tau_G = L/R = 4\,\mathrm{s}$, thus, $T_{a,G} = 5\tau_G = 20\,\mathrm{s}$
- **Objectives:** *reference tracking*
  $y^*(t \geq 1\,\mathrm{s}) = 3\,\mathrm{A}$ and $T_{a,F} = 1\,\mathrm{s}$, thus, $\tau_F = T_{a,F}/5 = 0.2\,\mathrm{s}$

The aim of this exercise is to design a *current control loop*, i.e., to control the current flowing into the winding $i(t)$ according to a given step-wise reference $i^*(t)$, by using a PI controller while achieving a final settling time of $T_{a,F} = 1\,\mathrm{s}$. In particular, the reference changes from $0\,\mathrm{A}$ to $3\,\mathrm{A}$ (i.e., $Y_{\mathrm{ss}} = 3$) at $t = 1\,\mathrm{s}$, holding this value until the end of the simulation.

According to Figure 6.1, $v(t)$ is the voltage of a variable voltage source (e.g., a controllable DC voltage supply), while R and L are the circuit parameters. Applying the Kirchhoff's voltage law, it follows that:

$$v(t) - Ri(t) - L\frac{\mathrm{d}i(t)}{\mathrm{d}t} = 0 \tag{6.1}$$

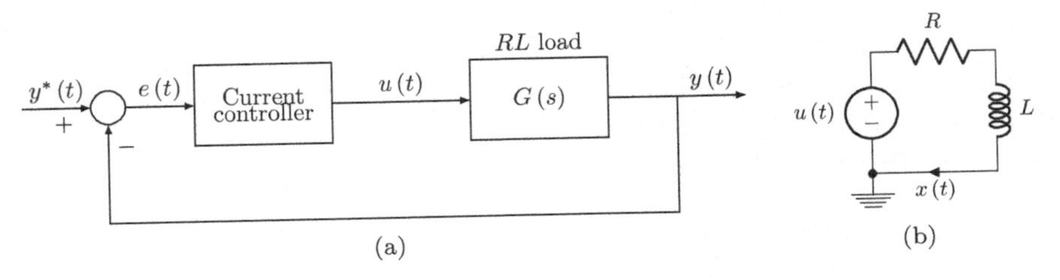

Figure 6.1 *RL*-circuit under investigation.

DOI: 10.1201/9781003196938-6

which is a differential equation describing a 1st order linear system. The voltage $v(t)$ is the controllable variable, i.e., $u(t) = v(t)$, which allows to increase/decrease the current flow according to the system parameters. Hence, such current control is based on the manipulation of the input voltage. By considering $x(t) = i(t)$ and $y(t) = x(t)$, the state-space representation shown in equations (5.1) and (5.2) is achieved considering $A = -R/L$, $B = 1/L$, and $C = 1$.

Then, the transfer function of the continuous-time linear system is defined as follows:

$$G(s) = \frac{i(t)}{v(t)} = \frac{y(s)}{u(s)} = \frac{1}{R + sL} = \frac{1/R}{1 + s\,(L/R)} = \frac{1/L}{s - (-R/L)} \tag{6.2}$$

where the time constant of the uncontrolled system is $\tau_G = L/R = 4\,\mathrm{s}$ and, thus, the settling time is $T_{a,G} = 5\tau_G = 20\,\mathrm{s}$. Note that τ_G and $T_{a,G}$ characterize the intrinsic response of such system (i.e., without the presence of any controller).

A PI controller is introduced. Its design depends on the choice of its performances, that is:

1. **The closed-loop settling time $\Longleftrightarrow$ bandwidth (cut-off frequency) ω_c**

   In this example, the controller should speed up the response of the resulting closed-loop system against reference transients (i.e., being faster than the uncontrolled system). According to the desired characteristics reported previously, it results $T_{a,F} = 1\,\mathrm{s} < T_{a,G} = 20\,\mathrm{s}$ and $\tau_F = T_{a,F}/5 = 0.2\,\mathrm{s} < \tau_G = 2\,\mathrm{s}$. Thus, the controlled system is going to result twenty times faster than the open-loop one.
   Remember: if a desired characteristic such as $T_{a,F}$ is not clearly specified, the choice of a suitable time response must be the first decision of the designer.

2. **The robustness level $\Longleftrightarrow$ phase margin ϕ_m**

According to the chosen design parameters, the *command-tracking performances* (i.e., how the resulting system is able to be fast and close to the given reference) or the *disturbance-rejection performances* (i.e., how the resulting system is able to cope the effects of a disturbance acting on it) may be emphasized. Note that, in general, high command-tracking performances results in poor disturbance-rejection one.

Given $R(s)$ of the same kind as that one reported in equation (5.8), the open-loop transfer function is:

$$L(s) = \frac{k_p(1 + sT_i)}{sT_i}\frac{1/R}{1 + s\,(L/R)} \overset{T_i = L/R}{\Longrightarrow} L(s) = \frac{k_p/R}{sT_i} \tag{6.3}$$

where a pole/zero cancellation is imposed by setting $T_i = \tau_G = L/R$. Since explicit formulas for k_p and k_i based on ω_c can be defined starting from the requirement $T_{a,F} = 1\,\text{s}$, it follows that $\tau_F = T_{a,F}/5 = 0.2\,\text{s}$, which leads to:

$$T_{a,F} \rightarrow \tau_F = \frac{T_{a,F}}{5} \rightarrow \omega_c = \frac{1}{\tau_F} = 5\ \text{rad/s} \rightarrow \begin{cases} k_p = \omega_c L & = 0.50\,\Omega \\ k_i = \omega_c R & = 0.125\,\Omega/\text{s} \end{cases} \tag{6.4}$$

Therefore, the computed k_p and k_i parameters ensure the command-tracking performances and, given $L(s)$ as a minimum phase system (mps), the phase margin ϕ_m is about 90° (i.e., high robustness). It must be noted that k_p and k_i have precise unit values, which in this case are Ω and Ω/s, respectively. The resulting closed-loop transfer function is:

$$F(s) = \frac{L(s)}{1 + L(s)} = \frac{1}{1 + sT_i} = \frac{1}{1 + s\,(L/R)} = \frac{1}{1 + s\,(k_p/k_i)} \tag{6.5}$$

At this point of the design procedure, all the characteristics of $F\,(s)$ are known. The next step is to numerically simulate the system.

6.1 Simulink® Simulation

In this book, the simulations of both uncontrolled and controlled system behaviors are carried out with MATLAB® and Simulink.®[1] This helps power electronics control engineers to gain an insight into the interaction of digital control algorithms and conversion systems. Simulink® allows to quickly model both the digital control algorithms and the analog circuits of a power electronic systems. According to the complexity and/or the requirements of the needed control scheme, the model can just focus on the system dynamics, i.e., focusing on the transfer functions and the transient response, or introduce more complex behavior at the component level, i.e., using detailed block representing resistors, inductors, capacitors, semiconductors, ... and their physical limitations, thus, covering a multi-domain modeling using built-in library blocks.

Simscape™ enables a quick modeling of physical systems within the Simulink® environment. In particular, Simscape™ Electrical™ (formerly SimPowerSystems™ and SimElectronics®) provides component libraries (e.g., semiconductors, motors, and converters) with the integration of mechanical,

[1]Note that Simulink® supports both floating-point and fixed-point computations, as well as continuous-time (analog), discrete-time (digital), hybrid (mixed-signal), and multi-rate systems.

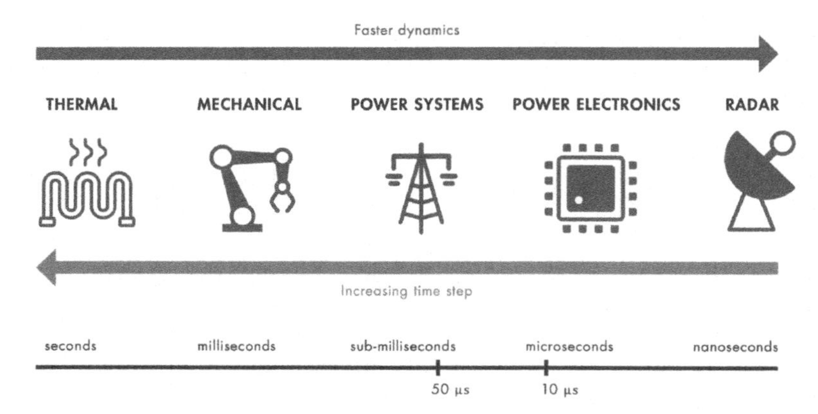

Figure 6.2 Relationship between system dynamics and time step [32].

thermal, and other physical aspects for modeling and simulating applications such as electromechanical actuation, drives, smart grids, and renewable energy systems. This book mainly refers to blocks from the *Simscape™ Electrical™ Specialized Power Systems* library, which can be found by clicking in the Simulink® library browser: Simscape → Electrical → Specialized Power Systems. For further information, the reader is referred to the MathWorks® website [30].

Using both standard blocks or Simscape™ components, analog (passive or active) and digital components can be modeled and simulated to validate the design of closed-loop control algorithms as well as the supervisory logic through real-time simulations.

Trade-off between *Simulation Speed* and *Model Complexity*:
The speed at which a controller must respond to ensure stable/safe operation of the system is a key point of Model-Based Design. Figure 6.2 shows a range of physical systems and an order of magnitude of their corresponding time constants; note that power electronics controllers are typically sampling in the order of µs or faster. In general, a model described by simple dynamics (e.g., transfer functions) will simulate really fast, though accuracy will not be maximized. Instead, a model with more complex dynamics (e.g., built in using elements from the Simscape™ library) will simulate slower, but will provide a more accurate description of the true power electronics behavior than the simplified representation. In the case of the considered *RL* load, the variable voltage source may be provided by a power converter stage. In practice, the latter returns a squared waveform output voltage, but it can be approximated by its average value. Therefore, the simulation will not capture the effects of switching behavior, but it will be suitable for tuning the control algorithms and testing the control under nominal operating conditions. The *RL* branch can be modeled as a set of linear differential equations with lumped parameters. Because it uses a simple mathematical representation of the

system behavior, the simulation will run fast. Assuming that the same system has to be installed in a real harsh environment (e.g., the power train of an electric vehicle) which is sensitive to temperature variations, the model should include a detailed power electronics switching behavior, effects of temperature on electrical efficiency, as well as the nonlinear model of the inductor, resulting to be mathematically more complex than before. As a result, the simulation will run really slow.

Before performing real-time simulations, the very first task is the initialization of all the needed parameters. In this book, this procedure is done using MATLAB® code by creating a script/source file (.m or m-file) that is called in Simulink® from the Model Properties → InitFcn callback (the .m file is executed every time the Simulink® model is running) or as PostLoadFcn callback (the .m file is executed only once). Nevertheless, the .m file could be even executed manually, thus, without creating such automatic routine in Simulink®.

Be careful: The book refers to one of these procedure in every exercise, since they allow easy initialization and edit of the simulation parameters, reducing the risk of manual change inside the blocks, which can lead to a difficult debug.

For the considered case study, the m-file which includes the code with all the definitions of the system parameters, the computation of the controller gains, and the transfer functions is the code reported here in the following:

```
%% define the Laplace operator
s       = tf('s');                 % then just call s
%% system parameters and transfer function:
R       = 0.025;                   % ohm
L       = 0.1;                     % H
tauG    = L/R;                     % ohm*s / ohm = s
wg      = sqrt(1/(R^2)-1)/tauG;    % rad/s
TaG     = 5*tauG;                  % s
G       = 1/(R + s*L);             % G(s)
%% closed-loop requirements
TaF     = 1;                       % s
tauF    = TaF/5;                   % s
wc      = 1/tauF;                  % rad/s (approximation)
%% PI parameters
% pole/zero cancellation procedure
kp      = wc * L;                  % ohm
ki      = wc * R;                  % ohm/s
Ti      = kp / ki;                 % s - equal to L/R
Rc      = kp + ki/s;               % R(s) - PI controller tf
Ts      = 1e-3;                    % sample time/discretization
%% transfer functions
Lc      = Rc*G;                    % L(s) - open-loop tf
```

```
Fc      = Lc/(1+Lc);                   % F(s) - closed-loop tf
%% voltage constraint
sat     = 0.5;                         % power supply limit
```

Then, it is enough to recall the variable names into a specified block in Simulink® to automatically assign the defined values, e.g., if R is used inside a Gain block, the value 0.025 is automatically updated. Given the parameters shown previously, the corresponding bode diagram of $G(s)$, $L(s)$, and $F(s)$ are reported in Figure 6.3, together with time constants and angular frequencies indications. Note that, $F(s)$ behaves exactly as a low pass filter as expected.

In order to show how to use both approaches, i.e., using standard or Simscape™ blocks, a Simulink® scheme is created such that:

- The controller is implemented using the standard block *PID Controller* from the library Continuous;
- The RL load is emulated through a transfer function (tf), i.e., a standard block, or building an equivalent circuit with blocks available in the Simscape™ library. Thus, the user has the possibility to select which one of the two modeling strategies should be included in simulation and then run it.

The full scheme is shown in Figure 6.4. The possibility to switch between the two load emulations is ensured by the manual switch `Sw2` which is a standard block from the Simulink® library. Every double-click done on the block changes its connection. Note that Figure 6.4 includes a further stage enabled through `Sw1`, which can add a saturation block (standard block) to bound the actuation signal $u(t)$. This will be discussed in details later in Section 6.2 referring to the windup effects. Thus, it is anticipated in Figure 6.4 just for

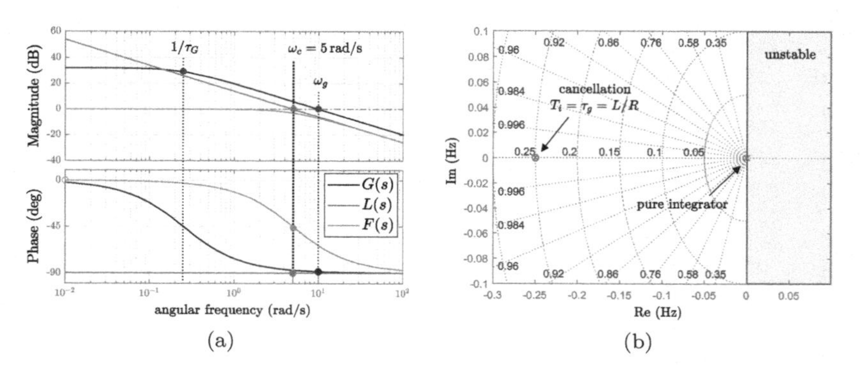

Figure 6.3 Bode diagrams for the RL-load case study.

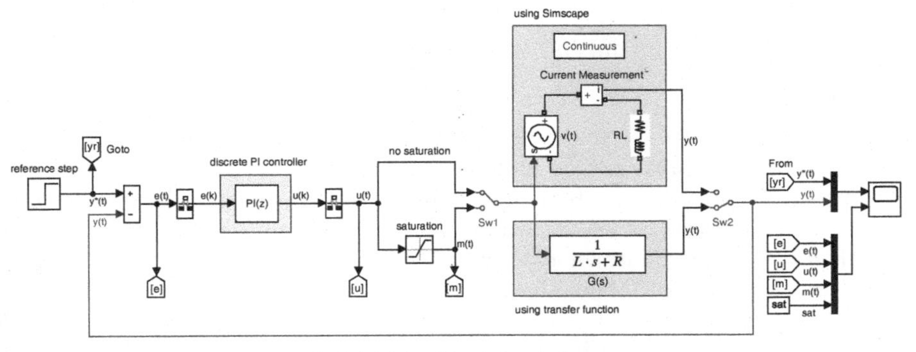

Figure 6.4 Simulink® scheme of a current control for an RL load using both standard and Simscape™ blocks. Rate Transition blocks are inserted to mimic the behavior of the A/D and D/A converters (see Figure 5.6 (b)).

the sake of completeness.

Remark: by using both standard or Simscape™ blocks to emulate the RL load, the control performances and the *averaged* system behavior are exactly the same. Thus, the user could equally adopt one of the two to evaluate the controller design, while the Simscape™ approach is preferable to emulate a more *accurate* behavior of the system at the component level.

The Simulink® file (`.slx`) related to Figure 6.4 can be downloaded from the editor web page, while the file settings are further detailed in the following sections.

6.1.1 Use of standard blocks (continuous/discrete)

Referring to the lower blue area reported in Figure 6.4, the closed-loop control of the RL load is created by using a **Transfer Fcn** block, which includes the transfer function $G(s)$, whereas a **PID Controller** block inside the pink region contains the parameters k_p and k_i of $R(s)$. More specifically:

- **Transfer Fcn block** (Simulink®)
  This block models a linear system through a transfer function in the Laplace-domain variable s. It can model single-input single-output (SISO) and single-input multiple-output (SIMO) systems. A generic transfer function $G(s)$ must be entered in the form:

$$G(s) = \frac{\text{num}(1)s^{n-1} + \text{num}(2)s^{n-2} + \cdots + \text{num}(n)}{\text{den}(1)s^{m-1} + \text{den}(2)s^{m-2} + \cdots + \text{den}(m)} \tag{6.6}$$

where num and den are matrices that contain the coefficients of the numerator and denominator ordered in descending powers of s with order n and m, respectively. For SISO systems (as in this example), these scalar coefficients have to be entered as vectors: numerator coefficients: $[1] \rightarrow \text{num}(1) = 1$ with $n = 1$; denominator coefficients: $[L\ R] \rightarrow \text{den}(1)s = L\,s$, $\text{den}(2) = R$ with $m = 2$; zero initial condition.

- **PID Controller block** (Simulink®)
  The PID Controller block can implement PID, PI, PD, P only, or I only controllers. The block can be specified for continuous-time or discrete-time domain operations. In both cases, the block output is, in general, a weighted sum of the input signal (P action), its integral (I action) and its derivative (D action). The block supports several controller types and structures. Configurable options in the block include:

  - Controller type (PID, PI, PD, P, I);
  - Controller form (serial or parallel);
  - Time domain (continuous or discrete);
  - Initial conditions and reset trigger;
  - Output saturation limits and built-in anti-windup mechanism;
  - Signal tracking for bump-less control transfer and multi-loop control.

  The internal structure of the block changes by activating different variant subsystems depending on how those options are edited.
  According to the requirements of the considered exercise, PI should be selected as Controller type. Then, the parameters should be set considering the correspondence P $= k_p$ and I $= k_i$, which are the controller parameters designed in continuous-time domain. However, this exercise foresees the implementation of a *discrete-time PI controller* as a final goal. Thus, the Discrete-time option in the Time-domain settings must be selected with Sample time T_s. All the other settings should be kept with their default values.

- **Step and Goto/From blocks** (Simulink®)
  The Step block provides a step between two definable levels at a specified time. When the simulation time is less than the Step time, the block output is the set equal to the Initial value. Otherwise, the output is the set on the Final value.
  Instead, the Goto block passes its input to its corresponding From block without any physical connection. Thus, their usage is manly targeted to achieve a more readable scheme without too many connection lines. The

input can be a real- or complex-valued signal or a vector of signals of any data type. A Goto block can pass its input signal to more than one From block, although a From block can receive a signal from one Goto block only. Goto/From blocks are matched through univocal tags/label.

Note that Rate Transition blocks (standard blocks) are used to interface the continuous-time scheme with the discrete-one (i.e. A/D and D/A). They are set either with Sample Time T_s or -1.

Case study settings:
Start time: `0(s)`; Stop time: `4(s)`; Solver type: `Variable-step`; Solver: `ode23t`; Solver details: `default`. This is a first example of simulation in which the main focus is the understanding of the closed-loop dynamics. Therefore, a `Variable-step` solver is used. Neverthless, in the next chapters a `Fixed-step` solver is preferred, being closer to operations in a embedded environment.

6.1.2 Use of Simscape™ (specialized power systems)

Referring to the upper blue area shown in Figure 6.4, the closed-loop control of the RL load is now created combining the PID Controller block (i.e., using a standard block) and the Simscape™ blocks, which are used to model the electrical system in a more physical manner. Specifically, all these new components are described here in the following:

- **Controlled Voltage Source block** (Simscape™— Specialized Power Systems)
  The Controlled Voltage Source block converts a Simulink® input signal into an equivalent voltage source. The generated voltage is driven by the input signal of the block which is associated to a control signal (e.g., coming from a control loop). Positive/negative terminals are associated to electrical ports. Thus, voltage signals cannot be directly connected to standard Simulink® block, e.g., scope, logic operators, gains. In this case, the Source type is set on DC, the Initial amplitude on 0, whereas no measurements are foreseen.

- **Series RLC Branch block** (Simscape™— Specialized Power Systems)
  The Series RLC Branch block implements an impedance as a series combination of R, L, C elements. At the specified frequency, the load exhibits a constant impedance value. By double clicking on the block, it is possible to select which element is to be used, e.g. if the C it is not used we select the RL only. According to the aforementioned design, the adopted settings are Branch type: RL, then Resistance (Ohms): R and Inductance (H): L. Do not select Set the initial inductor current checkbox and it is up to the reader set any measurement.

- **Powergui block** (Simscape™— Specialized Power Systems)
  The powergui block is necessary to simulate any Simulink® model containing Simscape™ Electrical™ Specialized Power Systems blocks. It stores the

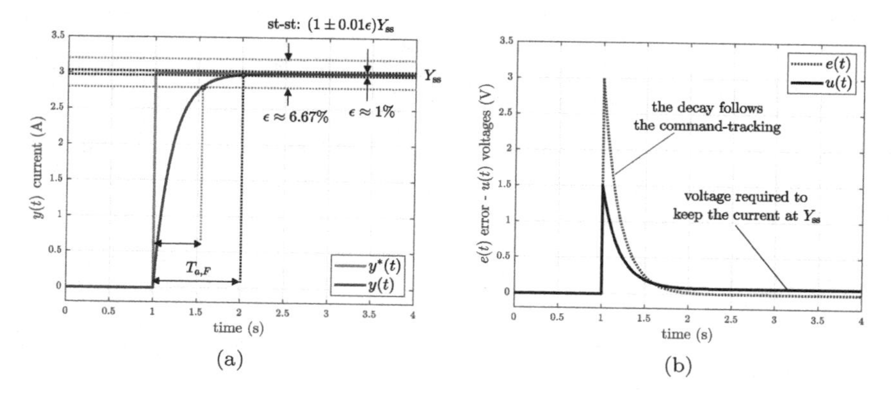

Figure 6.5 Performances of the current control of an *RL* load.

equivalent Simulink® circuit that represents the state-space equations of the resulting scheme and it allows to choose the solver for this circuit (which could be different from that one used for the reaming part of the scheme based on standard blocks). The available solvers are: Continuous, which uses a variable-step size; Discrete, which uses fixed time steps; Phasor. The powergui block also provides tools for steady-state analysis and for advanced parameter design. For this exercise, the recommended Simulation type is Continuous. No other settings should be edited.

In addition, a Current Measurement block must be inserted in the scheme to measure the instantaneous current flowing in the electrical connection line. Its output provides a data signal that can be used by other Simulink® blocks, thus, moving from Ampere to a numerical data (e.g., 1 A is different to numerical 1).

Generally speaking, the stator windings in electrical machines behave as *RL* loads plus the effects of back-emfs. Therefore, the control design proposed in this chapter directly links to the torque regulations for electrical machines reported in the next exercises.

6.1.3 Controller performances

Given the design and the settings of the aforementioned case study, the corresponding closed-loop scheme is simulated in MATLAB® and Simulink®. The resulting performances are shown in Figure 6.5. Figure 6.5 (a) shows the current flowing into the winding $y(t) = i(t)$ (blue line), which is correctly following the step-wise reference $y^*(t) = i^*(t)$ (red line). The set point changes from 0 A to 3 A (i.e., $Y_{ss} = 3$) at $t = 1$ s. Different $T_{a,F}$ could be defined according to the value of the error ϵ. Namely, Figure 6.5 (a) shows two different $T_{a,F}$ computed as the time interval to reach: 2.79 A for $\epsilon \approx 6.67\%$ (green dotted line) with $T_{a,F} \approx 0.6\,\text{s}$; 2.97 A for $\epsilon \approx 1\%$ (black dotted line) with $T_{a,F} \approx 1\,\text{s}$. Therefore, the definition of the settling time in terms of accuracy is necessary to classify the performances of the systems and to compare them, e.g., the comparison among two systems must refer to the same degree of accuracy to understand which behaves better. From now, the proposed procedure refers to the design guidelines reported in Chapter 5, in which $\epsilon \approx 1\%$ is suggested as reference accuracy, thus, entering in the region $[(1-0.01\epsilon)Y_{\text{ss}}\ (1+0.01\epsilon)Y_{\text{ss}}]$, where $T_a = 4 \div 5\tau$ holds. It results that $T_{a,F} = 1\,\text{s} < T_{a,G} = 20\,\text{s}$ and $\tau_F = T_{a,F}/5 = 0.2\,\text{s} < \tau_G = 2\,\text{s}$, so the closed-loop control system is going to be ten times faster than the open-loop one.

The effectiveness to achieve the desired steady-state value Y_{ss} can be verified via the final value theorem. Since the Laplace transform of a step-wise signal in time-domain is μ_s/s, where μ_s is the amplitude value (i.e., 3 A), by multiplying $F(s) = y(s)/y^*(s)$ with μ_s/s it results $Y_{\text{ss}} = F(s)\mu_s/s$. Namely:

$$\lim_{s\to 0} sY_{\text{ss}} = \lim_{s\to 0} sF(s)\frac{\mu_s}{s} = \lim_{s\to 0}\frac{\mu_s}{1+sT_i} = \mu_s = 3\,\text{A} \tag{6.7}$$

Note that the results holds for every considered value of T_i, that is, for every design choice.

Based on Figure 6.5 (b), the command-tracking performances are achieved thanks to the sudden change of the input voltage $u(t) = v(t)$ at $t = 1\,\text{s}$. In particular, the error $e(t)$ and the control variable $u(t)$ are following a decay which is equal (but mirrored) with respect to $y(t)$, as shown in Figure 6.5 (a). At steady-state, $e(t) = 0$ while $u(t) \neq 0$ since the system changed its operating point and a certain voltage is necessary to assure the required current flow.[2] This is still a consequence of the Kirchhoff's law, where at steady-state the current derivative can be neglected while computing the voltage value,[3] that is:

$$v(t) - Ri(t) - L\frac{\mathrm{d}i(t)}{\mathrm{d}t} = 0 \quad \rightarrow \quad U_{\text{ss}} = V_{\text{ss}} = RI_{\text{ss}} = RY_{\text{ss}} = 0.075\,\text{V} \tag{6.8}$$

Given the aforementioned parameters and performances, the designed PI controller achieves all the specifications.

After this first design, the controller parameters are modified to evaluate the potential issues due to a wrong design. Two scenarios are considered and the performances of the new controllers are compared to the previous one (the correctly designed one). The plots shown in Figure 6.6 reports the behaviors of all the design controllers.

[2] Note that a voltage drop is always present among the inductor (and its internal series resistance) and the resistance.

[3] The steady state quantities are denoted as uppercase letter with subscript ss.

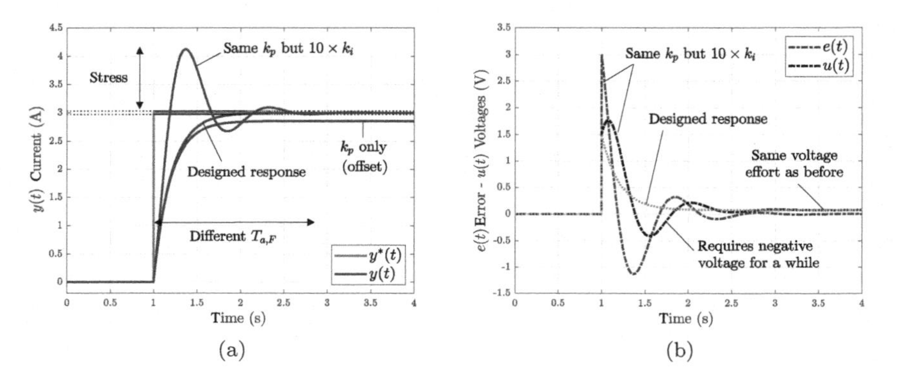

Figure 6.6 Performances of different current controllers.

More specifically, the two new controller are designed such that:

1. Integral action is removed, i.e., $k_i = 0$.

   It is worth noting that a simple P controller keeps a permanent offset at steady-state. The k_p action contributes to quickly increase the voltage only, bringing the output close to the current reference. The distance between $y(t)$ and $y^*(t)$ depends on the value of k_p. In case of a P controller, no pole/zero cancellation is possible. The open-loop and closed-loop transfer functions simply result:

$$L(s) = k_p \frac{1}{R + sL} \quad F(s) = \frac{k_p}{k_p + R + sL} \tag{6.9}$$

   Then, the cutoff frequency can be computed by imposing $|L(s)|_{s=j\omega_c} = 1$. It can be easily proved that this equality leads to $\omega_c = \sqrt{k_p^2 - R^2}/L$. If the considered bandwidth is again $\omega_c = 5\,\text{rad/s}$, the resulting k_p is obviously different. Moreover, the application of the final value theorem on the controlled system leads to:

$$\lim_{s \to 0} sF(s)\frac{\mu_s}{s} = \lim_{s \to 0} \frac{k_p \mu_s}{k_p + R + sL} = \frac{k_p \mu_s}{k_p + R} \tag{6.10}$$

   where the result is k_p-dependent. An increase in k_p might improve the reference tracking, while the offset persists since it could be zeroed only if $k_p \to \infty$. This also implies that a lower voltage effort $u(t)$ than before might be used to keep a (wrong) steady-state value $i(t)$.

2. k_p is kept equal to that one of the previous controller, while k_i is chosen ten time larger, i.e., $10 \times k_i$.

By increasing the integral part so significantly, the current response shows an oscillating behavior. Figure 6.6 (a) reports that, by changing the integral action, a faster transient response might be achieved at the cost of an higher stress on the system due to the overshoots. As an example, if the circuit cables see their insulation designed for a specific current peak and a certain temperature limit/heat (usually provided as maximum values of the cable class and diameter), the repetitive current overshoots could get them closer to those limits. Plastic insulation and jackets that surround the wires are usually rated to withstand up to $90 - 100\,°\mathrm{C}$, but temperatures that approach this limit are not recommended anyway. Thus, oscillating response of $i(t)$ might be not acceptable for this reason.

Hence, any physical constraints must be considered at the early design stage. Even if the transient response provided by this new controller is faster than before, the settling time is longer due to the oscillating behavior of $i\,(t)$, thus, leading to a late entry in the safety region $\epsilon = \pm 1\%$.

Oscillations also may imply a further stress on the power supply. Both $e(t)$ and $u(t)$ become negative for certain periods of time. Therefore, the energy is also flowing back to the power supply (which provides the actual value of DC voltage $u(t)$) and not only toward the RL load. This reversed power flow is only possible if the power supply is bidirectional, i.e., if it is able both to receive and dissipate/store energy, otherwise, another physical constraint must be considered during the controller design.

Even in this case, the final value theorem ensures $Y_{ss} = 3\,\mathrm{A}$.

6.2 Derive an Anti-Windup PI Controller Scheme

A linear controller (e.g., PI) is simple to design, but its performances are good as long as the dynamics remain close to what is foreseen by the linear theory. Nevertheless, real systems are affected by non-linearities and physical limitations. For instance, electrical drives suffer of magnetizing phenomena, skin effects, cable length issues, both electro-thermal and electromagnetic interference. In addition, power supplies are limited. In particular, these latter are among the easiest cases to treat since they represent static non-linearities. By considering the control variable $u(t)$ bounded by the energy flow capabilities of a power supply, it follows that:

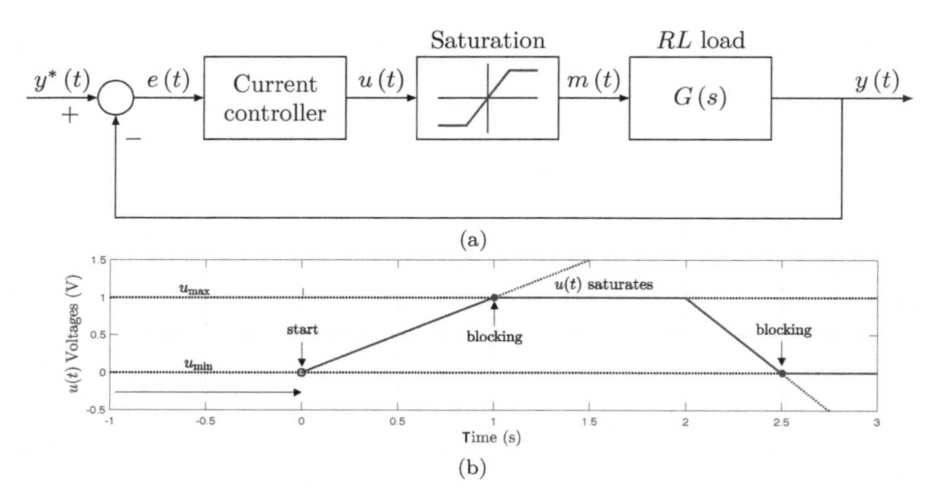

Figure 6.7 Closed-loop control scheme with saturated actuation (a) and example of saturated output (b).

$$m(t) = \text{sat}\,(u(t)) \begin{cases} u_{\min} & \text{if} \quad u_{\min} \geq u(t) \\ u(t) & \text{if} \quad u_{\min} < u(t) < u_{\max} \quad \text{(linear region)} \\ u_{\max} & \text{if} \quad u(t) \geq u_{\max} \end{cases} \tag{6.11}$$

where $m(t)$ is the actuation variable. Since the control variable is a voltage level, $u_{\min}$ and $u_{\max}$ are minimum and maximum allowed voltages that the DC power supply can output, that is, $v_{\min}$ and $v_{\max}$. For this case study, a (fictitious) unidirectional DC power supply which is able to provide a variable voltage from $u_{\min} = 0\,\text{V}$ to $u_{\max} = 0.5\,\text{V}$ is assumed to be available. If these limitations are introduced in the proposed model as reported in equation (6.11), the closed-loop scheme changes as shown in Figure 6.7 (a). The saturation block receives $u_{\min}$ as input and it outputs $u_{\max}$. There is a Simulink® block that can mimic this behavior among the standard ones under the name `saturation`. From now on, a discrimination is introduced between:

- $u(t)$, that is, the control variable returned by $R(s)$ as results of the control design (i.e., output of the PI controller);
- $m(t)$, that is, the actuation variable seen by the *RL* load, where $m(t)$ could saturate depending on the value assumed by $u(t)$.

From the controller perspective, note that the previous design of $R(s)$ does not include any information regarding the saturation of $u(t)$.

Indeed, Figure 6.8 (a) reports the transient response of the previous control scheme without saturation (blue dotted line), the related command-tracking

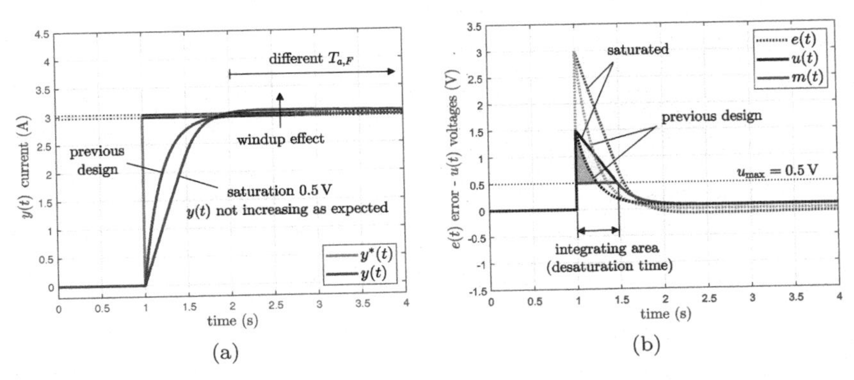

Figure 6.8 Comparison between unsaturated and saturated transient responses.

performances and the evolution of the same system including saturation effects (blue solid line). For this latter, the output $y(t)$ takes long to reach the steady-state condition due to the *windup* effect of the integrator, which also generates an overshoot.

Moving to Figure 6.8 (b), since $R(s)$ operates regardless any saturation limit, the voltage is required to be $u(t) = 1.5\,\text{V}$ at time instance $t = 1\,\text{s}$ to bring $y(t)$ closer to $y^*(t) = Y_{ss} = 3\,\text{A}$. However, the DC power supply is able to feed only $m(t) = u_{\max} = 0.5\,\text{V}$. Thus, the system is fed by this maximum voltage value. At the following time instant, the output $y(t)$ do not increase as expected due to a lower actuation. This limitation implies a larger error $e(t)$ compared to the unsaturated case, which shows an higher $y(t)$ due to an higher actuation signal (which has a peak value exactly equal to $u(t)$). The effects of this saturation on $u(t)$ and $y(t)$ are more clear as long as $u(t)$ returns below $u_{\max}$, which enables the system to re-enter into the linear region. Nevertheless, the effects due to the presence of a large error are still present. Thus, even if $m(t) = u(t)$, the system requires a long time to reach the steady state and it shows an overshoot as well.

This is a well-known problem which has a really practical explanation. More specifically, the windup effect is mainly due to the integrator part of the PI controller, which for $u(t) > u_{\max}$ keeps integrating the tracking error $e(t)$ producing $m(t) \neq u(t)$. When $m(t) = u(t)$, the controller re-enters into the linear region behavior. This means to wait for the integrator discharge (also called integral discharge time or de-saturation time), which depends on the integration area defined by the unsaturated $u(t)$ (yellow area in Figure 6.8 (b)).

This effect is particularly clear in Figure 6.9, which includes a further simulation of a system with a stricter voltage limitation, i.e., $u_{\max} = 0.3\,\text{V}$.

The larger the integrating area, the longer and the stronger the windup effect. So, the case with saturation set at 0.3 V implies a longer windup and

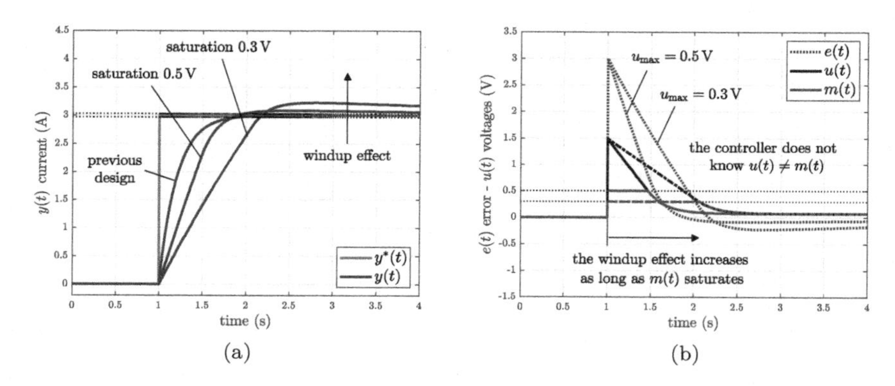

Figure 6.9 Comparison between different transient responses and the related windup effects.

a quite larger settling time $T_{a,F}$, keeping $\epsilon = 1\%$. To mitigate the windup problem, the controller must be augmented according to one of the many anti-windup schemes available in the literature. In this book, the back-calculation principles are considered, which lead the compensation schemes reported in Figure 6.10.

By exploiting the block schemes theory, it can be proven that the diagrams reported in Figure 6.10 (a) and (b) are equivalent, creating both an anti-windup compensation. The back-calculation method uses a feedback loop to discharge the integrator when the controller hits specified saturation limits, avoiding to enter in nonlinear operations. The back-calculation coefficient k_{aw} is computed as $k_{aw} = 1/T_i = k_i/k_p$. This time, constant T_i determines how quickly the integrator of the PI controller is reset by the anti-windup loop. The regulator reported in Figure 6.10 (a) is subjected to the continuous-time equations reported here in the following:

$$u(t) = \text{sat}\left(p(t)\right) \begin{cases} u_{\min} & \text{if} \quad u_{\min} \geq p(t) \\ k_p e(t) + \frac{1}{1+sT_i} u(t) & \text{if} \quad u_{\min} < p(t) < u_{\max} \\ u_{\max} & \text{if} \quad p(t) \geq u_{\max} \end{cases} \tag{6.12}$$

where $p(t)$ is an auxiliary variable used to keep the notation short. Therefore, the key aspects of an anti-windup scheme can be summarized as:

- No effect when the actuator is not saturating, i.e., within the linear region controllers with and without anti-windup scheme behave the same;
- Just one additional parameter $k_{aw} = 1/T_i$ is required;
- If the saturation limits are not clearly specified, a mathematical model of the actuator should be used.

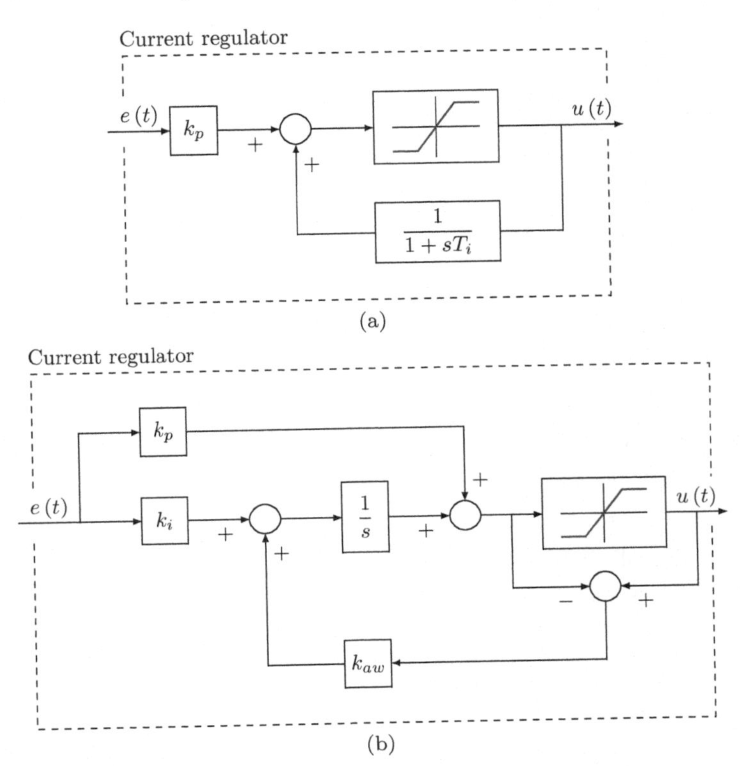

Figure 6.10 Example of back-calculation anti-windup scheme to update $R(s)$.

Both schemes reported in Figure 6.10 share the idea to include the saturation limits as thresholds inside a suitable saturation model. This latter may be just implemented by looking at the data sheet of the DC power supply or derived by using some identification techniques, especially in case where those limits are uncertain. Note that, in this case, it is quite easy to define the saturation limit. However, this is not always the case, e.g., accurate describe magnetizing phenomena or skin effects are not always available.

Simulink®: to include a PI controller with anti-windup configuration in simulation, insert a PID block keeping the settings shown previously and click on the Output Saturation tab in the block dialog. There, select Limit output and enter the saturation limits. Then, select back-calculation from the Anti-windup method menu and specify the Back-calculation coefficient Kb equal to k_{aw}. Hence, it is only necessary to change few settings into the PID block to use an anti-windup scheme, which makes their implementation quite simple even for the implementation into MCU proposed in the following Chapters.

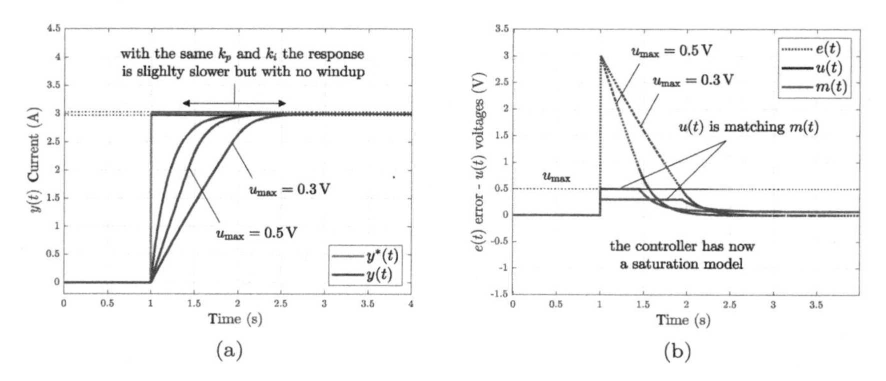

Figure 6.11 Performances of the current control of an RL load including anti-windup compensations.

The results of simulations including anti-windup strategies are summarized in Figure 6.11. In practice, the transient response reported in Figure 6.11 (a) do not present any overshoot. This is true even if stricter limitation are introduced, like $u_{\text{max}} = 0.3\,\text{V}$. Looking at Figure 6.11 (b), it can be noted that $u(t) = m(t)$ holds for all simulations, i.e., the controller "knows" the saturation limits. Nevertheless, the settling time is now longer than $T_{a,F} = 1\,\text{s}$, which was computed with the previous (unsaturated) design. This is obviously a consequence due to the limit introduced on $u(t)$ (limited actuation/limited energy) which cloud not make $i(t)$ increasing as expected from the previous design. Therefore, the transient takes longer to reach the steady-state condition, while assuring no overshoots. In case none of them is foreseen, the $T_{a,F}$ difference is negligible. However, if the introduction of an anti-windup scheme lead to violations in the settling time constraint, the design parameters k_p, k_i might be also updated accordingly.

It must be remembered that if saturation phenomena are neglected in the design phase, they can lead to closed-loop instability, especially if the process is open-loop unstable.

6.3 Design Summary

The procedure to analyze a system, design a closed-loop controller, and make it suitable to be implemented into a microcontroller can be summarized through the following key points:

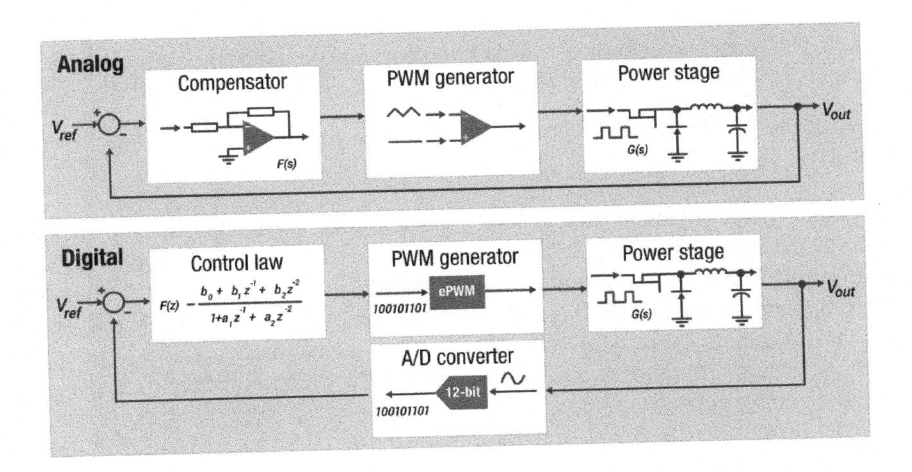

Figure 6.12 Management of a power control topology using integrated circuits vs discrete analog components. The term digital is used to denote that all external signals are converted from their native analog state to a digital state such they can be operated on by the main controller in the system [8].

- Analyze and describe the uncontrolled system dynamics/natural dynamics.

  This is necessary to understand the order of $G(s)$ as well as defining τ_G;

- Select the output $y(s)$ and input/control $u(s)$ variables.

  Given $G(s)$, then it is possible to define a closed-loop control scheme;

- Select the controller type in continuous-time domain.

  In this book, linear controllers are considered only, such as the PI ones. The Laplace operator is exploited to derive $R(s)$;

- Consider all the requirements to compute the controller parameters.

  Once the problem requirements are set, use one of the criteria shown in Chapter 5 to compute k_p and k_i;

- Discretize the controller.

  A discretization time T_s must be chosen as well as a suitable integration method according to the location of the system poles to map $R(s) \leftarrow R(z)$. If necessary, the system transfer function can be discretized too;

- Check if relevant limitations are present in the system.

  For instances, it should be verified possible actuation limits due to real devices capabilities. Then, saturation should be included in the control scheme;

- If necessary, adjust the controller with an anti-windup scheme.

  Each of the methods shown in this Chapter can be used to eliminate windup effects which may lead to damage or aging of system components;

- Select which elements of the control scheme have to be implemented on the MCU.

Note that the controller design has been started in continuous-time toward discrete-time domain. This is necessary to meet the working principle of embedded platforms, i.e., systems with discrete nature. Nowadays, this technology represents the main industrial standard for power electronic-based applications (and many more) instead of using the more historical analog control systems. Their difference is shown in Figure 6.12 which consider a closed-loop voltage control scheme.

The proposed discrete-time domain approach led to a Simulink®-based scheme suitable for the automatic code generation features. More details on this topic are reported in the upcoming chapters.

7

Manipulate the Variables Format: Data Types

The main objective of this book is to empower the reader with the ability to work and coding with standard MCU. As already discussed in the previous sections, due the discrete/digital nature of embedded systems, tuning and simulations of power electronic setups are carried out in the discrete-time domain. Nevertheless, real implementations of control algorithms could not be possible without considering even the hardware structure of a MCU. Clock frequency, maximum sampling time, memory storage, peripherals resolution, computational capabilities are just some examples of hardware-based requirements and specifications. In particular, the computational capability of a MCU is defined by the nature of the variable which may be involved in the computations and the type of operations they are asked to perform. The underlying concept is also called data-type analysis.

7.1 Fixed Point vs Floating Point Representation

The TMS320F28069M microprocessor adopted on the LAUNCHXL-F28069M operates with 32 bit word length, for which numerical representations can be either in *fixed point* or *floating point* format.

Fixed point: fixed point format exploits the two's complement representation to allow the computations between rational numbers. The positive and negative extreme values differ by one least-significant bit (LSB). Keeping a 32 bit word length, the bit format is:

$$\underbrace{\mathrm{SI}}_{\text{2 bit integer}}\;.\underbrace{ffffffffffffffffffffffffffffff}_{\text{30 bit quotient(fraction)}} \tag{7.1}$$

which is also called 2Q.30 (or just IQ30), where the 30 "f" represent the number to the right of the radix point.

Instead, the two remaining bits on the left side of the radix point represent the integer part of the number. Namely "S" is the sign bit, where 0 is allocated for positive numbers and 1 for negative ones, and one single bit "I". Hence, the maximum positive value of the integer part of such numbers is decimal

DOI: 10.1201/9781003196938-7

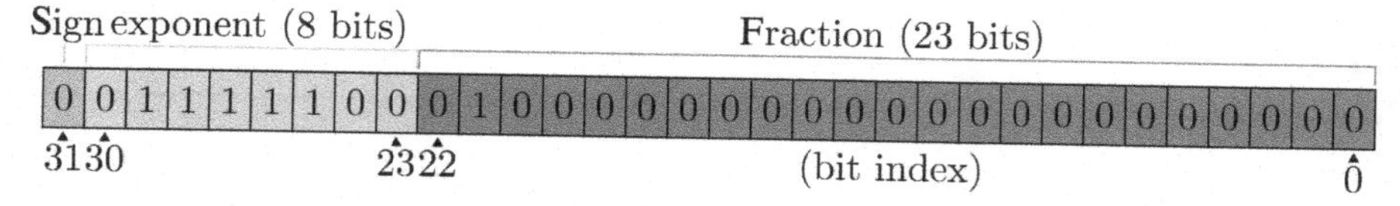

Figure 7.1 General representation of a floating point format number.

"1" (i.e., integer bits $S = 0, I = 1$) and the minimum decimal value is "-2" (i.e., integer bits $S = 1, I = 0$).

Floating point: The bit format for floating-point numbers is

$$\mathrm{S}\ \underbrace{eeeeeeee}_{\text{8 bit exponent}}\ .\underbrace{fffffffffffffffffffffff}_{\text{23 bit quotient(fraction)}} \tag{7.2}$$

which refers to the structure reported in Figure 7.1.

Floating point format works with a sign-magnitude representation. The sign of the number is represented the bit S, where 0 identifies positive numbers and 1 negative numbers. The 8 "e" bits represent the exponent of the floating point magnitude, in the range $-127\ldots126$. Instead, the quotient or fraction contains 23 "f" bits which represent the digits at the right of the radix point.

Hence, the real value assumed by a given 32-bit data with a given biased sign, exponent e, and a 23-bit fraction is computed as:

$$(-1)^{b_{31}} \times 2^{(b_{30}b_{29}\ldots b_{23})_2-127} \times (1.b_{22}b_{21}\ldots b_0)_2 \tag{7.3}$$

which yields to the following compact form:

$$(-1)^{\mathrm{S}} \times 2^{e-127} \times \left(1 + \sum_{i=1}^{23} b_{23-i}2^{-i}\right) \tag{7.4}$$

The 32 bit floating point representation allows to reach $\pm 2^{(255-127)}(1-2^{-23}) \approx \pm 3.4028 \cdot 10^{38}$. Thus, numbers within a very wide given range can be defined. Note that the sign-magnitude representation adopted in the floating point format defines a range within $\pm$ the maximum magnitude.

Example

As an example, Figure 7.2 shows how the value $x = 0.15625$ is represented in floating point format. Namely, x is split in terms of bits as follows:

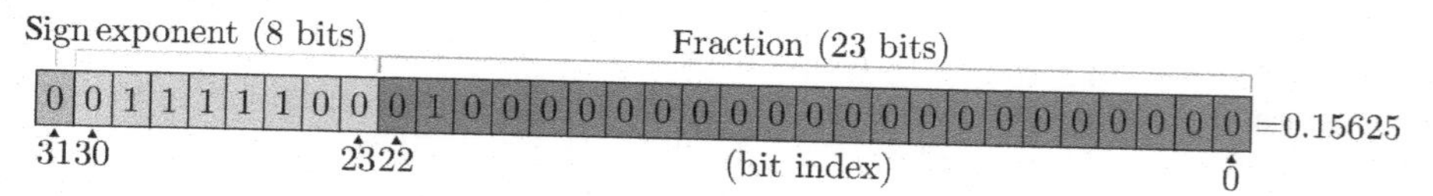

Figure 7.2 Representation of $x = 0.15625$ in floating point format.

Table 7.1 Comparison of 32 bit floating and fixed point representations: ranges and resolutions.

	Floating point	**Fixed point (IQ30)**
Max. decimal value	$\pm 2^{(255-127)}(1-2^{-23}) \approx 3.4028 \cdot 10^{38}$	$(2-2^{-30}) \approx 1.9999999991$
Min. decimal value	$\pm 1.1755 \cdot 10^{-38}$	-2.000000000
Relative resolution	$2^{-24} \approx 5.9605 \cdot 10^{-8}$	$2^{-32} \approx 2.3283 \cdot 10^{-10}$
Absolute resolution	$2^{-24} \cdot 2^{-127}$	$2^{-30} \approx 9.3132 \cdot 10^{-10}$

- Sign: $S = b_{31} = 0$
  $(-1)^S = (-1)^0 = +1 \in \{-1, +1\}$
- Exponent: $e = (b_{30}b_{29} \ldots b_{23})_2 = \sum_{i=0}^{7} b_{23+i}2^{+i} = 124 \in \{1, \ldots, (2^8-1)-1\} = \{1, \ldots, 254\}$
- $2^{(e-127)} = 2^{124-127} = 2^{-3} \in \{2^{-126}, \ldots, 2^{127}\}$
- Fraction: $1.b_{22}b_{21} \ldots b_0 = 1 + \sum_{i=1}^{23} b_{23-i}2^{-i} = 1 + 1 \times 2^{-2} = 1.25 \in \{1, 1 + 2^{-23}, \ldots, 2 - 2^{-23}\} \subset [1; 2 - 2^{-23}] \subset [1; 2)$

Thus:

$$x = (+1) \times 1.25 \times 2^{-3} = +0.15625$$

Note that:

- $1 + 2^{-23} \approx 1.000\,000\,119$
- $2 - 2^{-23} \approx 1.999\,999\,881$
- $2^{-126} \approx 1.175\,494\,35 \times 10^{-38}$
- $2^{+127} \approx 1.701\,411\,83 \times 10^{+38}$.

To the aim of providing a better understanding of the floating and fixed point representations, they are compared in Table 7.1.

The floating point format range is much wider than the fixed point one, but this advantage comes with the penalty of a less favorable relative resolution, for which the eight bits used for the exponent are not available for resolution.

For both floating and fixed point formats, the radix point can be moved to any location within the 32-bit word to manage the trade off between range and resolution.

Table 7.2 `Double` data type.

Bits	Usage
63	Sign (0 =positive, 1 =negative)
62 to 52	exponent, biased by 1023
51 to 0	Fraction f of the number $1.f$

Remark

A signed 32-bit integer variable has a maximum value of $2^{31} - 1 = 2147483647$, whereas a 32-bit base-2, floating-point variable can reach numbers up to $(2 - 2^{-23}) \times 2^{127} \approx 3.402823 \times 10^{38}$.

It is important to note that all integers with six or fewer significant digits as well as any number that can be written as 2^n (with $n \in [-126, 127]$) can be converted into floating-point values without any loss of precision.

7.2 Single vs Double Precision

As already mentioned in the previous section, a floating point representation can describe a wider range of numbers than a fixed point one, if the same bit width is considered at the cost of a loss in precision. Even if the previous section refers to 32-bit long words, when high accuracy is required, a *double precision* format could be used by enlarging the length of the representation reported previously up to 64 bit. More precisely, in the IEEE 754 (1985) standard [1], the 32-bit base-2 format is officially called `single`, while the additional floating-point format of 64-bit base-2 is denoted as `double`.[1] Note that one of the first programming languages to provide single-and double-precision floating-point data types was Fortran. Nevertheless, depending on the software in use, the command name to identify those data types may change.

Both MATLAB® and Simulink® represent numbers in floating-point notation unless otherwise stated (i.e., explicitly declared). By default, the data type is double precision, but it is possible to convert any number in single precision (or other data types) with a simple conversion function. According to the IEEE Standard, any value stored as a double requires 64 bits formatted as shown in Table 7.2.

The values that can be represented with `double` numbers lay in the negative range $[-1.79769 \times 10^{+308}, -2.22507 \times 10^{-308}]$ and in the positive range $[2.22507 \times 10^{-308}, 1.79769 \times 10^{+308}]$.

[1]Note that a base-10 representations was introduced more recently.

Table 7.3 **Single** data type.

Bits	Usage
31	Sign (0 =positive, 1 =negative)
30 to 23	exponent, biased by 127
22 to 0	Fraction f of the number $1.f$

In floating-point single-precision representations, any stored value requires 32 bits, which are formatted as shown in Table 7.3. Note that, `single` variables require less memory storage than `double` ones, since they are using 32 bits in spite of 64. However, because they are stored with fewer bits, numbers of type single are represented to less precision than numbers of type double.

Because the default numeric type for MATLAB® is the `double` one, the value of variable `x = 26.09` is spread over 64 bit. The function `isfloat()` can be used to verify that `x` is a floating-point number. If this is the case, this function returns logical 1 (true), logical 0 (false) otherwise. Note that, numeric data, characters or strings, and logical data can be converted in floating-point single- and double-precision representations using the MATLAB® functions `single()` and `double()`, respectively.

Creating Floating-Point Data

Floating-point numbers can be used to store values greater than $-3.4 \approx \times 10^{38}$ or lower than $\approx 3.4 \times 10^{38}$. For numbers that lie between these two limits, both double- or single-precision representation can be equally used, but the single one requires less memory. Here in the following, a simple way of defining numbers with several data types in MATLAB® is explained.

Double Precision

Since the default numeric type for MATLAB® is `double`, any double-precision number can be declared with a simple assignment statement:

```
x = 25.783;
```

The function `whos` shows the information associated with variable `x`, that is a 1-by-1 array of type double which is stored in 8 bytes, as reported in the following:

```
>> whos x
Name          Size              Bytes  Class   Attributes
x             1x1                   8  double
```

Instead, the function `isfloat` can be used just to verify that `x` is a floating-point number. This function returns logical 1 (true) if the input is a floating-point number and logical 0 (false) otherwise.

```
>> isfloat(x)
ans=
    logical
    1
```

Other numeric data, characters or strings, and logical data can be converted to double precision using the function `double()`. The example reported here in the following converts a signed integer number to double-precision floating point one:

```
y=int64(-589324077574);      % 64-bit integer variable
x=double(y);                 % y is converted to double
>> x
x=
    -5.8932e+11
```

Single Precision

Conversion functions are needed in MATLAB® to create floating point single-precision variables starting from the default data type:

```
x = single(25.783);
```

The `whos()` function can be used to store the characteristics (i.e., name, size, bytes, class, attributes, as shown previously) of variable `x` in a structure. For example, the bytes field allows to verify that single variables require 4 bytes only:

```
xAttrib = whos('x');
>> xAttrib.bytes
ans=
    4
```

Instead, functions `double()` and `single()` can convert other numeric data, characters or strings, and logical data to single precision variables. The example reported here in the following converts a signed integer number to single-precision floating point one:

```
y=int64(-589324077574);      % 64-bit integer variable
x=single(y);                 % y is converted to single
>> x
x=
    -5.8932e+11
```

Table 7.4 Summary of some data types available in MATLAB®.

Signed 8-bit integer	-2^7 to 2^7-1	$-128 \div 127$	`int8`
Signed 16-bit integer	-2^{15} to $2^{15}-1$	$-32768 \div 32767$	`int16`
Signed 32-bit integer	-2^{31} to $2^{31}-1$	–	`int32`
Signed 64-bit integer	-2^{63} to $2^{63}-1$	–	`int64`
Unsigned 8-bit integer	0 to 2^8-1	$0 \div 255$	`uint8`
Unsigned 16-bit integer	0 to $2^{16}-1$	$0 \div 65535$	`uint16`
Unsigned 32-bit integer	0 to $2^{32}-1$	–	`uint32`
Unsigned 64-bit integer	0 to $2^{64}-1$	–	`uint64`

Other Data types

MATLAB® provides several commands to directly define or convert data in a specified base and precision. Table 7.4 summarizes the data types which are mainly used in this book.

7.3 Use of *Scaling* in Fixed Point Representation

The theoretical procedure to represent real numbers using integer variables starts by selecting a suitable *base value*. In digital theory, numbers are typically computed through powers of 2. Furthermore, it is recommend to make an intermediate step while doing the conversion, that is, moving from a real value in SI units X_{SI} to a per-unit (p.u.) representation x_{pu}. Numbers like x_{pu} can assume values in the range $[0,1]$. Namely, they are computed as follows:

$$x_{\mathrm{pu}} = \frac{X_{\mathrm{SI}}}{X_{\mathrm{b}}} \tag{7.5}$$

where X_{SI} is the real value in SI units to be translated and X_{b} is the base (reference) value of the p.u. representation.

Then, x_{pu} is translated as:

$$\#x_{(N_{\mathrm{bit}})} = x_{\mathrm{pu}} 2^{N_{\mathrm{bit}}} = \frac{X_{\mathrm{SI}}}{X_{\mathrm{b}}} 2^{N_{\mathrm{bit}}} \tag{7.6}$$

where N_{bit} is number of bits of the selected base value and $\#x_{(N_{\mathrm{bit}})}$ is the value of X_{SI} translated in integer numbers. In other words, the representation through scaled integer numbers defines a correspondence with values ranging in one unit.

Example

Given $I_\mathrm{b} = 20\,\mathrm{A}$ as base value and $N_\mathrm{bit} = 6\,\mathrm{bit}$, translate the current value into scaled integers considering $i(t) = 17.5\,\mathrm{A}$ and $10.2\,\mathrm{A}$.

$$\#i_6 = \frac{I}{I_\mathrm{b}} 2^{N_\mathrm{bit}} = \frac{17.5}{20} 2^6 = 56 \tag{7.7}$$

$$\#i_6 = \frac{10.2}{20} 2^6 = 32.64 \approx 32 \tag{7.8}$$

Note that the data conversion is not exact in the second case. Thus, an approximation error is introduced. Such error is inversely proportional to the selected number of bits for N_bit.

Regarding the value adopted for N_bit, it is possible to determine the maximum and the minimum numbers that can be represented. Thus, for $N_\mathrm{bit} = 6\,\mathrm{bit}$:

$$\texttt{int6}: \ \#x_6 \in [-32,\ 31] \qquad \texttt{uint6}: \ \#x_6 \in [0,\ 63] \tag{7.9}$$

If a 6-bit base value is selected and a 8-bit data type is used for storing the translated numbers, there is a margin not to have any overflow in case of a wrong choice of X_b.

In addition, the choice not to saturate the bits of the processor is appropriate while making mathematical computations. If a processor working with signed `int16` (i.e., $\#x_{16} \in [-32768,\ 32767]$), and a base values of unsigned 12-bit integers (`uint12`, i.e., $\#x_{\mathrm{u}12} \in [0,\ 4095]$) are considered, the following computations might be carried out:

- **Sum:** in the worst case scenario, where both addends assume the maximum allowed value, it follows that $4095 + 4095 = 8190 < 32767 \rightarrow$ no overflow.
- **Product:** this operation behaves differently with respect to the sum. To the aim of providing a better understanding to the reader, an example is exploited starting directly from p.u. representation:

$$\left.\begin{aligned} x_a &= 3.56790 \\ x_b &= 1.27383 \end{aligned}\right\} \rightarrow x_c = x_a \cdot x_b = 4.54490 \tag{7.10}$$

Moving into integer numbers, it follows:

$$\left.\begin{aligned} \#x_{a,12} &= 3.56790 \cdot 2^{12} = 14614 \\ \#x_{b,12} &= 1.27383 \cdot 2^{12} = 5217 \end{aligned}\right\} \rightarrow \#x_{c,12} = 4.54490 \cdot 2^{12} = 18615 \tag{7.11}$$

Instead, performing the same computations using the converted values:

$$\#x_{c,12} = \#x_{a,12} \cdot \#x_{b,12} = 76241238 \neq 18615 \tag{7.12}$$

Thus, the two results are inconsistent. This ambiguity is a consequence of the *properties of exponents*:

$$\#x_{c,12} = x_a \cdot 2^{12} \cdot x_b \cdot 2^{12} = x_c \cdot 2^{12 \cdot 2} \rightarrow \#x_{c,24} \tag{7.13}$$

Namely, the base value *moves* from a 12 to 24 bit representation. Thus, the right way to avoid such issue is to use a `shift` operator. Namely:

- **Right shift ($\gg$):** division by a power of two, e.g., right shift of 3 bit $\rightarrow 1/2^3$
- **Left shift ($\ll$):** multiplication by a power of two, e.g., left shift of 3 bit $\rightarrow 2^3$

Consequently, the previous multiplication can be performed as follows:

$$\#x_{c,12} = (\#x_{a,12} \cdot \#x_{b,12}) \gg 12 = 18613 \tag{7.14}$$

that is, the shift is done over a number of bit equal to that one of the base and the approximation in the result $18613 \approx 18615$ is the consequence of the loss of some digits in the shift operation.

Note that the choice to use `uint12` as base value for processors working with `int16` allows a tolerance of 4 bit to cover model uncertainties. However, the partial result of the multiplication $\#x_{a,12} \cdot \#x_{b,12}$ needs a storage in a 32-bit variable (e.g., `int32`) since the partial result has a base value of 24 bit.

7.4 Converting from Decimal Representation to Single Format

Another interesting representation for embedded systems is the decimal (or base 10) one. This is a representation of non-negative real numbers r in the form of a series, traditionally written as the sum reported here in the following:

$$r = \sum_{i=0}^{\infty} \frac{a_i}{10^i} \tag{7.15}$$

where a_0 is a non-negative integer, and a_1, a_2, ... are integers satisfying $0 \leq a_i \leq 9$, which are called the digits of the decimal representation.

To convert a base 10 real number into a `single` (32-bit base-2) format the outline reported here below should be used:

1. Consider a real number with an integer and a fraction part, such as $r = 12.375$;
2. Convert and normalize the integer part into binary;
3. Convert the fraction part;
4. Sum the two results and adjust them to produce a proper final conversion.

Conversion of the fractional part: consider 0.375, i.e., the fractional part of 12.375. To convert it into a binary fraction, multiply it by 2 first, then re-multiply the new fraction by 2 until a zero fraction is found or until the precision limit is reached (which is 23 digits for `single` format). Thus:

- $0.375 \times 2 = 0.750 = 0 + 0.750 \rightarrow b_1 = 0$

  The integer part is the binary fraction digit. Therefore, 0.750 must be multiplied again by 2 to proceed;

- $0.750 \times 2 = 1.500 = 1 + 0.500 \rightarrow b_2 = 1$;

- $0.500 \times 2 = 1.000 = 1 + 0.000 \rightarrow b_3 = 1$

  That is, the fraction part is $= 0.000$.

Hence, $(0.375)_{10}$ can be exactly represented in binary as $(0.011)_2$. Not all decimal fractions can be represented in a finite digit binary fraction. For example, 0.1 does not have any exact representation, i.e., only an approximated translation can be performed. Coming back to the example under investigation, it follows that:

$$(12.375)_{10} = (12)_{10} + (0.375)_{10} = (1100)_2 + (0.011)_2 = (1100.011)_2$$

Since `single` data type requires real values to be represented in the format reported in equation (7.3), 1100.011 is shifted to the right by 3 digits to become $(1100.011)_2 \times 2^{-3}$. Thus, $(12.375)_{10} = (1100.011)_2 \times 2^{-3}$, from which exponent and fraction can be determined:

- The resulting exponent is 3, that is the shift performed previously. Therefore, $e = 130 = 10000010$ to obtain $130 - 127 = 3$ in the biased form;

- The fraction is 100011.

From these outcomes, the resulting `single` representation of 12.375 is determined as:
$0 - 10000010 - 10001100000000000000000 = 41460000_H$

Be careful: if 68.123 is to be converted into `single` representation using the above procedure, $42883EF9_H$ is expected as the final result, being the last 4 equal to 1001. However, due to the default rounding behavior of the IEEE 754 format, the result is actually $42883EFA_H$, whose last 4 bits are 1010.

Example

A value of 0.375 is consider. Knowing that $0.375 = (1.1)_2 \times 2^{-2}$, the exponent is -2 and in the biased form it is $127 + (-2) = 125 = 0111\ 1101$. The fraction is 1 by looking at the right part of binary sequence 1.1 ($1 = b_{22}$). Finally, the resulting `single` representation of the real number 0.375 is:

$$0 - 01111101 - 10000000000000000000000 = 3ec00000_{\mathrm{H}}$$

7.5 Processing the Data: Implementation Hints

Floating point MCU based controllers are more expensive then fixed point ones. Furthermore, a fixed point controller algorithm can be executed on a floating point MCU, but not vice-versa. The software complexity of industrial applications is related to the available hardware capability. The latter could be dependent on, e.g., cost decisions, footprints, and harsh environment. Assuming to implement a given control algorithm, if a powerful microprocessor with a large memory allocation is available, the complexity of the computations is not an issue. Therefore, the best possible accuracy (i.e., floating point variables) is achieved. On the other hand, cheap platform foresee the employment of ad hoc or underestimated memory. Then, both the architecture and even the computations may be revisited to speed up the processing (i.e., exploitation of integer variables). Some typical drawbacks of the digital implementation into microcontrollers are related to the management of data types and computations among the variables (mathematical operations). In general, it is possible to build a firmware based either with double or single precision. However, microcontrollers have always limited computational power. Hence, complex variable structure such as `float` and `double` have slow computation time. It is important to remember that `single` variables requires less memory than the previous ones. Finally, to speed up the processor operation (and memory usage) further, variables can be translated into integers numbers. Note that, given a firmware implementation, possible loss of accuracy while doing operations with floating-point arithmetic may be caused by the limitations of the adopted hardware. For example, this latter could not have enough bits to represent the result.

Summary

Computational efforts required by the design of firmware running on an embedded platform is dependent on the type of required operations and by the nature of the variable involved in such calculations. Due to the reduced availability of computational power, it is a best practice to avoid the introduction of complex numerical structures like `float` or `double` in the code. The best solution to optimize computational speed consists in representing all the variables with integer numerical structures (`int8`, `int16`, `int32`) or unsigned integer (`uint8`, `uint16`, `uint32`) wherever it is possible. For product and/or more complex operations (e.g., embedded functions), the complexity should be limited to `single` variables. It is quite handy to manage data type conversion in Simulink® by using the data-tye conversion block (it is a standard block). Its usage is exploited in the next chapters while studying practical cases study.

Part III

Real-Time Control in Power Electronics: Peripherals Settings

Introduction

Every time a firmware foresees the employment of several peripherals, the user must take care of their scheduling, with particular attention to their specific settings, timing (e.g., sample time) and the synchronization between them. For instance, ADC and ePWM peripherals, which are two of the most important ones for any motor control/power electronic-based firmware, can be coordinated to be executed according to a specific order/priority. This is done by using synchronization event (e.g., one ePWM module may trigger the Start of Conversion of several ADC blocks) and/or interrupt service routines. Note that, even if a control scheme/firmware is well designed, its behavior is not ensured without a proper internal timing.[1]

Nevertheless, several peripherals need to be investigated in order to understand how a closed-loop scheme can be implemented and which debugging tools are possible. Simulink® Coder™ generates standalone C and C++ code from Simulink® schemes. This part describes how to set the Simulink® environment, the peripherals and the target hardware (i.e., LaunchXL F28069M Piccolo™) in order to achieve automatic code generation. In all the examples reported in this book, *.c files are generated and built into executable files (which can run on the adopted board) automatically from Simulink®. Such procedure dramatically speeds up design testing and prototyping. In the next chapters, the peripherals reported in the following table will be presented:

Block Name	**I/O**	**Firmware / Testing (.slx)**
Serial Send	O	Testing
Serial Receive	I	Testing
SCI XMT	O	Firmware
SCI RCV	I	Firmware
GPIO DI	I	Firmware
GPIO DO	O	Firmware
ADC	O	Firmware
ePWM	I	Firmware
eQEP	O	Firmware

[1]The *Timing and Synchronization between Peripherals* is a topic as important as the design of the controller parameters.

DOI: 10.1201/9781003196938-7

8

Basic Settings: Serial Communication COM and Hardware Target

Before using the F28069M LaunchPad™, the communication channel between the board and the host PC should be set as well as the Simulink® environment. In particular, the latter settings should cover the compiler selection as well as hardware targeting for automatic code generation.

8.1 Virtual Serial Communication through COM port

In case of a first time connection between the F28069M LaunchPad™ and the PC, the steps reported in the following describe how to configure the virtual COM port of the board:

- Connect the board to the PC through the USB cable provided in the TI package;
- Open the device manager window (see Figure 8.1 (a) and (b)):
  - Open the **TI XDS100 Class Auxiliary Channel** (or **TI XDS100 Channel B** in previous Windows releases, see Figure 8.1 (b)) properties and tick **Enable VCP** checkbox;
  - Disconnect and reconnect the board;
  - A new COM port **XDS100 Class USB Seral Port (COMx)** should appear in section **Port (COM and LTP)**.

It is possible to redefine the COMx number (e.g., COMx → COM8) by clicking on the **Advanced...** button in the **Port Settings** pane. Note that the *baud rate*[1] inside the COMx settings must not be changed. This value can be adjusted directly in the Simulink® environment if required.

[1]**Baud rate** refers to the number of signal or symbol changes that occur per second. To obtain the actual Bit/Rate the formula is: $BitRate = BaudRate \times nBit$ where nBit is the number of bit required to represent that symbol.

DOI: 10.1201/9781003196938-8

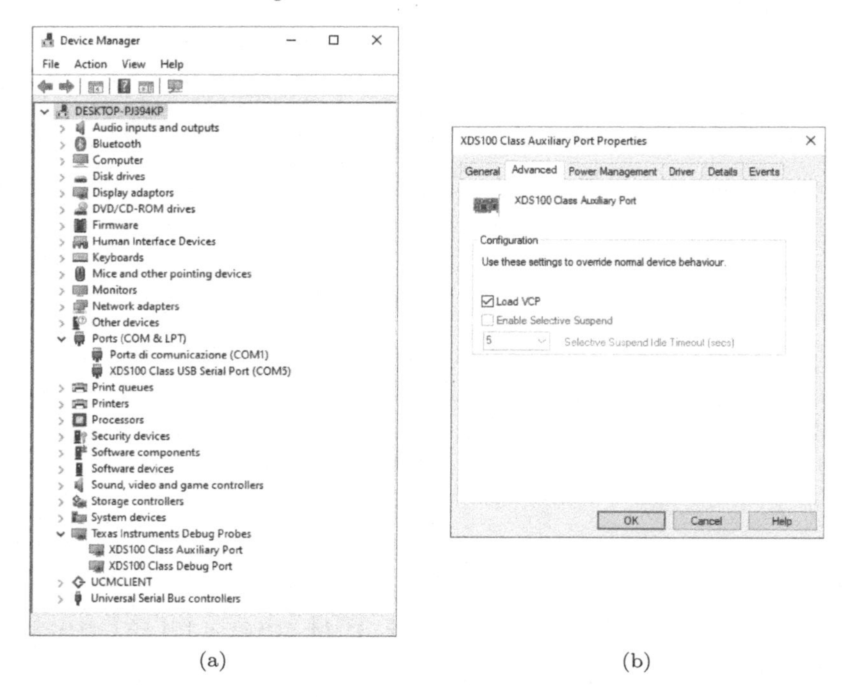

Figure 8.1 Device Manager window (a) and port properties window for XDS100 Class Auxiliary Port (b).

The virtual COMx is not only used for communication (data exchange) purposes, but as programming interface too, i.e., letting the JTAG[2] Debug Flow. Indeed, many C2000 MCU boards have a JTAG emulator implemented on-board and they do not require an external one. Unless there is an application requirement, TI suggests using the onboard emulator for development purposes. The XDS100 and the XDS110 are two target emulators that are found on TI C2000 Evaluation Modules. For completeness, note that those emulators are physically connected to:

- XDS100v1 → FTDI FT2232
- XDS100v2 → FTDI FT2232
- XDS110 → TI MCU TM4C1294NCPDRI3R

For the JTAG emulator to communicate the PC, the driver files need to be installed. Their installation typically occurs co-incident to the installation of

[2]The use of JTAG in embedded systems aims to development, debugging, and testing. JTAG uses a 5-pin implementation in most systems: TDI (Test Data In), TDO (Test Data Out), TCK (Test Clock), TMS (Test Mode Select), TRSTn (Test Reset).

Code Composer Studio. To check that the drivers are successfully installed, it is enough to connect the PC to the JTAG emulator and power it up.

Remark

The Target Configuration File (.ccxml) contains the information necessary to connect the PC to target device and the JTAG emulator being used. To view the current target configurations, select Target Configurations under the "View" tab in CCS.

9

Simulink® Configuration

Every time a blank Simulink® project is opened, the following steps must be implemented/verified to assure a correct automatic C-code generation and firmware deployment on the target microcontroller:

- Open the **model configuration parameters** menu by clicking on the Model Settings icon in the Modeling bar[1] of Simulink® window;
- Open the **code generation** panel:
  - Select **Texas Instruments™ Code Composer Studio (C2000)** as **Toolchain** setting as shown in Figure 9.1;
  - Set the **System target file** as **ert.tlc** and **C** as **Language**.
- Open the **hardware implementation** pane:
  - Select **TI Piccolo F28069M LaunchPad** as **Hardware board**;

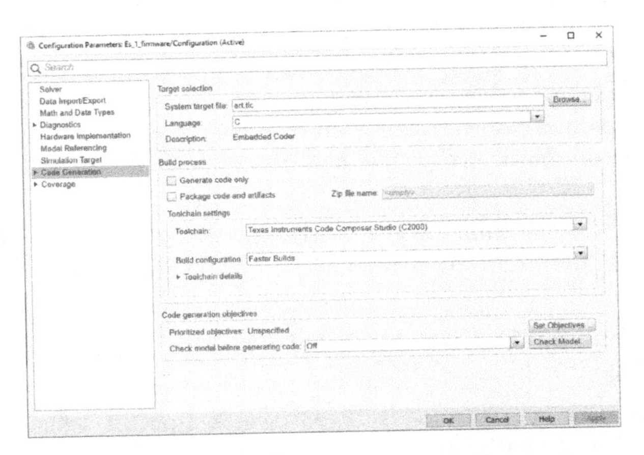

Figure 9.1 Simulink® settings: Code Generation.

[1]In releases which are earlier than the 2019a one, this icon is in the main menu bar of the Simulink® window.

DOI: 10.1201/9781003196938-9

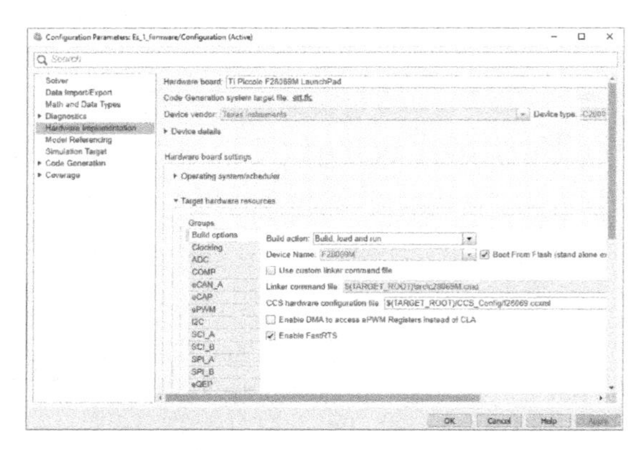

Figure 9.2 Simulink® settings: Hardware Implementation for TI LaunchXL F28069M boards.

- Enlarge the **Target hardware resources** menu;
- Select **Build options** from the target hardware resources **Groups**:
  * set **Build action** on **Build, load and run**, as shown in Figure 9.2;
  * thick **boot from flash**[2] checkbox;
  * verify the linker command file:
    $(TARGET_ROOT)\src**c28069M.cmd**
  * verify CCS hardware configuration file:
    $(TARGET_ROOT)\CCS_Config**f28069.ccxml**

  where $(TARGET_ROOT) automatically set the directory path according to the path of the MATLAB® installation. It must be said that, if a different linker command file and/or CCS hardware configuration file are needed, the full directory path should be specified, which then becomes a local setting. In this case, the user must remember to use custom paths every time new Simulink® files are created.

- Select **SCI_A** from **target hardware resources** menu (see Section 3.1.4 for the hardware arrangement that enables this channel). Moreover:
  * define the **Desired baud rate in bits/sec** for the SCI_A: the default value is set to 5.625×10^6 bit/s;
- Repeat this last step for **SCI_B** (see Section 3.1.4 for the hardware arrangement that enables this channel).

[2]This option allows the board to store and load the firmware on the flash memory onboard the LaunchPad™. Otherwise, the code is temporarily stored in the RAM memory of the board not allowing any stand alone executions.

9.1 Simulink® Environments: Firmware vs Testing

Automatically generated C code from Simulink® models greatly improves design, testing and prototyping speed for any project.

Simulink® Embedded Coder™ generates standalone C/C++ code from Simulink® models for a wide variety of applications. This Section describes how to set the Simulink® environment to achieve automatic code generation. Basically a **.c* file is generated and built into executable HEX file (alternatively, binary files can be adopted) which is loaded on the Flash or RAM of the board through the COMx port,[3] and then executed on the MCU.

9.1.1 Overview

The Simulink® environment is used both to design the control system firmware and to test it giving inputs from the host PC to the microcontroller. In order to differentiate better the code running inside the MCU from the simulation test code, two different .slx files are built:

1. **Firmware.slx:** this is the Programming Environment for the MCU, where the internal algorithm is built. It defines the programmable logic that the input/output peripheral has to follow (see Figure 9.3 (a)). It is mainly subejct to the samplig time T_s. By means of automatic code generation, this Simulink® file is translated into executable implementation through the following steps:

    - C compiler from the Texas Instruments™ IDE Code Composer™ Studio is launched along with Simulink® Embedded Coder®;
    - *.c* and *.h* files are created and a report interface (Report Generation) is produced. There, the code is accessible and the compiler routine verified, as shown in Figure 9.4;
    - The resulting file is downloaded into the MCU memory (Flash or RAM) and, then, executed.

    The Simulink® support package that allows to program the peripherals onboard the LaunchPad™ is the **Embedded Coder Support Package for Texas Instruments C2000 Processors**. This add-on installs a library which is entirely dedicated to the **C2000**-based MCU boards.

2. **Testing.slx:** this is the Testing/Debugging Environment that Simulink® executes on the host PC to send/receive data to/from the LaunchPad™, while the loaded firmware is in execution. This

[3] UART peripheral, SCI, and COMx port refer to the same subject, i.e., the serial communication between the LaunchPad® board and the Host PC.

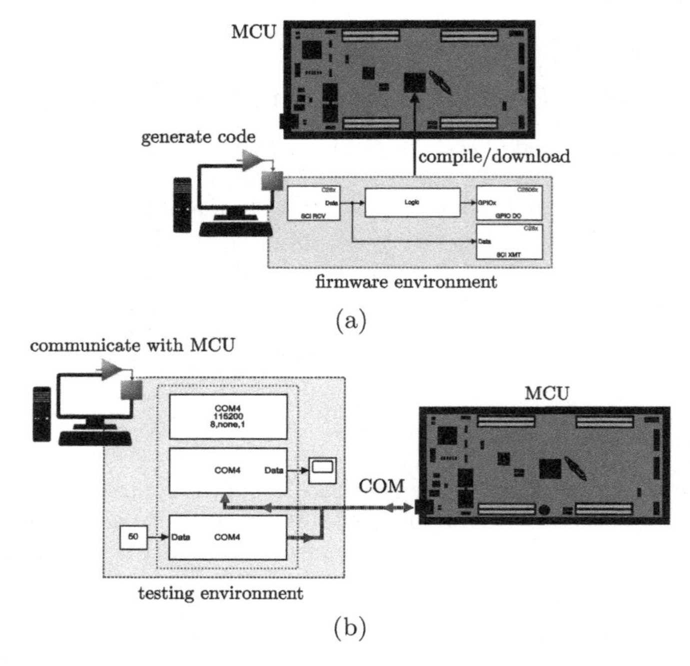

(a)

(b)

Figure 9.3 *Firmware* (a) and *testing* (b) environments in Simulink®.

file works as a simulation except that the signal are sent/received from the embedded board through the COM port (see Figure 9.3 (b)). Indeed, the scheme runs at the simulation step-size T_{sim}, while the speed of communication depends to the reception and transmission time T_{RX} and T_{TX}, respectively (see Section 10.4). Note that the Testing environment is not always necessary, i.e., a firmware may not require any serial COM debugging tool.

9.1.2 Execution in Simulink®

Due to this separation between the *firmware* and the *testing* code, these two .slx files are handled in two different ways:

- To load the C-code generated from firmware.slx into the LaunchXL F28069M, the button Build, Deploy & Start (Build, Deploy & Start, DEPLOY) in the Hardware bar should be clicked;
- To run the simulation file testing.slx, click on the Run button (Run) on the Simulation bar.

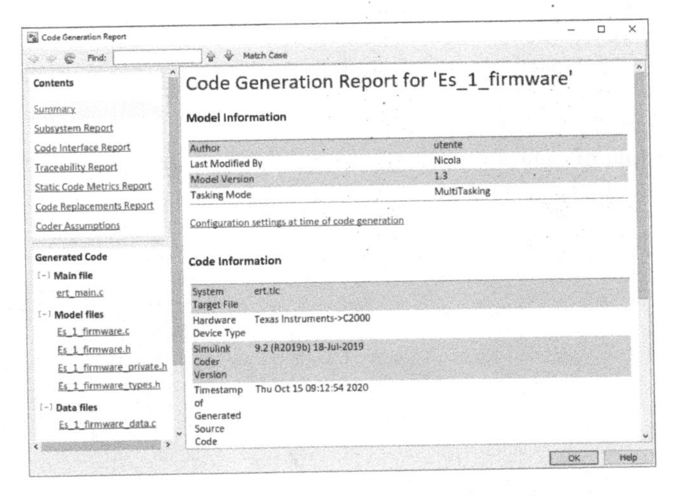

Figure 9.4 Code generation report.

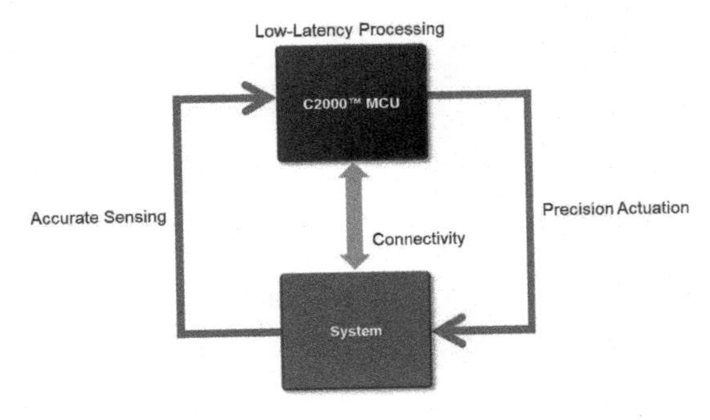

Figure 9.5 Example of a real-time control system [15].

9.2 MCUs and Real-Time Control with Simulink®

Since firmware.slx and testing.slx are two separate files, the Simulink® environment leads to build a real-time control system, where the MCU (which is loaded with the firmware file) sends/receives data with the host PC.

Such systems have time dependence on when they need the input to remain stable. A simple visualization of a type of real-time control system, which is suitable for the Simulink® environment, is reported in Figure 9.5. A fully realized controller would be capable of measuring key parameters of the system (sensing), applying control algorithms to the incoming data (processing), and, then, affecting the system to achieve the desired change given by the control algorithm (actuation). Simple exercises, such as single peripheral testing, will require only a F28069M LaunchPad™ board connected to the host PC to program the platform through the JTAG cable. The latter will be also used as a virtual serial port to exchange data and provide touting actions. More complex exercises requires the effective presence of both converter and load stages in addition to custom interface boards.

An example of complex control system is provided in Figure 9.6. Here in the following, the required hardware will be detailed at the beginning of every exercise.

Figure 9.6 Example of a motor control test bench based on the LaunchPad™ F28069M Piccolo™ programmed by the host PC (see Appendix B).

10

Serial Communication Interface (SCI) Peripheral

The main board communication is based on a Universal Asynchronous Receiver-Transmitter interface (UART, also known as SCI), which is a hardware device for asynchronous serial communication[1] in which the data format and transmission speed are configurable. This is used as a serial communication protocol over the peripheral port of the PC. The electric signal levels and methods are handled by an external driver circuit.
The UART takes bytes of data and transmits the individual bits in a sequential fashion. At the destination, a second UART re-assembles the bits into complete bytes. Each UART contains a shift register, which is the fundamental method of conversion between serial and parallel forms.

Serial transmission of digital information (bits) through a single wire or other medium is less expensive and complex than parallel transmission, which requires multiple wires.
For today personal computers (PC), the serial communication interface (SCI) can be virtually emulated over the USB link. This is the reason why the Enable VCP (Virtual COM Port) option must be set for the USB peripheral in the PC device manager. This operation is required for almost all the exercises presented in this book.

The LaunchXL F28069M Piccolo™ has two SCI modules: A and B. In particular, the receiver and transmitter of each of them can operate both in full-duplex[2] and half-duplex communication and they support the following data types: int8, uint8, int16, uint16, int32, uint32, single. The baud-rate that is chosen in the implementations reported in this book is 115 200 baud/s. This value can be raised up to 5.625 Mbaud/s, although the communication can be more prone to errors (data mismatch issues) due to an high volume of data to be exchanged (i.e., transferred/received) by the MCU within the sampling time T_x. For instance, if the data transmission from the MCU to the PC cannot be realized within T_x, the PC may receive fragmented data with

[1] **Asynchronous serial communication** is a form of serial communication in which the interfaces of the communicating endpoints are not continuously synchronized by a common clock signal. Instead of a common synchronization signal, the data stream contains synchronization information in form of start and stop signals before and after each unit of transmission.

[2] **Full-duplex communication** is adopted in point-to-point systems and it means that the data can flow in both direction simultaneously.

DOI: 10.1201/9781003196938-10

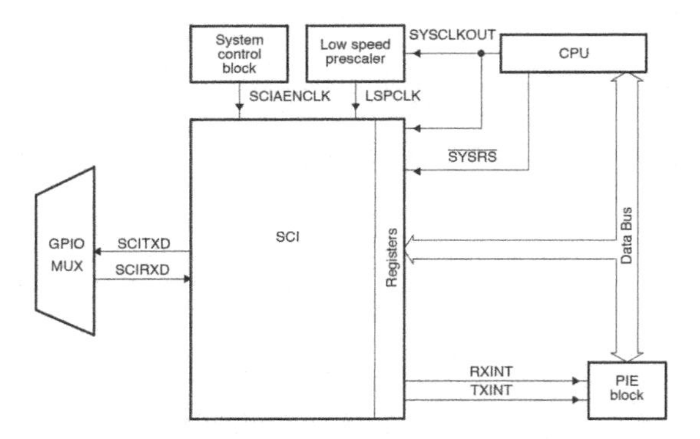

Figure 10.1 SCI-CPU Interface [17].

unreadable values. Hence, the transmission rate (both baud rate and sampling time) should be set considering the computational load given by the firmware.

10.1 Hardware Details

The serial communications interface (SCI) is a two-wire asynchronous serial port, commonly known as a UART. The SCI modules support digital communications between the CPU and other asynchronous peripherals that use the standard non-return-to-zero (NRZ) format. The SCI receiver and transmitter have a 4-level deep FIFO each for reducing servicing overhead, and they have their own separate enable and interrupt bits. Both of them can be operated independently for half-duplex communication, or simultaneously for full-duplex communication. As already mentioned, the communication bit rate can be set to different speeds through a 16 bit baud-select register. Finally, the SCI checks receive data for break detection, parity, overrun, and framing errors to specify data integrity. A picture representing the structure of the SCI interfaces is shown in Figure 10.1. The features of this module are listed here in the following:

- Two external pins:
  - SCITXD: SCI transmit-output pin;
  - SCIRXD: SCI receive-input pin.
- Both external pins can be used as GPIO if they are not used for SCI, thanks to a suitable arrangement of jumpers JP6 and JP7 (see Table 3.2).
- The baud rate is programmable up to 64000 different rates. However, 115 200 baud/s is the standard value adopted in all the exercises reported in this manuscript.

- Data-word format consists in:
  - One start bit;
  - Data-word length programmable from 1 to 8 bit;
  - Optional *even/odd/no parity* bit;
  - 1 or 2 stop bits;
  - An extra bit to distinguish addresses from data (address bit mode only).
- Four error-detection flags: *parity*, *overrun*, *framing*, and *break detection.*
- Two wake-up multiprocessor modes: idle-line and address bit.
- Half- or full-duplex operation.
- Double-buffered receive and transmit functions.
- Transmitter and receiver operations can be accomplished through interrupt-driven or polled algorithms with status flags.
- Separate enable bits for transmitter and receiver interrupts (except BRKDT).
- NRZ[3] (non-return-to-zero) format.
- **Enhanced features**:
  - Auto-baud-detect hardware logic;
  - 4-Level transmit/receive FIFO.

10.2 Firmware Environment: Send and Receive Data through Serial Communication

The two blocks reported in the following are used in the firmware environment to establish the serial communication between the F28069M LaunchPad™ board and the host PC to send and receive data (see Figure 10.2). They can be found in the *Embedded Coder Support Package for Texas Instruments C2000 Processors* library, inside the *C2806x* subset.

10.2.1 C2806x SCI receive

This block (see SCI RCV in Figure 10.2) corresponds to a code section which allows to receive data coming from the connected PC to the LaunchPad™ through virtual COM port. In the following, the main settings of this block are reported.

[3]In telecommunication, a **non-return-to-zero** (**NRZ**) line code is a binary code in which ones are represented by one significant condition, usually a positive voltage, while zeros are represented by some other significant condition, usually a negative voltage, with no other neutral or rest condition.

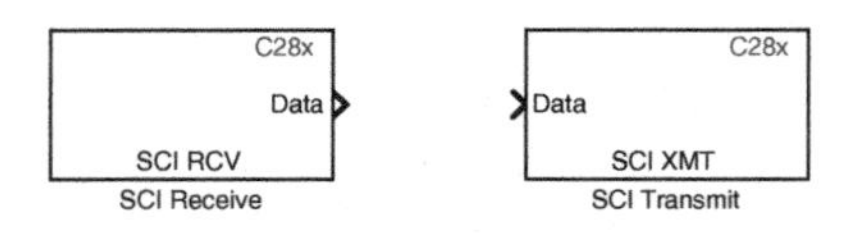

Figure 10.2 SCI Receive (SCI RCV) and Transmit (SCI XMT) blocks to for programming C2806x processors.

Block Parameters:

- **SCI Module**

  This label sets the SCI module which is adopted for communication. For the LaunchPad™ F28069M it can be A or B.

  Remark: this setting should match with the Serial Send block channel selected in the Testing Environment (i.e., added inside the *testing.slx* file, see Section 10.3.2).

- **Additional Package Header**

  This field specifies the data located before the received data package. It generally indicates the start of data. The additional package header must be an ASCII value. A string or number (0–255) can be used. Single quotes around strings must be entered in this field, but the quotes are not received or included in the total byte count. To specify a null value (no package header), enter two single quotes alone ''. Additional package headers specified here must match with that one entered in the Serial Send block placed in the *testing.slx* file (see Section 10.3.2).

- **Additional Package Terminator**

  This field specifies the data located at the end of the received data package. It generally indicates the end of data. The additional package terminator must be an ASCII value. A string or number (0–255) can be used. Single quotes around strings must be entered in this field, but the quotes are not received nor included in the total byte count. To specify a null value (no package terminator), enter two single quotes alone ''. Additional package terminators specified here must match with that one entered in the Serial Send block placed in the *testing.slx* file (see Section 10.3.2).

- **Data Type**

  This option specifies the data type of the output data. Available options are int8, uint8, int16, uint16, int32, uint32 and single.

- **Data Length**

  This value specifies the size of data that the block should receive at each time step. Numbers greater than 1 indicate that arrays are transferred. The data length must match the input data length, which is sent through the Serial Send block in the *testing.slx* file (see Section 10.3.2).

- **Initial Output**

  This parameters sets the output default value from the SCI Receive block. This value is used, for example, if a connection time-out occurs and the **Action taken when connection timeout field** is set to **Output the last received value**, but nothing has been received yet.

- **Action taken when connection times out**

  This label allows to specify the output in case a connection time-out occurs. If **Output the last received value** is selected, the outputs return the last received value. If no value is received, the SCI Receive block outputs return the Initial output value. If **Output custom value** is selected, the custom value is set through the parameter **Output value when connection times out**.

- **Output value when connection times out**

  Parameter that sets the output time-out custom value.

- **Sample Time**

  This field specifies the sample time T_s in seconds for input sampling of the block. To inherit sample time and to execute this block asynchronously, set it equal to -1.

- **Output receiving status**

  By enabling this feature, a status block providing the status of the serial communication is created.

- **Enable Receive FIFO interrupts**

  If this option is selected, an interrupt is posted when the FIFO is filled with a specified amount of data, allowing the subsystem to perform any action. For example, the C28x Hardware Interrupt block can be used for triggering the SCI Receive block to read the data as soon as it is received. If the option is not selected, the SCI Receive block is in polling mode and it checks the FIFO for data. If any data is present, the block reads and outputs it. If data is not present and blocking mode is enabled, the block waits until data is available. In non-blocking mode, the block goes on with the execution of the algorithm without waiting for any received data.

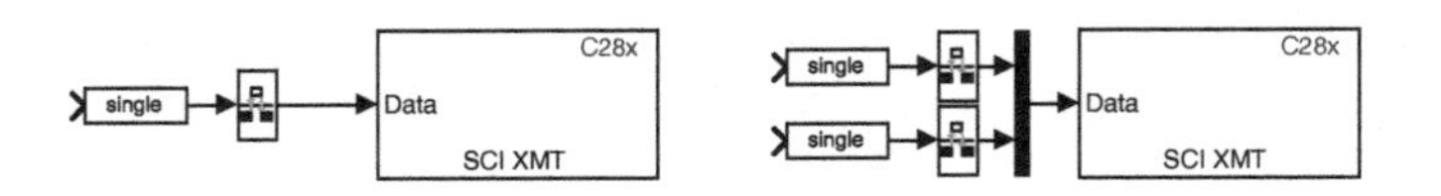

Figure 10.3 Examples of Rate Transition and Data Type Conversion blocks connected to a SCI Transmit block.

- **Receive FIFO interrupt level**

  This parameter appears only if **Enable Receive FIFO interrupts** is ticked. When it is enabled, the receive FIFO generates an interrupt when the number of received data bytes is greater than or equal to the value selected for this parameter.

10.2.2 C2806x SCI transmit

This block (see SCI XMT in Figure 10.2) allows the transmission of data from the board to the host PC Through SCI. For example, this communication can be used to visualize the values acquired by the ADC converters inside the onboard sensors in the Simulink® environment. In the following, the main settings of this block are reported without any further explanations. Indeed, they are the same as those described for the C2806x SCI Receive block (see Section 10.2.1).

Block Parameters:

The block parameters of C2806x SCI Transmit are less than C2806x SCI Receive, but with the same meaning. In particular they are: *SCI Module, Additional Package Header, Additional Package Terminator, Enable transmit FIFO interrupts, Transmit FIFO interrupt level.* See the previous section for further details.

Remarks

The transmission rate as well as the trasnmitted data format/length are not internally set by this SCI blocks. These have to be (hardly) fixed through specific Simulink® blocks, such as Rate Transition and Data Type Conversion, according to the firmware execution. This is summarized in Figure 10.3, which shows a couple of possible arrangements related to data exchange required in the following exercises. Note that, multiple data/variables can be sent exploiting Mux blocks (data is organized in $[1\ n]$ arrays). Double-precision floating point data is not recommended due to its large size. This is particularly critical when more than one variable have to be transmitted/received.
Finally, note that no COM port is specified in SCI RCV or XMT, since those blocks are translated into code which is downloaded directly on the hardware board.

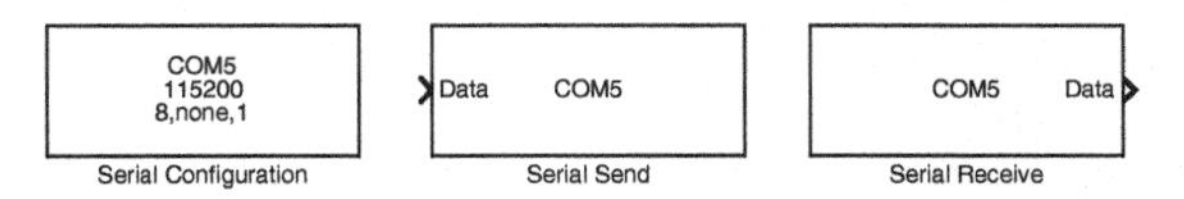

Figure 10.4 Serial Configuration, Send, and Receive blocks for setting serial communications between the *testing.slx* file and the LaunchPad™ F28069M.

10.3 Testing Environment: Send/Receive Data through Serial Communication

The set of blocks listed here in the following (and shown in Figure 10.4) are used in the Testing Environment to allow interaction between the host PC and the F28069M LaunchPad™ board through the Simulink® software. Moreover, debugs of the generated code deployed on the board can be done in a separate simulation file. Such blocks are available in the *Instrument Control Toolbox™* library, as already mentioned in Section 4.3.

10.3.1 Serial configuration

This master block (see Figure 10.4) configures the parameters for the serial communication, which is carried out together with the Serial Receive and Serial Send blocks over COMx port. Its main settings are reported here in the following.

Block Parameters:

- **Communication Port**

  This setting specifies which available **COMx** port is exploited for serial communication. This choice must match with the COM port number selected in the Serial Send and Serial Receive block as well as with the VCP generated in the PC Device Manager.

- **Baud rate**

  This option allows to set the rate at which signals are transmitted for the serial interface. The default value is 9600 baud/s. This number should match the one specified in the Model Configuration Parameter window.

- **Data bits**

  This label specifies the number of bits of a single chunk of data (i.e., baud) that are transmitted over the serial interface. The default value is 8 and other available values are 5, 6, and 7.

- **Parity**

  This field allows to specify the desired check of parity bits in the data bits transmitted through serial port. This is set to **none** by default. The other available choices are:

- **Even**: parity bit is set equal to 0 if the number of ones in a data set of bits is even;
- **Odd**: parity bit is set equal to 1 if the number of ones in a data set of bits is odd;
- **Mark**: parity bit is always set equal to 1;
- **Space**: parity bit is always set equal to 0.

- **Stop bits**

  This option specifies the number of bits used to indicate the end of the chunk of data. The selected number of data bits impacts on this parameters since it determines the choices available for the stop bits. If data bits is set equal to 6, 7 or 8, then, the default value is 1 and the other choice at disposal is 2. Otherwise, the only available number is 1.5.

- **Byte order**

  This label sets the byte order. If byte order is **LittleEndian** (default value), then the instrument stores the least significant byte in the first memory address. Conversely, if byte order is set to **BigEndian**, the most significant byte is placed first in memory and the least significant one last.

- **Flow control**

  This field specifies the process of managing the rate of data transmission on the serial port. Flow control can be set to **none** or **hardware**, depending if the user wants to have no flow control or to let the hardware determine it.

- **Timeout**

  This setting specifies the amount of time that the model waits for the data during each simulation time step. The default value is 10 s.

10.3.2 Serial send

This block (see Figure 10.4) is needed to send data from the host PC to the LaunchPad™ F28069M board via serial communications interface (SCI) through the COMx port. In the following, its main settings are reported.

Block Parameters:

- **Communication Port**

  This field allows to specify an available COM port. This choice must match with the COM port number selected in the Serial Configuration block (see Section 10.3.1).

- **Header**

  This label specifies additional data placed at the beginning of the information frame to be transmitted before sending it through serial port. By default, **<none>** or no header is specified. The header entered here must match with that one chosen in the SCI Receive block in the *firmware.slx* file (see Section 10.2.1).

- **Terminator**

  This label specifies the additional data placed at the end of the information frame to be transmitted after sending it over the serial port. By default, **<none>** or no terminator is specified. The terminator entered here must match with that one chosen in the SCI Receive block in the *firmware.slx* file (see Section 10.2.1).

- **Enable blocking mode**

  This flag allows simulation halting while sending data. Namely, the Serial Send do not send any other data until a signal is received. This option is selected by default. It is cleared in case it is not desired to have write operations that should pause the simulation.

Remark:

A Data Type Conversion block should be connected before the Serial Transmit block to avoid the transfer of any double-precision floating point data. The setting of this additional block should match the data type selected in the option panel of the SCI Receive block in the Firmware Environment (see Section 10.2.1).

10.3.3 Serial receive

This block (see Figure 10.4) is intended to receive data to the host PC via COM port from the MCU. Its main settings are reported here in the following. Some parameters are not explained since they are the same as those described for the Serial Send block (see Section 10.3.2).

Block Parameters:

Some parameters of Serial Receive block are the same as those reported in the previous Section, i.e., *Communication Port, Header, Terminator*. The other parameters are:

- **Data Size**

  Through this label, it is possible to specify the size of the transmitted data. Both 1-D vectors and matrices are accepted, e.g., [1 1] corresponds to a scalar value, while [1 3] corresponds to an array of size 3.

- **Data Type**

  This label specifies the type of the received signal (i.e., int8, uint8, int16, uint16, int32, uint32 and single). This setting should be coherent with the corresponding one specified in the Data Type Conversion block connected to the SCI XMT block in the firmware environment (see Section 10.2.2). It is important to note that the board F28069M can handle single-precision floating-point numbers, but this choice implies an heavier transmission and computational load. Hence, this kind of transmission likely overloads the communication buffer. To avoid this problem, int8/uint8 or int16/uint16 data types are preferable in case many signals have to be transferred.

- **Action when data is unavailable**

  This option can be set choosing among one of the following actions: **Output last received value**, **Output custom value** or **Error**. It is very similar to the setting reported in Section 10.2.1, i.e., Action taken when connection times out for firmware environment (see Section 10.2.1).

- **Custom value**

  This parameter is enabled only if **Action when data is unavailable** is set on **Output custom value**. It specifies a value to output when no data is received. The default value is 0.

- **Block sample time**

  This setting specifies the sample time of Serial Receive during simulation. Namely, it indicates the rate at which the block is executed during simulation. This value must be specified in seconds.

Remarks

Firmware and Testing Environments have common features: while both SCI RCV/Serial Receive blocks must specify *Sample Time* and *Data Type*, SCI Transmit and Serial Send do not have these options available and they must be checked and possibly adapted to be synchronized with other blocks by employing the Data Type Conversion block, the Rate Transmission block, and Mux block in case of multiple data transmission.

10.4 Time Variable Settings (Sample Rates)

While designing both the firmware and testing environment for control systems, the reader comes across different sample rates/times options which must be set correctly (they are worth of much attention too). At this point of the book, a brief summary of those times is reported in the following:

- **CPU Clock** - T_{ck}

  Typically fixed by the internal or external board oscillator; for the LaunchPad™ F28069M it is equal to $T_{\text{ck}} = 11\,\text{ns}$, which means a clock frequency of $f_{\text{ck}} = 90\,\text{MHz}$.

- **Integration Time** - T_s

  This value can be edited by the user in the **Configuration Parameters** window in the *firmware.slx* file. This number represents the minimum integration time used to discretize the system and to perform the computation. In this book, it is mainly identified with T_s. This value must be set according to the complexity of the code running on the MCU. Given that T_s defines the minimum, other sampling times must be multiple of T_s. Indeed, an algorithm might have parts of the code running at different rate, from the slowest (not critical) up to the fastest (highly critical).

- **Sample Time Tx** - T_{TX}

  This parameter represents the time rate at which data are sent to the Serial Send block and stored in the TX Serial Buffer during the transmission of signals to the LaunchXL F28069M board. The same occurs when data is sent from the MCU to the host PC via the SCI XMT. In both cases, it is set through Rate Transition blocks. This parameter must match with the SCI Receive sample time. In this book, $T_{\text{TX}} = 0.01\,\text{s}$ is mainly used .

- **Sample Time Rx** - T_{RX}

  It represents the Time Rate at which data are sampled from the RX Serial Buffer during the reception of signals from the LaunchPad™. This parameter can be set in the Serial Receive block and it must match the SCI Transmit sample time. According to the firmware requirements, the time range could vary from $T_{\text{RX}} = 0.01\,\text{s}$, $T_{\text{RX}} = 0.001\,\text{s}$, ..., up to $T_{\text{RX}} = T_s$.
  In particular, the most significant consequences of a wrong choice for $T_{\text{TX}}/T_{\text{RX}}$ (or data type) settings are mismatches such as:

  - Overloading of the transmission buffer causing data spikes and/or wrong values;
  - Simulation time much slower than real time (depending on the capabilities of the host PC), causing large delay between sent and received data within a given time span.

- **Fixed Step-Size** - Δt

  Regarding the firmware environment, every *firmware.slx* relies in a discrete framework. Thus, the book case studies will focus on Simulink® schemes which adopt a fixed-step solver option with a discrete solver such as ODE4 (Runge-Kutta) and a fixed-step size T_s, i.e., $\Delta t = T_s$.
  In the simulation environment (i.e.,*testing.slx*), the simulation starts at $t = 0$ and ends at $t = T_{\text{end}}$. The simulation time step T_{sim} may be fixed or vary

Figure 10.5 Connection between the F28069M LaunchPad™ board and a host PC.

during each run according to the selected solver options in Simulink®, i.e., discrete- or variable-step size, respectively. Since the firmware is designed in a discrete-time domain, the same is done for the testing environment, e.g., keeping $\Delta t = T_{\text{sim}} = T_s$ or $\Delta t = T_s/10$ (to increase the visualization resolution). Note that, every time the *enabling blocking mode* option is enabled in the Serial Send block, the simulation environment waits for the data acquisition before increasing the Simulation Time by one step (i.e., $t_{i+1} = t_i + \Delta t$), resulting in simulation times which are totally different from the real time. Namely, the more the testing environment receives data, the longer the simulation will take to run.

10.5 Examples on Serial Communication

In order to understand the difference between firmware and testing environments, two simple exercises are proposed here in the following.

Example 1

Build a firmware aimed at generating an internal signal from the board F28069M and send it to the PC through the serial port

A step-by-step procedure is reported here below to guide the reader through this first exercise. The LaunchPad™ F28069M is programmed in such a way to transmit an internally generated signal to the host PC through the virtual serial port COMx. SCI module A is used. Thus, the right jumper configuration should be realized on the board (see Table 3.2).

Firmware Environment

c28069_internalSin_F.slx (solver: fixed step - ODE4, step size: $T_s = 100\,\mu s$)

Open a new blank Simulink® project and configure the environment as shown in Chapter 9. Then:

- Open the **Model Configuration Parameters** window:
  - Select these two settings in the **solver selection** sub-menu:
    * type = fixed step;
    * solver = discrete (no continuous states);
  - Set the **fixed-step size** equal to $T_s = 100\,\mu s$.
- Place a **SCI Transmit** block in the file and double click on it to modify its parameters:
  - Select SCI module: A;
  - Remove both additional package header `'S'`→'' and terminator `'E'`→''.
- Insert a **Sin Wave** block to generate the internal signal. Double click on it:
  - Set sine type: sample based;
  - Set time: use simulation time.
    Namely, the implemented sine wave is $y(k) = A \sin\left(2\pi \frac{k}{n_s}\right) + b$.
  - Set the gain equal to $A = 0.5$;
  - Set the bias equal to $b = 0.5$;
  - Set the number of samples per period $n_s = 100$;
  - Set number of offset samples = 0;
  - Set sample time $T_{sig} = 1000T_s = 0.1\,s$.
- Insert a **Pulse Generator** block to have a possible alternative to the sine-wave at disposal. Double click on it:
  - Set pulse type: sample based;
  - Set time: use simulation time;
  - Set the gain equal to $A = 0.5$;
  - Set the number of samples per period $n_s = 100$;
  - Set the pulse width 50 → 50 %, i.e., equally spaced square-wave;

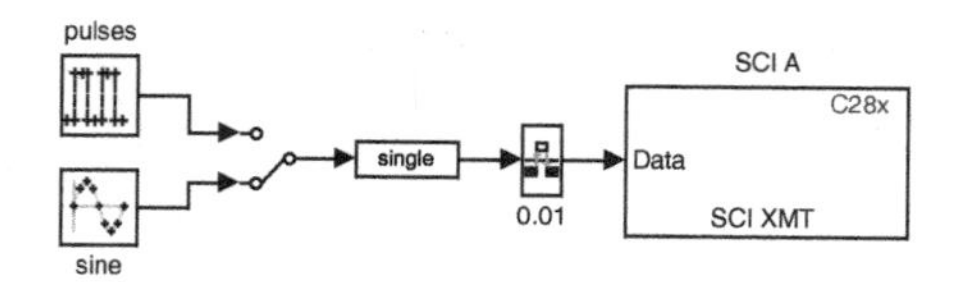

Figure 10.6 Simulink® scheme included in c28069_internalSin_F.slx.

- Set phase delay = 0;
- Set sample time $T_{sig} = 0.1\,\text{s}$.

- Insert a **Manual Switch** block to change the source input.

All system parameters have to be defined and initialized by a m-script before the firmware deploy. This can be done inside Simulink®, e.g., in *Model Properties/Main/InitFcn* or as a separate MATLAB® code. In this exercise, most of the parameters are manually defined in the each block settings, while only T_s and T_{sim} require a global assignment. For simplicity, both of them will be defined directly in Simulink® for either firmware and software environment.

The first internal signal is a sample based sine-wave. Namely, the value of k is increased every T_{sig} returning the actual value of $y(k)$ at each time step, until the reset condition is reached $k = n_s$. Hence, the sine-wave period is equal to $T = T_{sig} \cdot n_s = 10\,\text{s}$ and its corresponding frequency is $f = 0.1\,\text{Hz}$. Note that T_{sig} is a mutiple of T_s. Since the F28069M LaunchPad™ board includes a floating-point unit, single data type could be used to send floating-point signals through the COMx. This assignement is done via the **Data Type Conversion** block. A **Rate Transition** block is used to fix the sample rate of the SCI XMT to $T_{\text{TX}} = 100T_s = 0.01\,\text{s}$. Finally, connect the aforementioned blocks as shown in Figure 10.6. To translate the file c28069_internalSin_F.slx into C-code language and to run it on the board F28069M, this Simulink® scheme should be compiled, built, deployed and run by clicking on **Build, Deploy & Start** 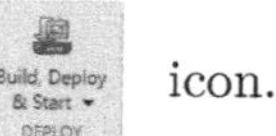 icon.

By double clicking on the manual switch icon, the input signal can be changed to the pulse generator. This block works analogously to the sine-wave one, thus, its explanation is not repeated here.

Be careful: the sine-wave can be edited any time by acting on the parameters of the Sin Wave block. To make these changes effective on the board, the translation in C-code and the deploy to hardware must be repeated every time. The same holds for the Pulse generator block.

Testing Environment

c28069_internalSin_T.slx (solver: fixed step - ODE4, step size: $T_{\text{sim}} = T_s$)

Open a new blank Simulink® project:

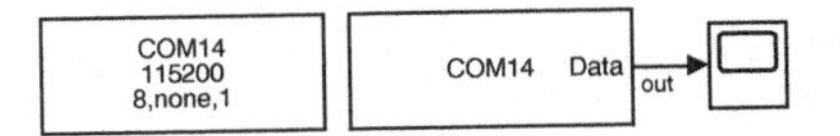

Figure 10.7 Simulink® scheme included in c28069_internalSin_T.slx.

- Open the **Model Configuration Parameters** window:
  - Select these two settings for the **solver selection** sub-menu:
    * type = fixed step;
    * solver = discrete (no continuous states);
  - Set the **fixed-step size** equal to $T_s = 100\,\mu s$;
  - Set the **stop time** equal to inf.
- Insert a **Serial Configuration** block. Double click on it:
  - Select the available COMx port as **communication port**;
  - Define the **baud rate** equal to 115200. This value must match with that one defined in the Model Configuration Parameters in the c28069_internalSin_F.slx file (see Chapter 9);
  - Set **data bits**: 8;
  - Set **stop bits**: 1;
  - Set **byte order**: LittleEndian;
  - Set **timeout**: 10.
- Insert a **Serial Receive** block. Double click on it:
  - Select the right COMx as **serial port**;
  - No **header** or **terminator** should be specified;
  - Set **data size** [1 1], since one signal is sent, received and visualized only;
  - Set **data type**: single;
  - Set **sample time**: 0.1 s.

Note that the data type sent from the board to the PC is single. Thus, `single` is set as data type in the Serial Receive parameters without any need for additional data conversion. The resulting sine-wave can be visualized through a Scope block. Finally, connect these blocks as shown in Figure 10.7. In particular, this scheme can also include a **Dashboard Scope** with signal associated to the output of the Serial Receive. This is a simple way to dynamically visualize signals over time. This approach is suggested for the next exercises included in the book which requires SCI as debugging tool. For simplicity reasons, Dashboard Scopes will be included in the testing environments, but not

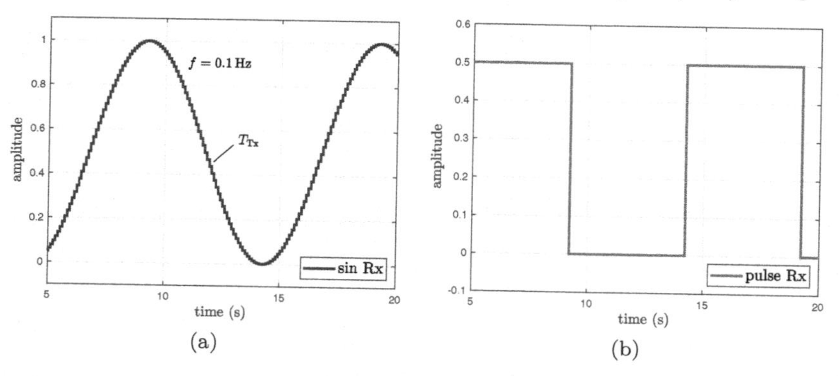

Figure 10.8 Data read from the SCI A when the internal signal is generated by Sine Wave (a) and Pulse Generator (b) blocks.

reported in any figure. Moreover, the Testing Environment does not need to deploy code to target. Indeed, this file should be run only as standard simulations to start sending data through the serial port. In order to continuously run the simulation, the stop time is set equal to inf (i.e., infinity). It is important to note that the serial configuration is the "main" block which sets the parameters of the COMx communication, possibly overwriting previous settings edited in the PC device manager. The results of this exercise are shown in Figure 10.8.

Example 2

> Build a firmware aimed at processing a signal received from serial communication by the MCU and, then, send it back to the host PC

In this example, the F28069M LaunchPad™ board must be able both to read an external command and send it back through the virtual serial COMx port. SCI A module is used. Thus, the right jumper configuration should be realized on the board (see Table 3.2).

Firmware Environment

c28069_externalSin_F.slx (solver: fixed step - ODE4, step size: $T_s = 100\,\mu s$)

Open a new blank Simulink® project and configure the environment as shown in Chapter 9. Then:

- Open the **Model Configuration Parameters** window and use the same settings adopted in **Example 1**.

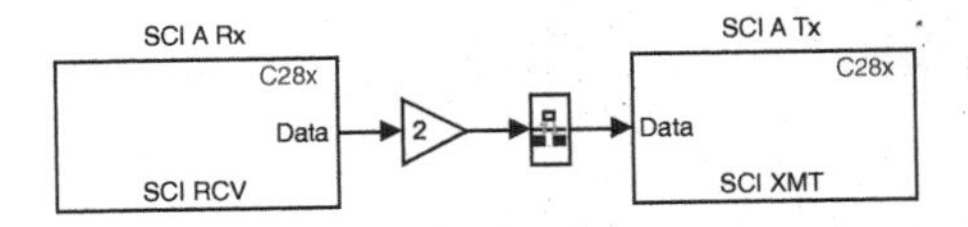

Figure 10.9 Simulink® scheme included in c28069_externalSin_F.slx.

- Place a **SCI Transmit** block in the file and and use the same settings adopted in **Example 1**. The Rate Transition block is set with an Output port sample time equal to 0.01 s.
- Insert a **SCI Receive** block in the file and double click on it to modify its parameters as follows:
  - Select SCI module: A;
  - Remove both additional package header `'S'`→'' and terminator `'E'`→'';
  - Set data type: single;
  - Set data length: 1;
  - Set initial output: 0;
  - Set sample time: 0.1 s.
- Add a **gain** block and set its value equal to 2. Moreover, set the output data type as single in the Signal Attributes tab.

Finally, connect these aforementioned blocks as shown in Figure 10.9. In this example, the board F28069M reads an external signal from the COMx port, multiplies it through a gain and sends it back to the PC. The signal coming from the COMx port is bounded between $[0, 1]$ and it is in single-precision floating point format. A data type conversion is not necessary because the SCI RCV is already set to handle `single` data, while the sample time is set to 0.1 s, which is equivalent to 10 Hz communication frequency (sample & hold effect on the processed signal).

Testing Environment

c28069_externalSin_T.slx (solver: fixed step - ODE4, step size: $T_{\text{sim}} = T_s$)

Open a new blank Simulink® project:

- Open the **Model Configuration Parameters** window and use the same settings adopted in **Example 1**;
- Place a **Serial Configuration** block and use the same settings adopted in **Example 1**;
- Insert a **Serial Send** block. Double click on it:

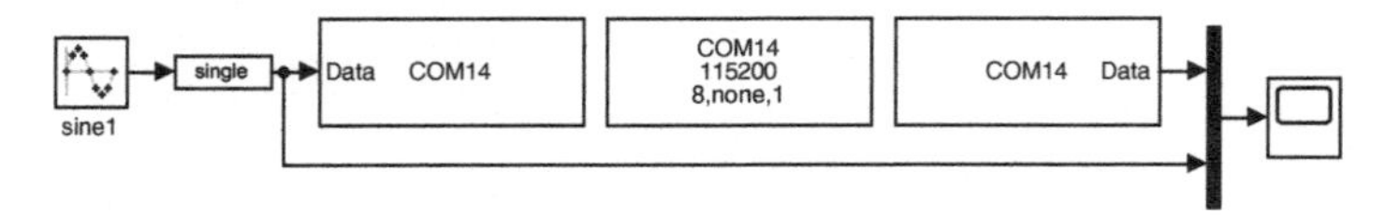

Figure 10.10 Simulink® scheme included in c28069_externalSin_T.slx.

- Select the right COMx port as serial port;
- Do not insert any header or terminator.

- Add a **Serial Receive** block and use the same settings adopted for **Example 1**;
- The external signal is generated by a **Sin Wave** block with the same parameters set in **Example 1**

In this example, the communication through COMx involves single-precision floating point numbers and it is full-duplex. Hence, both the Serial Receive and the SCI Receive blocks are set to work with single data type without the need for any additional data conversion. The input and output sine-waves can be plotted in the same Scope block through a multiplexer (MUX) block. Finally, connect these blocks as shown in Figure 10.10. The resulting waveforms are reported in Figure 10.11.

Computational delay: it is important to note that there is a slight shift between the two waveforms. In general, this computational delay is due to both the finite processing time of the microcontroller (T_{ck}) and the sample time of the serial communication (T_{RX}). In this case, the latter is much larger

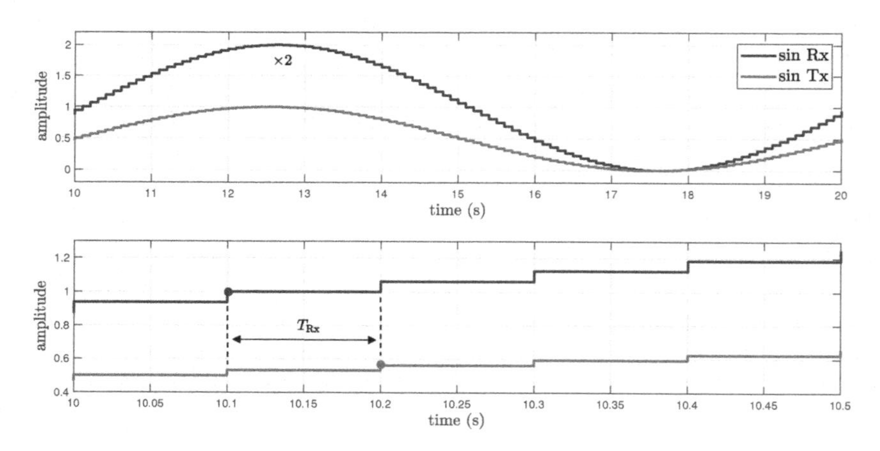

Figure 10.11 Transmitted/received sine-wave signals over time with a detailed view of the computational delay which causes the shift between them.

than the former. Hence, this computational delay is mainly due to the serial communication. This is detailed by a zoom view reported in the lower side of Figure 10.11, allowing to visualize T_{RX}.

11

GPIO Peripheral—Digital Input/Output

This Chapter describes the general-purpose input/output (GPIO) peripheral and how to deal with digital input/output pins. Indeed, the GPIO peripheral has general-purpose digital state pins that can be configured as both inputs and outputs. The peripheral has a resolution of 16 bits and its internal circuits (and logic) work in the voltage range of $0 - 3.3\,\mathrm{V}$.

When configured as an output, an internal register can be written to control the state driven on the corresponding pin. When configured as an Input, it is possible to detect its level by reading the state of an internal register. Both reading and writing operations for those registers can be easily done in Simulink® through suitable blocks.

The most common settings foresees the assignment of different pins of the LaunchPad™ board (i.e., among the 80 pins accessible on top of it) to each GPIO channels. For instances, a GPIO peripheral may be used also for internal purposes to control some communication features (e.g., CAN, SPI, etc..). Moreover, some GPIO channels share the external pin of the LaunchXL F28069M with other peripherals (e.g., ePWM, eQEP, eCAP). Hence, a multiplexing mechanism is adopted to select the right function on the shared pin. The assignment of some pins can be reconfigured in Simulink® according to specific necessities. The default pin-out of the adopted board is reported in Figure 3.4 and 3.5.

11.1 Hardware Details

The GPIO pins are multiplexed by the GPxMUXn registers, which selects the operation to be performed on the shared pins. On this board, they are labeled with the GPIO signal to which they are connected, that is, GPIO0-GPIO58.

Moreover, the LaunchPad™ board has three I/O ports:

- **Port A** which includes GPIO0-GPIO31 signals;
- **Port B** which includes GPIO32-GPIO58 signals;
- One analog port, which includes AIO0-AIO15 signals.

DOI: 10.1201/9781003196938-11

In particular, the pins of the analog port are connected to the channels of the analog-to-digital converter peripheral ADCA0-7 and ADCB0-7. Further details on this peripheral are reported in the next Chapter.

These pins can be operated as digital I/O (referred to as GPIO) or connected to one up to three I/O peripheral signals via the GPxMUXn registers. If digital I/O mode is selected, the pin "direction" (i.e., input or output) is configured via the GPxDIR registers. It is also possible also to qualify the input signals to remove unwanted noise via the GPxQSELn, GPACTRL, and GPBCTRL registers. Note that most of the pins are re-configurable (i.e., they are associated to many registers). However, Simulink® provides a limited access to them. Indeed, only some of them can be edited through the dedicated block library. For a more flexible customization of the code, the reader is suggested to use Code Composer Studio™.

Every time a reconfiguration of the I/O pins is needed, it is suggested to follow these steps:

1. **Organize a pin-out table.**
2. **Enable or disable internal pull-up resistors.**

   To enable or disable the internal pullup resistors, write to the respective bits in the GPIO pullup disable (GPAPUD and GPBPUD) registers. For pins that can operate as ePWM output pins, the internal pullup resistors are disabled by default. All the other GPIO-capable pins have the pull-up resistor enabled by default. The AIOx pins do not have internal pull-up resistors. It must be noted that the pin register access in its full extent is available only through Code Composer Studio™.
3. **Select input qualification.**

   In case a pin is used as an input, it may be needed to specify its input qualification. This information should be set in the GPACTRL, GPBCTRL, GPAQSEL1, GPAQSEL2, GPBQSEL1, and GPBQSEL2 registers. By default, all the input signals are synchronized by **SYSCLKOUT**.
4. **Select pin function.**

   Configure the GPxMUXn or AIOMUXn registers such that each pin to be re-configured is connected to one of the available peripheral. By default, all GPIO-capable pins are configured at reset as general purpose input pins.
5. **For digital general purpose I/O, select direction of the pin.**

   Every time the pin is configured as a GPIO, the direction of the pin should be specified as either input or output in the GPADIR, GPBDIR, or AIODIR registers. By default, all GPIO pins are inputs. To change the pin from input to output, first load the output latch with the value to be driven by writing the appropriate value to the GPxCLEAR, GPxSET, or GPxTOGGLE (or AIOCLEAR, AIOSET, or

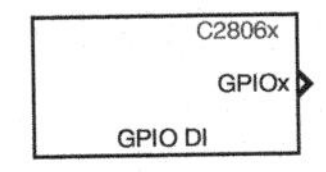

Figure 11.1 C2806x GPIO Digital Input block.

AIOTOGGLE) registers. Once the output latch is loaded, change the pin direction from input to output via the GPxDIR registers. The output latch for all pins is cleared at reset.

6. **Select low power mode wake-up sources**:

   Specify which pin is able to wake the device up from HALT and STANDBY low power modes. The pins are specified in the GPIOLPMSEL register.

7. **Select external interrupt sources.**

   Specify the source for the XINT1 - XINT3 interrupts. For each interrupt, one of the port A signals can be specified as the source. This is done by specifying the source in the GPIOXINTnSEL register. The polarity of the interrupts can be configured in the XINTnCR register.

11.2 Firmware Environment: GPIO Peripherals

In this Section, the two GPIO blocks available in the *Embedded Coder Support Package for Texas Instruments C2000 Processors* library (C2806x subset) are presented.

11.2.1 C2806x GPIO digital input (GPIO DI)

This block (see Figure 11.1) allows to configure a GPIO pin as a digital input. Namely, it sets the GPxMUXn on digital input (DI). Once a pin is configured as GPIO DI, it becomes unavailable for digital output or other peripheral operations. The GPIO peripheral acquires a voltage signal from the corresponding pin V_{pin} and it is interpreted as high (True) or low (False), according to the voltage thresholds reported previously.

The main settings of this block are reported here in the following.

Block Parameters:

- **GPIO Group**

  GPIO signals are organized into groups. This setting allows the selection of the group of pins to view or configure. Changing group, different pins can

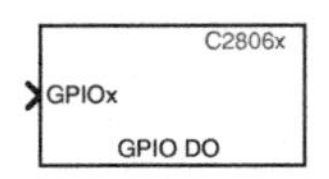

Figure 11.2 C2806x GPIO Digital Output block.

be assessed by ticking the corresponding check-boxes. Only pins that belong to the same GPIO group can be selected simultaneously. The reader should refer to the pin-out reported in Figure 3.5 for the pin/GPIO correspondence.

- **Sample Time**

  This field specifies the time interval that elapses between two consecutive output samples. Set this parameter equal to -1 to inherit sample time from the adopted solver.

- **Data Type:**

  This parameter sets the data type of the input, which is read as a 16-bit integer and, then, it is cast to the selected data type. Valid data types are auto, double, single, int8, uint8, int16, uint16, int32, uint32 or boolean. Regardless of the data type, high states are translated as the logic state 1 and low ones as logic state 0. Indeed, the state of the pins can be either high or low, depending on the voltage applied on it V_{pin}:

  – Voltage inputs between 0 and 1.3 V lead to LOW output values, i.e., logic 0;
  – Voltage inputs over 1.5 V lead to HIGH output values, i.e., logic 1;
  – Voltage inputs in the range $[1.3, 1.5]$ V lead to random output values, i.e., logic 0 and 1.

Remark

Since voltage inputs in the range $[1.3, 1.5]$ V lead to random logic states, it is recommended to keep V_{pin} quite far from the high/low voltage threshold.

11.2.2 C2806x digital output (GPIO DO)

This block (see Figure 11.2) allows to configure a GPIO pin as a digital output. Namely, it sets the GPxMUXn on digital output (DO). Once a pin is configured as GPIO DO, it becomes unavailable for digital input or other peripheral operation. Digital output GPIO blocks receive an input signal in Simulink® and it sends an output voltage on the corresponding pin:

- 0 (low/false signal): $V_{\text{pin}} = 0\,\text{V}$ (pin grounded);
- 1 (high/true signal): $V_{\text{pin}} = 0\,\text{V}$ (pin pulled high).

Every digital output GPIO has a toggle option that inverts the output signal on the pin. The main settings of this block are reported here in the following. One parameter is not explained since it is used to set the same function described for the C2806x GPIO Digital Input block.

Block Parameters:

- **GPIO Group**
- **Regular- and Toggle mode**

  For each selected pin, regular- (default) or toggle mode can be selected to drive the GPIO signal. In regular mode, true inputs pull the GPIO pin high. False ones ground the pin. In toggle mode, true values switch the pin output level continuously. Thus, a true input switches the output voltage from high to low and vice-versa. Instead, if the input is set to false, the pin output level is unaffected. Namely, the last logic value is kept constant as long as the the input is kept equal to 0.

  Remark
  The switching frequency in toggle mode is limited by the 20 MHz rate of the pin buffers. Regardless of the input data type, every time a numerical signal is applied to a GPIO in regular-mode, this value is interpreted as:

  - Low if the signal applied to the GPIO is a 0 numerical value;
  - High if the signal applied to the GPIO is any numerical value $\neq 0$.

11.3 Examples with GPIO blocks

Example 1

> Build a firmware which is able to drive the right GPIO signal to turn the blue led mounted onboard the LaunchPad™ board on/off

The led turn on/off is the "hello world" for micro-controller progrmming. This simple exercise is essential to understand the electrical characteristics of the microcontroller in use. The schematics of the LaunchXL F28069M board as well as its pin-out can be found in the corresponding User's guide [16]. From this file, it can be noted that this board includes two programmable leds which are physically connected to two digital pins: a blue (GPIO39) and red (GPIO34) one. An initialization script can be used to load some constants, e.g., in *Model Properites/Callbacks/InitFcn*.

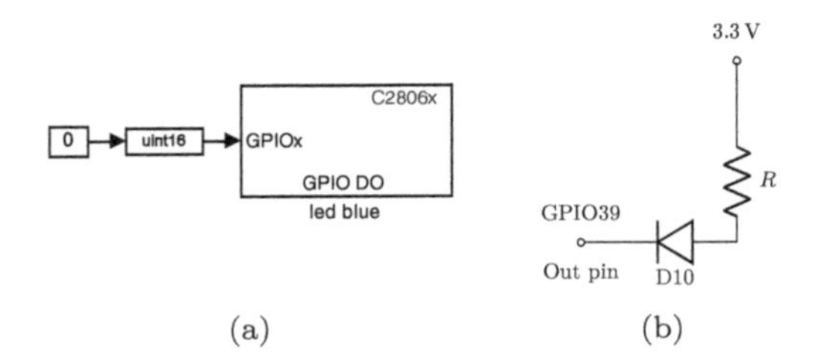

Figure 11.3 Simulink® scheme included in c28069_ledB_F.slx (a) and physical connection between the diode D10 and the corresponding output pin (b).

Firmware Environment

c28069_ledB_F.slx (solver: fixed step - ODE4, step size: $T_s = 100\,\mu s$)

Open a new blank Simulink® project and configure the environment as shown in Chapter 9. Then:

- Open the **Model Configuration Parameters** window and use the same settings adopted for **Example 1** in Section 10.5;
- Insert a **Digital Output** (GPIO DO) block and double-click on it:
  - Select GPIO Group **GPIO32~GPIO39**;
  - Flag **GPIO39**;
  - It is possible to test the firmware without enabling **Toggle mode** and then with it to appreciate how the output is affected. Note that once every time a pin flag is modified, the build and deploy operation must be repeated.

A constant block is used to produce an internal command signal which drives the GPIO39 with 0 or 1 values. The sample time of the constant block is set equal to $T_{sig} = 1000T_s = 0.1\,s$. The led is turned on/off accordingly. Since the resolution of GPIO is 16 bit, a conversion using a Data Type Conversion block must be performed to obtain an uint16 format. Note that even a different data type might be used. Finally, all the aforementioned blocks should be connected as shown in Figure 11.3 (a). It is worth noting that Figure 11.3 (b) highlights that the circuit works with a *negative logic*. Namely, the led anode is connected to $V_{cc} = 3.3\,V$ through a resistor, while the cathode is in contact with GPIO39. As a consequence, a high value for GPIO39 implies an high pin voltage, which keeps the led off since no difference in potential (voltage) is present . On the other hand, a zero value implies a zero pin voltage, which turns the led on. Figure 11.4 shows a typical voltage waveform that can be measured on the GPIO pin every time it is driven in toggle mode and its input is set equal to 1.

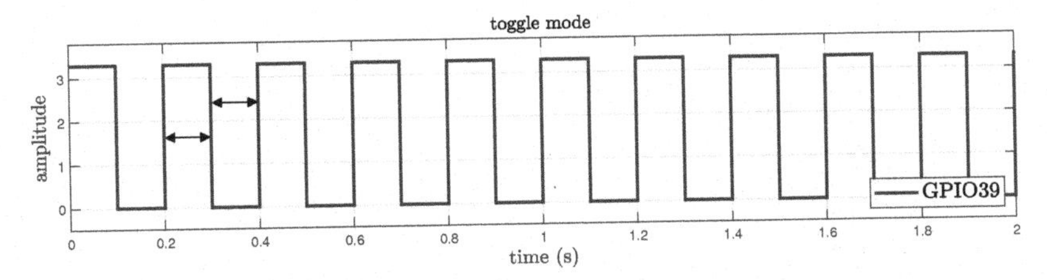

Figure 11.4 Example of GPIO39 output voltage over time in toggle mode.

Remark

The input value of GPIO39 can be edited by changing the value of the constant block. The translation in C-code and the deploy to hardware must be repeated at each modification of this value.

Example 2

> Build a firmware which is able to drive the right GPIO signals to to make the red and the blue leds blink alternately

In this exercise, the discrete time Delay block is exploited to set a certain delay of N_d time steps for the driving signal of one led. The time delay in seconds is computed as $N_d T_{sig}$, where $T_{sig} = 1000 T_s$ and T_s is the fixed step size of the solver.

In order to make the leds blink, the toggle mode must be enabled in the GPIO DO block parameters window. The interleaved blinking is achieved by selecting an odd time delay (e.g., $N_d = 1$ which imply one step delay)

Firmware Environment

c28069_ledBR_F.slx (solver: fixed step - ODE4, step size: $T_s = 100\,\mu s$)

Open a new blank Simulink® project and configure the environment as shown in Chapter 9. Then:

- Open the **Model Configuration Parameters** window and use the same settings adopted for **Example 1** in Section 10.5; the constant block value must be equal to 1 in order to use the toogle mode.
- Insert a **Digital Output** (GPIO DO) block and double click on it:
  - Select GPIO Group **GPIO32∼GPIO39**;
  - Flag **GPIO39**;

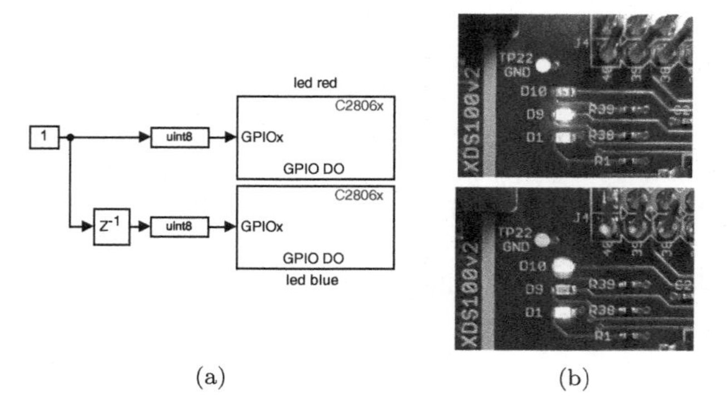

(a) (b)

Figure 11.5 Simulink® scheme included in c28069_ledBR_F.slx (a) together with the red (D9) and blue (D10) leds blinking accordingly (b).

  - Enable **Toggle mode**.

- Insert a new **Digital Output** (GPIO DO) block and double-click on it:
  - Select GPIO Group **GPIO32∼GPIO39**;
  - Flag **GPIO34**;
  - Enable **Toggle mode**.

  The blinking speed depends on the sample time T_{sig} set in the constant block.

- Insert a **Delay** block (it can be found in the Discrete library) and:
  - Set **Sample time** equal to 0.1 s.
  - Set **Delay Length** equal to 1, i.e., one step delay referred to T_{sig}.
- Insert two **Data Type Conversion** blocks to feed the two GPIO DO blocks with uint16 or uint8 data.

Finally, connect these blocks as shown in Figure 11.5 (a). The blinking effect would be directly visible on the board as reported in Figure 11.5 (b).

Example 3

Build a firmware which is able to drive the GPIO channels to turn the leds on/off by reading an external command signal through serial port

For this exercise, two Simulink® files must be created. One is needed to program the MCU board and another one to test it by sending data through COM port. SCI module A is used for communication purposes. Thus, a suitable jumper configuration should be realized on the board (see Table 3.2).

Firmware Environment

c28069_SCIledBR_F.slx (solver: fixed step - ODE4, step size: $T_s = 100\,\mu s$)

Open a new blank Simulink® project and configure the environment as shown in Chapter 9. Then:

- Open the **Model Configuration Parameters** window and use the same settings adopted for **Example 1** in Section 10.5.
- Insert two **Digital Output** (GPIO DO) blocks and double-click to set:
  - Select GPIO Group **GPIO32~GPIO39**;
  - GPIODO1: flag **GPIO34** (disable toggle mode);
  - GPIODO2: flag **GPIO39** (disable toggle mode).
- Insert a **SCI Receive** block and double-click on it:
  - Select SCI module A;
  - Remove both additional package header `'S'`→'' and terminator `'E'`→'';
  - Set data type: uint8;
  - Set data length: 1;
  - Set initial output: 0;
  - Set sample time: $T_{\text{sig}} = 0.1\,\text{s}$.
- Place an **Add** and a **Constant** block to invert the signals fed into GPIOs. Set the signs of the add block equal to $-+$ and the constant block sample time equal to 0.1 s. This arrangement is such that every time GPIO34 turns on, then, GPIO39 turns off and vice-versa.

In this case, the input of the SCI module is `uint8`. Thus, an input signal bounded between $[0, 1]$ is converted into a integer ranging in $[0, 255]$.
A sample time equal to 0.1 s (meaning a working frequency of 10 Hz) is set for the serial communication to ensure good data transfer. The serial communication was tested by the authors up to 5 kHz. However, this is not its upper limit, which depends by the firmware optimization. Finally, connect the aforementioned blocks as shown in Figure 11.6 (a).

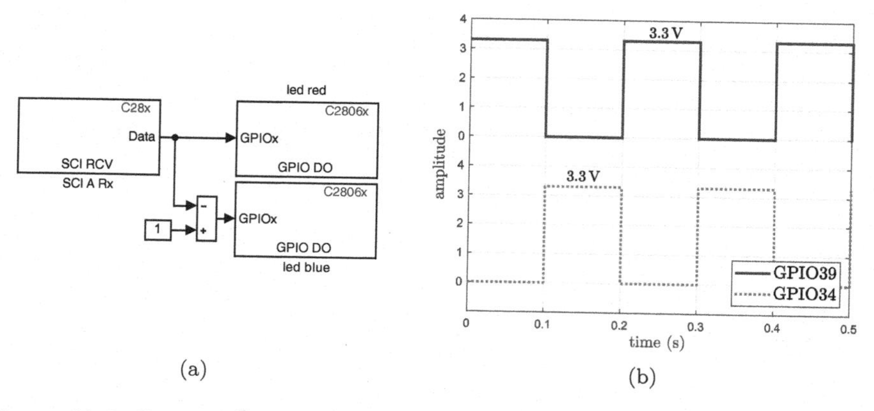

Figure 11.6 Simulink® scheme included in c28069_SCIledBR_F.slx (a) and expected outputs (condensed plots) on the corresponding GPIO pins (b).

Testing Environment

c28069_SCIledBR_T.slx (solver: fixed step - ODE4, step size: $T_{\text{sim}} = T_s$)

Open a new blank Simulink® project:

- Open the **Model Configuration Parameters** window and use the same settings adopted for **Example 1** in Section 10.5.
- Place a **Serial Configuration** block and use the same settings adopted for **Example 1** in Section 10.5.
- Insert a **Serial Send** block and use the same settings adopted for **Example 2** in Section 10.5.

This exercise consists in sending a constant command signal to drive GPIO34 and GPIO39 through the SCI A. Hence, a constant block is inserted into the scheme. Dashboard elements can be added to the testing environment scheme and connected to a constant block to easily perform live changes of the command signal ($0 \leftrightarrow 1$). Alternatively, a manual change of this value can be done by opening and modifying the constant block value (its sample time can still be settled as T_{sig}). Since the programming environment is set to receive uint8 data, a data type conversion must be performed before the Serial Send block through a **Data Type Conversion** block . Finally, connect these blocks as shown in Figure 11.7. Remember to edit an initialization script, e.g., in *Model Properites/Callbacks/InitFcn*, even for the testing environment.

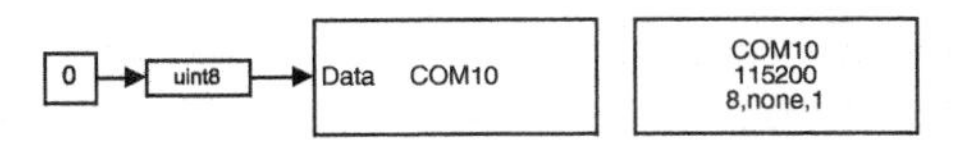

Figure 11.7 Simulink® scheme included in c28069_SCIledBR_T.slx.

Example 4

Build a firmware which is able to drive a PWM modulation of the blue led blinking. The modulation signal is read through COM port

If the led turn on/off is the "hello world" for microcontroller programming, the realization of a pulse-width-modulation (PWM) technique is the "hello world" for power electronics applications. In this example, a PWM logic is built from scratches using basic Simulink® elements. The Embedded Coder Support Package for Texas Instruments C2000 Processors library includes optimized PWM peripherals, but they are not used in this exercise. Indeed, this example is aimed at providing the reader more insight into the logic behind this modulation technique before to move in built-in peripherals.

The external command transferred thorugh COM port represents the modulation signal which is compared to a triangular carrier. The result of this operation is a train of pulses which drives the blinking of the blue led (GPIO39).

Firmware Environment

c28069_SCIledBpwm_F.slx (solver: fixed step - ODE4, step size: $T_s = 100\,\mu s$)

Open a new blank Simulink® project and configure the environment as shown in Chapter 9. Then:

- Open the **Model Configuration Parameters** window and use the same settings adopted for **Example 1** in Section 10.5.
- Insert a **SCI Receive** block and double-click on it:
  - Select SCI module A;
  - Remove both additional package header `'S'`→`''` and terminator `'E'`→`''`;
  - Set data type: `single`;
  - Set data length: 1;
  - Set initial output: 0;
  - Set sample time: 0.001 s.
- Insert a **SCI Transmit** block and set it as reported in Section 10.5.

- Place a **Rate Transition** block and set its sample time equal to $T_{\text{TX}} = 0.001\,\text{s}$ and a **Data Type Conversion** block set as `single` before sending the data via SCI XMT.
- Place another **Rate Transition** block and set its sample time equal to $T_s = 0.001\,\text{s}$. Namely, this block is to be placed aftera SCI RCV and all the computations done after it are executed every T_s.
- Add a **Digital Output** (GPIO DO) block and double-click on it:
  - Select GPIO Group **GPIO32∼GPIO39**;
  - Flag **GPIO39** (disable toggle mode).
- Place a **Saturation** block $[0, 1]$ to restrict the modulating signal in case a value greater than one is sent over SCIA.

The simplest approach to realize a PWM logic is using a **Relational Operator** block. The signal coming from SCIA represents the modulating signal, which is applied to the first input of the comparator. This is compared to a triangular carrier at fixed frequency, which is applied to the second input of such block. The desired behavior is such that:

- Every time the triangular waveform is greater than the modulating signal, the output of the comparator is low (0);
- Every time the modulating signal is greater than the triangular waveform, the output of the comparator is high (1).

This simple method produces an output square wave with a duty cycle that varies depending on the level of the modulating signal, thus, changing the blinking of GPIO39. The resulting switching frequency f_{sw} is equal to the carrier frequency f_c. For this exercise, let assume $f_{sw} = f_c = 20\,\text{Hz}$.

Nevertheless, in terms of implementation, the PWM principle can be realized in many (embedded) ways. For simplicity reason, this exercise will only discuss how implement the carrier signal. In particular, the Simulink® scheme include two different approaches allowing to switch between them through a **Manual switch**:

1. **Continuous-time strategy:**

   The simplest approach is to realize the carrier frequency as a repetitive sequence of values, as shown in Figure 11.8. This scheme comprises:

   - A **Repeating Sequence** used for building a symmetrical triangular carrier. Double-click on it:

     –Set time values: $[0\ 1/(2f_{sw}), 1/f_{sw}]$
     where $T_{\text{sw}} = 1/f_{\text{sw}} = 1/20 = 0.05\,\text{s}$;

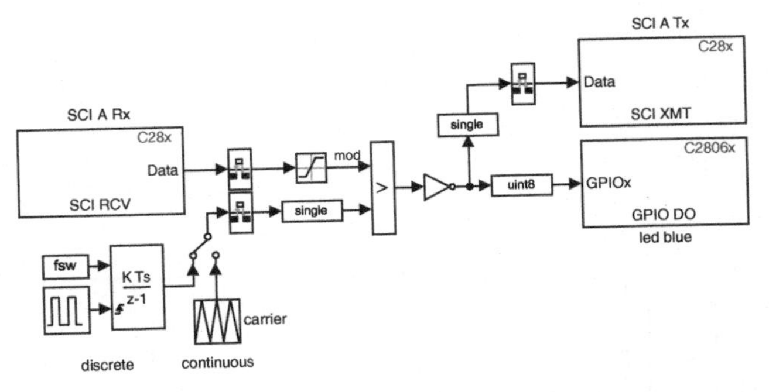

Figure 11.8 Simulink® scheme included in c28069_SCIledBpwm_F.slx.

–Set output values [0 1 0].

- A **Rate Transition** block after the carrier generation and set its sample time equal to $T_s = 100\,\mu s$, i.e., all the computations done after this block are be executed every T_s.
- A **Data Type Conversion** set to `single` values to be consistent with the modulating signal.
- A **Relational Operator**, which performs the comparison between modulation and carrier signals. Double-click on it:

  –Set the relational operator equal to $>$;

  –Set the output data type of this block as Boolean.
- A **Not** port, which allows to be consistent with the negative logic which drives the leds, that is:

  –PWM output $1 \to 0$

  –PWM output $0 \to 1$

  This strategy allows to obtain on the led connected to GPIO39 a slow blinking for low values of the modulating signal, i.e., mainly off within T_{sw}, while the contrary occurs for high value of the modulating signal, i.e., mainly on within T_{sw}.
- Another **Data Type Conversion** set to `uint8` is used to bound the GPIO39 input signal.

Finally, connect these blocks as shown in Figure 11.8. The resulting waveforms are reported in Figure 11.9. To use continuous-time blocks (i.e., the **Repeating Sequence** block), the feature for allowing continuous time operation of the C2000™ family is exploited. To enable this option, open the Simulink® **Configuration Parameters** window:

- Go in the **Code Generation**, **Interface** menu;

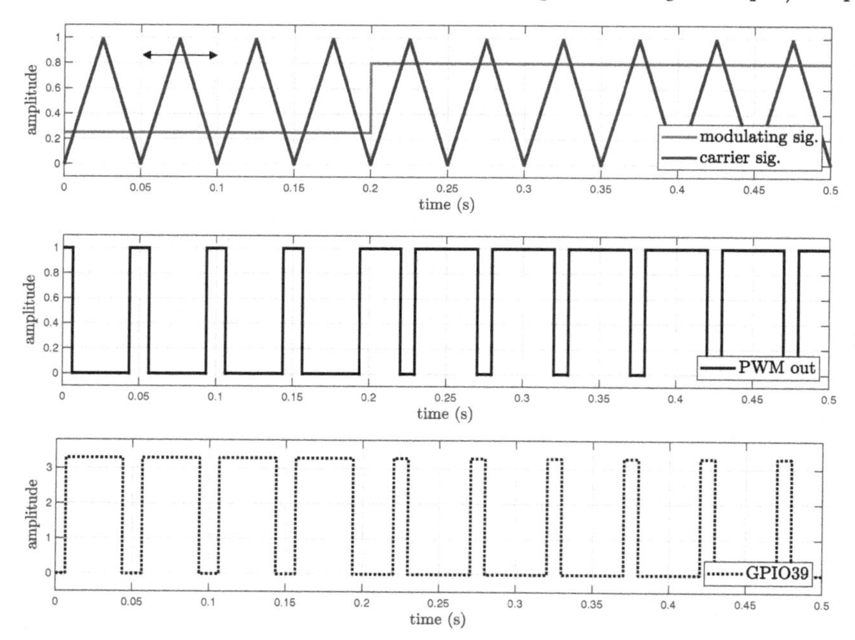

Figure 11.9 Resulting signals for a PWM logic which implements a carrier built up with continuous-time blocks.

- Flag the **continous time** checkbox in the **Software environment/Support** sub-menu.

2. **Discrete-time strategy:**

   Alternatively, the carrier can be built using a **Discrete-time Integrator** block. Indeed, the carrier signal can be approximated through a staired waveform. In this case, a saw-tooth carrier is built. Assuming to apply a Forward Euler discretization method, the discrete-time integrator block works according to the following equations:

$$\begin{cases} y(k) = y_0 & \text{if } k = 0 \\ y(k) = y(k-1) + K[t(k) - t(k-1)]u(k-1) & \text{if } k > 0 \end{cases}$$

   where $y(k)$ is the output of the block, $u(k)$ is the input of the block, k is the time step index, K is the gain of the integrator and y_0 is the initial condition. The parameters of the block are set as follows:

   - **Integrator method**: Integration: Forward Euler;
   - **Gain value**: 1;

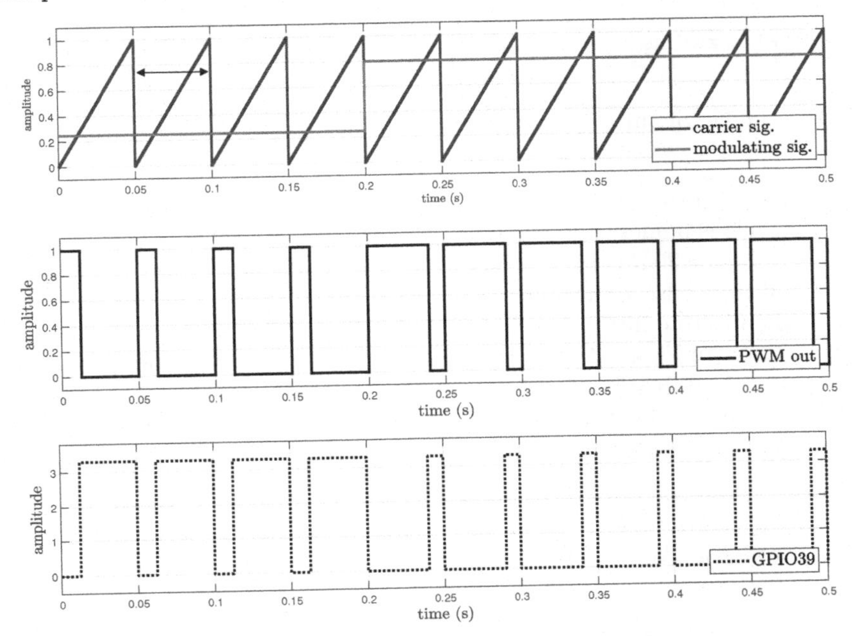

Figure 11.10 Resulting signals for a PWM logic which implements a carrier built up with a discrete-time integrator.

- **External reset**: Rising;
- **Initial condition**: 0;
- **Sample time**: T_s.

The external reset is needed to realize a saw-tooth waveform. Thus, a constant value is integrated until the external reset trigger commands the reset of the ramp. After that, the integration restarts from its initial condition. This external reset is realized through a **Pulse Generator** block set as follows:

- **Pulse type**: Time based;
- **Time (t)**: Use simulation time;
- **Amplitude**: 1;
- **Period (secs)**: $T_{sw} = 0.05\,\mathrm{s}$;
- **Pulse Width (% of period)**: 50;
- **Phase delay (secs)**: 0;
- Tick the checkbox **Interpret vector parameters as 1-D**.

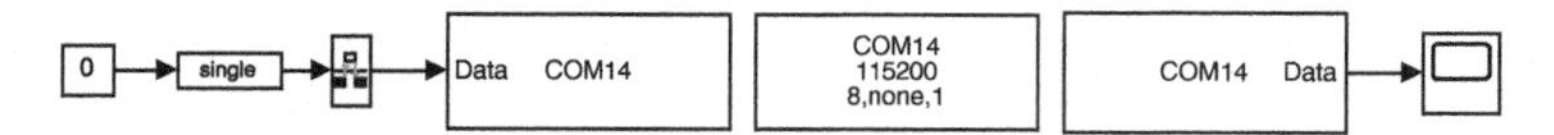

Figure 11.11 Simulink® scheme included in c28069_SCIledBpwm_T.slx.

> The Discrete-time Integrator block integrates a constant value which is numerically equal to the desired switching frequency, which is $f_{sw} = 20\,\text{Hz}$. In this case, the rising edge of the pulses created by the Pulse Generator block are enough for enabling the reset of the discrete integrator. The aforementioned blocks are arranged as shown in Figure 11.8. The resulting waveforms are reported in Figure 11.10. It is important to underline that to realize a given carrier frequency, the gain value of the integrator should be set equal to f_{sw}. In this case, the carrier peak value reaches 1 every $T_{sw} = 0.05\,\text{s}$.

In both cases, the initialization script (which can be created in *Model Properites/Callbacks/InitFcn* or in a separate m-file) now requires a global definition of f_{sw}. An example of such script is reported here below:

```
%% carrier frequency definition
fsw   = 20;
Tsw   = 1/fsw;
%% sampling time definition
Ts    = 100e-6;
```

Testing Environment

c28069_SCIledBpwm_T.slx (solver: fixed step - ODE4, step size: $T_{\text{sim}} = T_s$)

Open a new blank Simulink® project:

- Open the **Model Configuration Parameters** window and use the same settings adopted for **Example 1** in Section 10.5.
- Place a **Serial Configuration** block and use the same settings adopted for **Example 1** in Section 10.5.
- Insert a **Serial Send** block and use the same settings adopted for **Example 2** in Section 10.5.
- Place a **Constant** block. This value defines the modulating signal which has to be bounded between 0 and 1. A **Data type conversion** set to `single` and a **Rate Transition** with sample time equal to 0.01 s are included as well to prepare the data for transmission.
- Add a **Serial Receive** block and use the same settings adopted for **Example 1** in Section 10.5, apart from the fields **Data type**, which now should be set to `single`, and **Sample time**, that is equal to 0.01 s.

- Insert a Scope block to check if the control signal is set correctly.

Finally, connect all these blocks as shown in Figure 11.11. The waveforms that should be visualized in the scope are the those reported in the bottom plots of Figure 11.9 and 11.10 for the continuous and discontinuous carriers, respectively.

12

Analog to Digital Converter Peripheral

In this Chapter, the working principles and the main configuration of analog to digital converter (ADC) peripherals mounted on TI boards are briefly outlined. ADCs are fundamental elements to sample analog external signals and convert them into digital data to be processed by the MCU. In particular, the ADC peripherals installed on the F28069M LaunchPad™ board has 12-bit resolution (part SAR, part pipelined) with input pins operating in the range $0 \div 3.3$V and it able to work up to 3.46 MSPS. Since ADC peripherals includes many analog and digital circuits besides the analog-to-digital converter core, it is common to refer to them as ADC modules. The analog circuits include the front-end analog multiplexers (MUXs), sample-and-hold (S&H) circuits, the conversion core, voltage regulators, and other analog supporting circuits. Digital circuits include programmable conversions, interface to analog circuits and to device peripheral bus. Moreover, ADCs work at full system clock without the need of any prescaler.

The ADC core contains a single 12-bit converter fed by two sample-and-hold circuits (A and B). The S&Hs can sample both simultaneously and sequentially, and they are fed by up to 16 analog input channels (8 for S&H A and 8 for S&H B). The reader is reference to Figure 3.5 for more details on the pinout of the board. The ADC blocks available in Simulink® allow to set all the ADC modules options.

12.1 Operating Principle

The ADC peripheral allows the sampling and the conversion of analog external signals to discrete ones. In general, this process involves voltage signals and consists in two main steps:

1. Sampling and Holding;
2. Analog to Digital conversion.

In general, an analog signal changes in time continuously. Thus, its conversion into digital values is possible only if it is kept steady for a suitably long time. Namely, the voltage signal is held for a specific time interval and, then, it is converted. The sample and hold (S&H) circuit keeps the value stable for

DOI: 10.1201/9781003196938-12

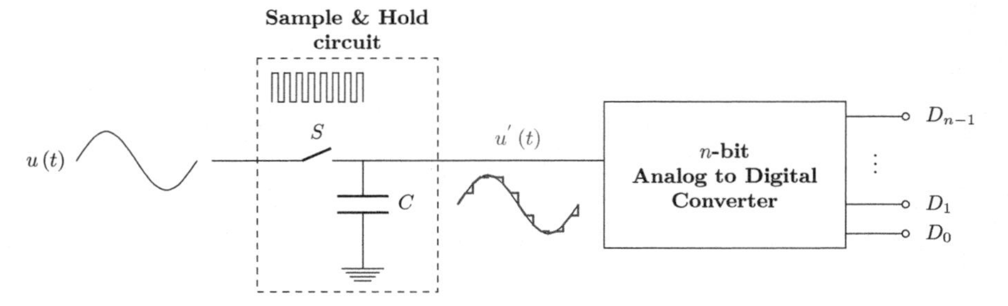

Figure 12.1 Operation principles of analog to digital conversions.

at least the time required for the digitizalization, which is performed by the ADC. This process is sketched in Figure 12.1.

12.1.1 Sample & hold

The S&H circuit has a switch which is normally kept open. The sampling process occurs every time this switch is closed, allowing the hold capacitor to charge (holding process). Ideally, when the measurement window is opened, the input signal charges the capacitor instantly. Actually, the charge is never sent on the capacitor immediately. Indeed, charging the hold capacitor takes a certain amount of time due to the dynamics of the circuit, which is typically a RC one. Once the capacitor is fully charged, its voltage is quantized by the analog to digital conversion stage.

12.1.2 Analog to digital converter

Once the input signal is held by the S&H, a numerical value is assigned to the capacitor voltage. Namely, the nearest value to the peak value of the held signal is searched. After this matching, the assigned number is encoded in the form of a binary number, which is represented over "n" bits, that is, the resolution of the ADC. The accuracy of this operation is strictly dependent on the resolution of the converter, i.e., high resolutions imply accurate conversions. It is possible to find standalone ADC chip (e.g., the TI ADS8661), however, the LaunchXL F28069M board integrates the 12-bit resolution ADC core and the related circuits inside the TMS320F28069M. Note that a 12-bit ADC translates external signals in integer numbers ranging from 0 to 4095.

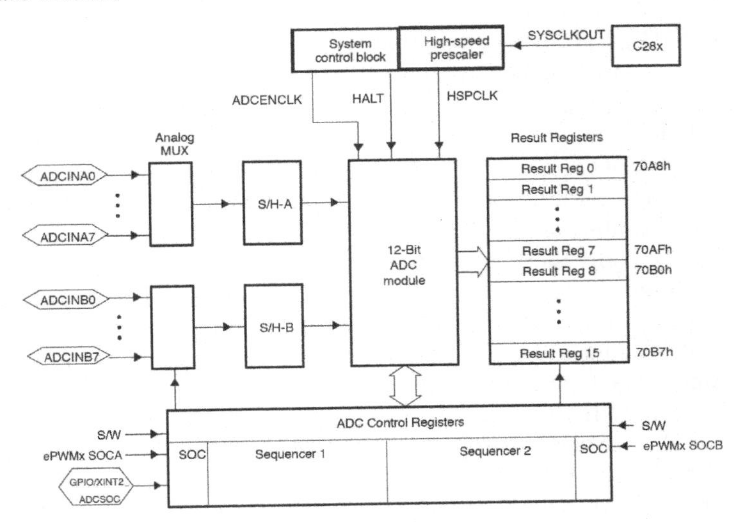

Figure 12.2 Block scheme showing the working principles of the ADC peripheral [19].

12.2 Hardware Details

A scheme of the 12-bit recyclic[1] ADC module installed on the LaunchXL F28069M board is shown in Figure 12.2. There, analog circuits such as the front-end analog multiplexers MUXs, sample-and-hold (S&H) circuits and the conversion core are represented as well as digital circuits such as programmable conversions, result registers, interface to analog circuits, interface to device peripheral bus and interface to other on-chip modules. It is important to note that part of this ADC module implementation is SAR [2] while another part is pipelined.[3] More specifically, the core of the ADC contains a single converter which is fed by two S&H circuits. These two circuits can sample signals simultaneously or sequentially up to 3.46 MSPS. The S&H circuits, in turn, are fed by 16 analog input channels by using two multiplexer.

The converter can be configured to run with an internal band-gap reference to create true-voltage based conversions or with a pair of external voltage

[1]Recyclic ADCs use the same module for both fine and coarse conversions of analog signals.

[2]SAR stands for successive approximation register. In this implementation, the signal is temporarily stored in a register and coded so that its most significant bit is equal to 1. Then, each bit of this code is compared with a reference voltage and converted step-by-step in the final digital signal.

[3]This implementation foresees the conversion through p stages. Each of them converts n bit at a time. This operation leads to some delays in the output since the converted signal is not available until all the stages complete the conversion. Then, the resulting signal is coded over np bit.

references VREFHI/LO to create full range ratiometric based conversions, that is, the signal is translated on a $0 \div 3.3$V voltage range.

This ADC peripheral has the ability to catch a series sequence of conversions, that is also called *simultaneous sampling mode.* This means that each time the ADC receives a start of conversion (SOC) request, it is able to schedule multiple conversions. The ADC can perform up to 16 sequential conversion. Their timing is specified in his SOCx register. Namely, the peripheral starts from the conversion specified in SOC0 followed by that one reported in SOC1, and so on. For each conversion, any of the available 16 input channels can be selected through the analog MUX. Input channel pairs can be simultaneously sampled on condition that the two coupled inputs have the same S&H offset. Namely, the two channels must have the same number, so that the pair ADCINA2-ADCINB2 is allowed, but ADCINA2-ADCINB3 is not. After conversion, the digital value of the selected channel is stored in the appropriate result register ADCRESULTn. Namely, there are 16 output registers; the first result is stored in ADCRESULT0, the second one in ADCRESULT1, and so on.

The basic operation principle of this peripheral is centered around the configurations of individual conversions, triggered at the start by a SOC event and at the end by an End Of Conversion (EOC) event. In particular, the 16 SOC registers can be configured by setting:

- Trigger source;
- Channel;
- Acquisition prescale window.

Moreover, multiple trigger sources are available:

- Software immediate start;
- ePWMx_ADCSOCA/ePWMx_ADCSOCB. Namely, they are interrupt signals generated by ePWM modules;
- XINT2_XINT2SOC. In this case, an external pin can be configured as a trigger;
- CPUx_TINTxn, i.e., CPU timers 0/1/2 interrupts;
- ADCINT1/2. Those ADC interrupts can be fed back to generate ADC sampling continuously. They can be set in the option *ADCINT will trigger SOCx.*

Finally, it is worth mentioning that the ADC peripheral contains 9 flexible PIE[4] interrupts, which can configure interrupt requests after any conversion.

12.2.1 Difference between acquisition window and sample time

As an advanced feature, external drivers vary in their ability to drive an analog signal quickly and effectively. Some circuits require longer time to properly transfer the charge into the sampling capacitor of an ADC. To account for this performance, the ADC peripheral supports control over the sample window length for each SOC configuration. Each ADCSOCxCTL register has a 6 bit field called ACQPS that determines the window size for the S&H. The value inserted in this field is one less than the number of cycles desired for the sampling window for that SOC. Thus, a value of 15 sets a 16 clock cycles sample time. The minimum allowed number is 7 clock cycles (i.e., ACQPS = 6). The total sampling time is found by adding the sample window size to the conversion time of the ADC, that is, 13 clock cycles. Hence, the overall sample time is:

$$(\text{ACQPS} + 1) + 13 = \text{ACQPS} + 14 \tag{12.1}$$

Instead, the Sampling Time register defines the rate of sampling from the ADC register. It can be chosen independently from the acquisition window, provided that it is greater than the observation interval. In this book, we are going to change directly the sample time defined as T_s.

12.3 Firmware Environment: ADC Peripheral

In this Section, the characteristics of the ADC block available in the *Embedded Coder Support Package for Texas Instruments C2000 Processors library* (C2806x subset) are briefly presented.

12.3.1 C2806x ADC

This block (see Figure 12.3) allows to set the channel of the ADC peripheral for acquiring an external analog signal. Usually, it is exploited for measurement purposes. Its main settings are reported here in the following and organized according to the tabs available in the Block Parameters window.

[4]The Peripheral Interrupt Expansion (PIE) is used to multiplex numerous interrupt sources into a smaller set. The PIE block can support up to 96 peripheral interrupts.

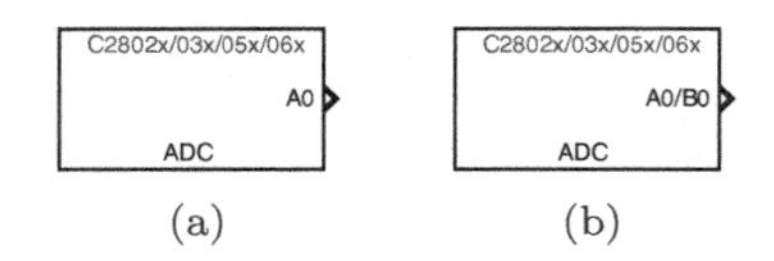

Figure 12.3 C2806x ADC block in two different configurations: (a) single sampling mode and (b) simultaneous sampling mode.

Block Parameters - SOC Trigger tab

- **Sampling Mode**

  This drop-down menu allows to select single or simultaneous sampling mode.

- **SOC Trigger Number**

  This setting allows to identify the SOC trigger adopted for each channel. In single sampling mode, one trigger can be selected only. In simultaneous sampling mode, paired signals (0-1, 2-3, . . . , 14-15) are available.

- **SOCx Acquisition Window**

  This parameter defines the length of the acquisition window in terms of ADC clock cycles. The value of this parameter depends on the SYSCLK and on the minimum ADC sample time. For the F28069M LaunchPad™ board, the minimum value that can be set here is 7.

- **SOCx Trigger Source**

  This drop-down menu allows to select the source that triggers the start of analog to digital conversions. As already mentioned in Section 12.2, the available entries are:

  – Software;

  – CPU0/1/2_XINT0/1/2n;

  – XINT2_XINT2SOC.
  XINT2SOC external pin should be specified inside the Configuration Parameters window, Hardware Implementation, Target hardware resources menu, ADC. This operation defines the GPIO channel connected to a pin that triggers the start of conversion;

  – ePWM1-8 _ADCSOCA/ePWM1-8 _ADCSOCB.

- **ADCINT will Trigger SOCx**

  Use the ADCINT1 or ADCINT2 interrupt to trigger a SOC after an EOC. Namely, those ADC interrupts act as feedbacks. These loops create a continuous sequence of conversions. The default selection, i.e., No ADCINT, disables this parameter.

- **Sample Time**

  This parameter specifies the time interval elapsed between two consecutive samples, e.g., T_s.

- **Data Type**

  This drop-down menu specifies the data type of the digital output data. The available options are double, single, int8, uint8, int16, uint16, int32, and uint32.

- **Post Interrupt at EOC Trigger**

  This checkbox enables post interrupts when the ADC triggers EOC pulses. When this option is selected, the dialog box displays the **Interrupt Selection** and **ADCINT# Continuous Mode** options.

- **Interrupt Selection**

  This allows to select which interrupt ADCINT# the ADC posts after triggering an EOC pulse. The following checkbox depends on the value specified here.

- **ADCINT# Continous Mode**

  Every time this checkbox is enabled, an ADCINT# is generated when the ADC generates an EOC signal.

Block Parameters - Input Channels tab

- **Conversion Channel**

  This drop-down menu allows to select the input channel to which the ADC conversion applies. It is possible to choose one (for Single sample mode) or two (for Simultaneous sample mode) channels.

12.4 Example with ADC block

In the following, the external board **extPot3** is used to manipulate up to three ADC inputs (see Appendix B for hardware details).

The extPot3 can be directly mounted on right-hand 40-pin socket of the LaunchXL F28069M board (see Figure 12.4) and it includes three linear 20 kΩ potentiometers connected to header J7, namely ADCB5 (pin 66), ADCA3 (pin67) and ADCB3 (pin68), which are used to create and vary input analog signals between $[0, 3.3]$ V. If the extPot3 is correctly mounted, a green led turns on. Figure 12.5 (a) shows the equivalent circuit of a linear potentiometer.

Figure 12.4 extPot3 mounted on right-hand 40pin socket of the F28069M LaunchPad™ board (see Appendix B for hardware details).

> Build a firmware aimed at reading an analog input from a potentiometer which modulates the blinking of the blue led

Referring to the previous PWM logic implementation (see Example 4 in Section 11.3), one of the external potentiometer is used now to vary the modulating signal. For instance, by reading ADCA3 (pin67), the F28069M board generates the pulse sequence depending on the modifications in the duty cycle which are related to the variations in the voltage level applied to the selected ADC input. Note that the digitalized measurement (from ADCA3) must be scaled between 0 and 1 to fit the PWM logic designed in the previous Chapter. Both the continuous- and discrete-time implementation of the carrier are still valid.

Firmware Environment

c28069_ADCledBpwm_F.slx (solver: fixed step-ODE4, step size: $T_s = 100\,\mu\text{s}$)

Open a new blank Simulink® project and configure the environment as shown in Chapter 9. Then:

- Open the **Model Configuration Parameters** window and use the same settings adopted for **Example 1** in Section 10.5.
- Insert a **GPIO DO** block and build the same **PWM logic** used in Chapter 11 by following the steps reported in **Example 4** in Section 11.3.
- The **ADC** block can be found in the C2806x subset of the library Embedded Coder Support Package for Texas Instruments C2000 Processors. Insert it in the scheme ans set its parameters as follows:

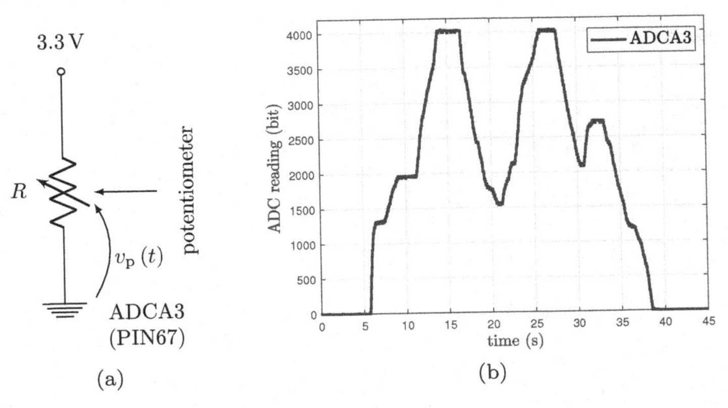

Figure 12.5 Equivalent circuit of a linear potentiometer (a) and example of 12-bit representation from ADCA3 before any scaling.

- Select ADCINA3 as **Conversion channel** in the **Input Channels** tab. Then, go back in the **SOC Trigger** tab;
- **Sampling mode**: Single sampling mode;
- **SOC trigger number**: SOC0;
- **SOCx acqusition window**: 7;
- **SOCx trigger source**: CPU0_TINT0n *or* Software;
- **ADC will trigger SOCx**: No ADCINT;
- **Sample time**: $T_{\text{ADC}} = 0.001\,\text{s}$;
- **Data type**: uint16;
- Flag **Post interrupt at EOC trigger** (optional);
- **Interrupt selection**: ADCINT1 (optional);
- Flag **ADCINT1 continous mode** (optional).

- Insert a **Data Type Conversion** block set to `int16` to deal with positive signal only.
- Place a **Gain** block for scaling the digitalized signal within the interval $[0, 1]$. This is fundamental because a 12-bit reading returns a value in a range from 0 ($0\,\text{V}$) to $2^{12} - 1 = 4095$ ($3.3\,\text{V}$, see Figure 12.5 (b)). Hence, set the gain as $1/(2^{12} - 1)$ or, to be more conservative, as $1/2^{12}$. In this last case, the representation range is not fully exploited.
- Insert a **SCI transmit** block to send the values returned by ADCA3 and the output of the PWM logic to the host PC via SCI A. the reader is suggested to use the same settings adopted in Chapter 11, i.e., `single` data type and transmission rating equal to 0.01 s.

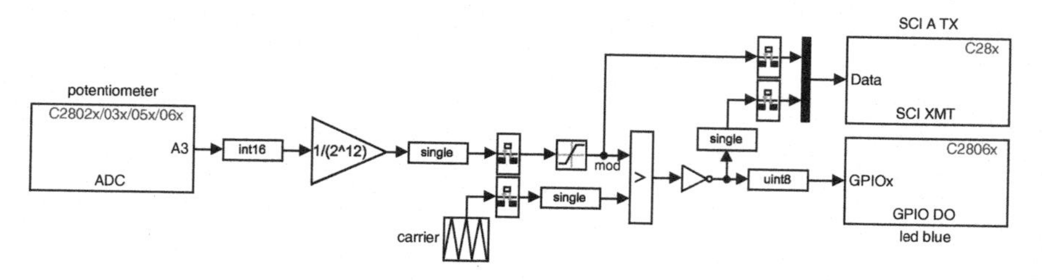

Figure 12.6 Simulink® scheme included in c28069_ADCledBpwm_F.slx.

The remaining part of the scheme directly refers to Example 4 in Section 11.3. Inside the ADC block, the sample time is set equal to $T_{\text{ADC}} = 0.001\,\text{s} \neq T_s$. In this exercise, the ADC reading does not represent a critical task in terms of timing, i.e., such peripheral is not synchronized with any other one. Therefore, the timing has been relaxed, i.e., $T_{\text{ADC}} > T_s$, to reduce the computational load on the processor. However, this is not always the case, and most of the time the ADC peripheral has to be synchronized with a particular signal as well as using T_s is necessary to improve plot resolution.

Note that **Rate Transition** blocks are placed to run the signal comparison every T_s. A continuous-time block is used to create the triangular carrier, resulting in a switching frequency of $f_{\text{sw}} = f_c = 20\,\text{Hz}$. This allow the reader to recognize the different led blinking during operation. The resulting scheme is reported in Figure 12.6.

The linear potentiometers mounted on the extPot3 board are connected to the 3.3 V supply, which is directly available on the LaunchXL F28069M board. A full rotation of a potentiometer leads to a voltage span from 0 V to 3.3 V or vice-versa, depending on the direction of rotation. The resulting modulation signal acquired by the ADC can be sent to the PC and visualized, as shown in Figure 12.7. Since the ADC peripheral has a 12 bit resolution, the output range is between 0 and 4095. This bit value can be nicely represented by `unit16` data type, which allows numbers between 0 and 65535. Note that, in the ADC settings does not exist a 12-bit data type, thus, 16-bit representation is the right choice to avoid data cast. Regarding global definitions, an initialization script an be created in *Model Properites/Callbacks/InitFcn* or in a separate `m`-file. An example of such script is reported here below:

```
%% carrier frequency definition
fsw   = 20;
Tsw   = 1/fsw;
%% sampling time definition
Ts    = 100e-6;
```

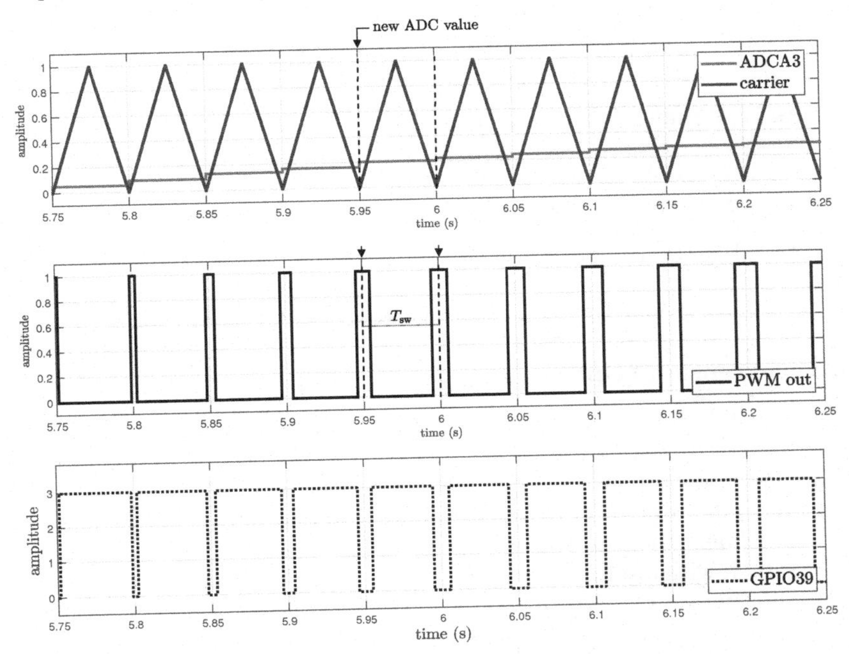

Figure 12.7 Signal treatment from ADC reading of a linear potentiometer input to GPIO39 output

Testing Environment

c28069_ADCledBpwm_T.slx (solver: fixed step - ODE4, step size: $T_{\mathrm{sim}} = T_s$)

Open a new blank Simulink® project:

- Open the **Model Configuration Parameters** window and use the same settings adopted for **Example 1** in Section 10.5.
- Place a **Serial Configuration** block and use the same settings adopted for **Example 1** in Section 10.5.
- Add a **Serial Receive** block and double click on it:
  - Select the right COMx as **serial port**;
  - No **header** or **terminator** should be specified;
  - Set **data size** [1 2], since one signal is sent, received and visualized only;
  - Set **data type**: `single`;
  - Set **sample time**: $T_{\mathrm{RX}} = 0.01\,\mathrm{s}$ (or $0.05\,\mathrm{s}$).

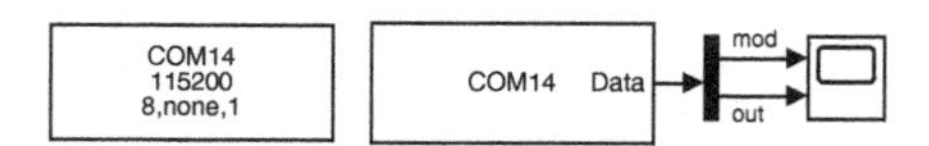

Figure 12.8 Simulink® scheme included in c28069_ADCledBpwm_T.slx.

- Insert a Scope block to check if the control signal is set correctly.
- Insert a Mux block to split the signals coming from serial communication.

The aforementioned blocks should be connected as shown in Figure 12.8.

Buzzer extension (Optional)

Since the extPot3 has a piezoelectric buzzer available, the manipulated value read by the ADC together with the PWM logic can be used to modulate the sound coming out of the buzzer itself. This latter can be directly connected to GPIO8 (pin78) or to ePWM5B (pin77) through a physical connection on the board. Then, the buzzer will produce a sound at every edge of the PWM output signal, with a sound which has an increasing frequency for high value of the modulating signal. Figure 12.9 reports the block arrangement for driving GPIO8 and a zoom on the extPot3 board showing the buzzer.

12.5 Synchronization between ADC modules

All the ADC modules developed for the family of C2000™ processors implement a conversion approach based on SOC signals. Every SOC instance allows a custom sampling for every channel, which can be edited in the **Block Parameters: ADC** window. SOCx numbers determine the priority of the chained conversion according the ascending number of the instances. Namely, SOC1 is processed before SOC2, SOC2 before SOC3, and so on. The **SOC trigger number** $(0 - 15)$ can be chosen in the **Block Parameters: ADC** window in the drop-down menu, while the **SOCx trigger source** option can be set as reported in Section 12.3.1.

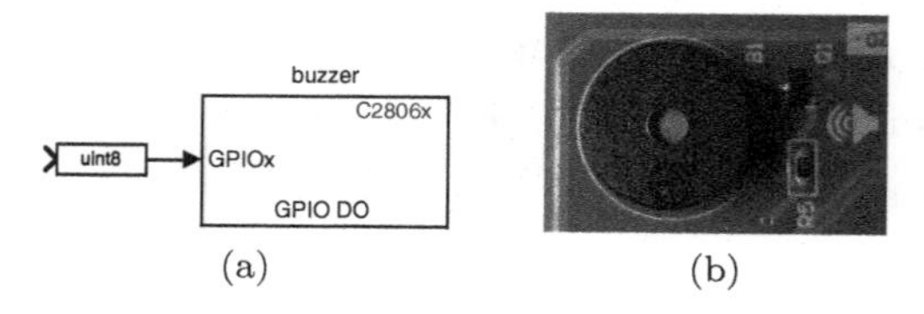

(a) (b)

Figure 12.9 Block arrangement for driving GPIO8 (a) and detail of extPot3 board showing the buzzer (b).

Moreover, it is possible to program the generation of an Interrupt signal at every EOC event. Interrupt signals are numbered from ADCINT1 to ADCINT9. In particular, ADCINT1 and ADCINT2 can be chosen for triggering other SOC instances only. To this aim, the check-box **Post Interrupt at EOC trigger** should be enabled. This will generate the appearance of the **Interrupt selection** frame in which an appropriate ADC interrupt signal useful for further synchronization could be selected.

Example

> Build a firmware aimed to synchronize two ADC modules in a master-slave chain configuration

This exercise is an extension of the previous one. Two ADC modules are exploited, ADCINA3 and ADCINB5. The main settings to create a master-slave chain configuration between ADC modules is presented.

Firmware Environment

c28069_ADCsync_F.slx (solver: fixed step - ODE4, step size: $T_s = 100\,\mu s$)

Modify the previous example by duplicating the ADC block and updating the settings as follows:

ADC1 settings:

- Select ADCINA3 as **Conversion channel** in the **Input Channels** tab. Then, go back in the **SOC Trigger** tab.
- **Sampling mode**: Single sampling mode;
- **SOC trigger number**: SOC0;
- **SOCx acqusition window**: 7;
- **SOCx trigger source**: CPU0_TINT0n *or* Software;
- **ADC will trigger SOCx**: No ADCINT;
- **Sample time**: $T_{\text{ADC}} = 0.001\,\text{s}$;
- **Data type**: uint16;
- Flag **Post interrupt at EOC trigger**;
- **Interrupt selection**: ADCINT1;
- Flag **ADCINT1 continous mode** (optional);

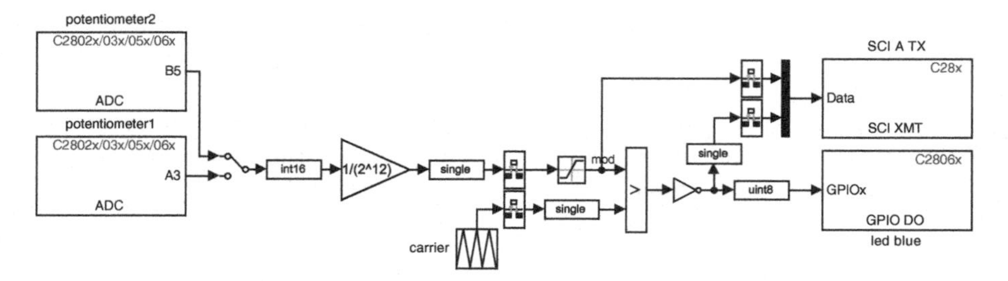

Figure 12.10 Simulink® scheme included in c28069_ADCsync_F.slx.

ADC2 settings:

- Select ADCINB5 as **Conversion channel** in the **Input Channels** tab. Then, go back in the **SOC Trigger** tab.
- **SOC trigger number**: SOC1;
- **ADC will trigger SOCx**: ADCINT1;
- Unflag **Post interrupt at EOC trigger**;
- Keep the other settings equal to those specified for ADC1.

In this case, ADC1 (ADCINA3) is the master that creates an interrupt signal ADCINT1 at its EOC. This signal is used to trigger the SOC of ADC2 (ADCINB5), which behave as slave.

Note that if only one ADC module is used, the EOC flag is not strictly necessary since no other modules requires such trigger; this also explain the settings reported in Section 12.4. The new scheme is shown in Figure 12.10.

Remark

Every ADC block exploits one SOC in **single sample mode** or two of them in **simultaneous sample mode**. It is forbidden to use the same SOC instance for different ADCs.

13

Pulse Width Modulator Peripheral

The enhanced pulse width modulator (ePWM) peripheral is a key element in controlling many of the power electronic systems found in both commercial and industrial equipment. These systems include digital motor control, switch-mode power supply control, uninterruptible power supplies (UPS), and other forms of power conversion. Each of the available ePWM modules generates signals through suitable switching patterns by means of a PWM logic, performing a digital to analog (DAC) function.

The F28069M LaunchPad™ board provides up to 8 ePWM modules with a total amount of 16 PWM channels, 8 of them high-resolution PWM (HRPWM), with 16-bit independent timer in each module. In particular, from such 16 channels, 12 of them are general purpose while 4 of them are filtered PWM, i.e., equipped with an RC output filter. Those channels performs the actual DAC function for the board.

Each ePWM module has two outputs, ePWMA and ePWMB. The signals available at the corresponding output pins can drive the gates of the upper and lower switches of a two-level converter topology, as shown in Figure 13.1. In this Chapter, the operation principles of the ePWM peripheral are briefly outlined as well as an explanation on how to configure the its settings in the Simulink® environment.

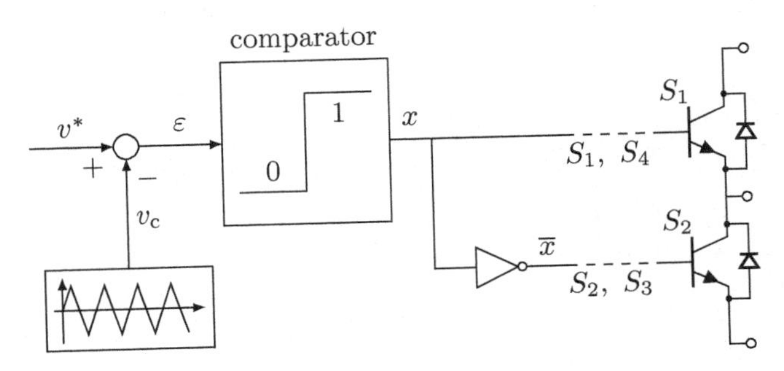

Figure 13.1 Actuation of a two-level voltage source converter (2L-VSC) with PWM logic.

DOI: 10.1201/9781003196938-13

13.1 Operating Principle

Since the working principle of PWM logic was already presented in Chapter 11, a specific example is considered here in the following. Namely, the actuation of a two-level voltage source converter (2L-VSC) which comprises two power switches S_1 and S_2 (it is not important to specify their technology) with anti-parallel diodes (see Figure 13.1). In a pulse width modulator, the reference voltage is compared to a triangular carrier (both of them are scaled over a $[0, 1]$ range) and this output is used to drive the converter switches in a complementary way. By considering a single-phase case, i.e., the drive of a single converter leg (phase A), a sinusoidal reference voltage $m(t)$ is compared to the triangular carrier signal generating a switching pattern $S(t)$ which has to be applied on the upper switch through the gate driver, i.e., $S_1(t) = S(t)$. Instead, the lower switch is actuated with the logic negation of the resulting switching pattern, that is, $S_2(t) = \text{not}(S_1(t))$. Therefore, S_2 operates in a complementary way with respect to S_1.

The operation of these switches given their switching patterns creates a pulsed voltage waveform at the output of the converter. This latter can be simply represented as $v_{\text{out}}(t) = S(t)V_{\text{dc}}$, i.e., the product between $S(t)$ and the supply voltage (dc-link) V_{dc}. Hence, the output voltage is bounded within two levels, i.e., 0 and V_{dc}. The fundamental component of this voltage is proportional to the reference voltage. Indeed, its harmonic spectrum includes the frequency of the modulation signal as fundamental component and higher frequency components around the multiples of f_{sw}, depending on the adopted PWM logic. The carrier signal can be implemented in different ways, e.g., realizing increasing saw-tooth, decreasing saw-tooth or triangular waveforms. The carrier frequency must be higher than the modulation one and, in case of a two-level converter, the resulting switching frequency equals the carrier one, that is $f_{\text{sw}} = f_c$. indeed, the comparator evaluates:

- If modulation signal < carrier signal → PWM signal high → $S(t) = 1$;
- If modulation signal > carrier signal → PWM signal low → $S(t) = 0$.

This mechanism allows to dynamically adjusts the duty cycle $d(t) = T_{\text{on}}/T_{\text{sw}}$ of the output square wave, i.e., by varying the control signal, and, hence, its average value.

Assuming $f_{\text{sw}} = f_c = 1\,\text{kHz}$, a sine wave at frequency $f_1 = 50\,\text{Hz}$ ($T_1 = 0.02\,\text{s}$) as modulation signal and $V_{\text{dc}} = 10\,\text{V}$, the corresponding resulting sinusoidal PWM modulation is shown in Figure 13.2. The first order harmonic computed for $v_{\text{out}}(t)$ is matching with the fundamental components f_1, which amplitude is dependent on V_{dc}. It is important to note that such modulating signals are commonly used in AC drives, due to the AC nature of many application. However, constant modulating signals are used in DC drives, and their value represents the duty cycle $m(t) = d(t)$ of the output

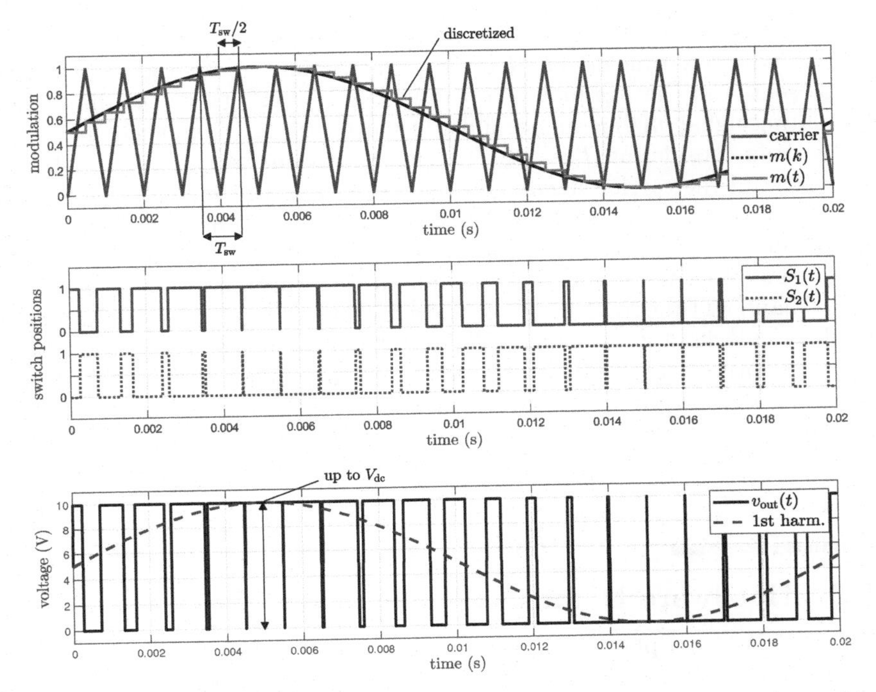

Figure 13.2 Results of a sinusoidal PWM implementation in terms of switching pattern and output voltage of the 2L-VSC.

PWM modulated signal. Therefore, this modulation technique is extremely useful in power electronics-based application because:

- Small motors often requires low power consumption and the use of (discontinuous) PWM signals allows to efficiently control a power converter by reducing the power losses in the switches (both due to conduction and switching) compared to other kind of linear controller.[1] Indeed, the switching devices foresees two states only: short and open circuit.
- The motor dynamics (which can be modeled as an RL circuit and a controlled voltage source, that is, the back-emf) act as a *low pass filter*. Hence, undesired high-order harmonics resulting from the PWM implementation might be rejected by the machine itself without the need for additional filter, being mostly sensitive to low frequency components such as the fundamental one.

[1] An old alternative was a simple circuit in which the current flow was linearly controlled by varying a resistance, causing much energy losses and, thus, heat.

It is essential to remind that the aforementioned approach has to be translated into discrete-time domain k for implementations on MCUs. This process is represented in the top plot reported in Figure 13.2, where the discretized modulating signal $m(k)$ is superimposed to $m(t)$. For instance, such discretization is done considering $T_s = T_{\rm sw}/2$. Thus, the value of $m(k)$ is updated at every peak of the carrier signal. By using $T_s = T_{\rm sw}$, the value of $m(k)$ is updated every time the carrier equals zero. The choice of this sampling time is quite important for embedded implementation and for the resulting voltage spectrum as well, which may now include further harmonics.

In Chapter 11 the implementation of a PWM logic realized by using simple Simulink® blocks was presented. This operation is conceptually fine, but it is not an optimal solution for embedded implementation. Indeed, the PWM processing logic can be optimized to achieve high-resolution at low latency rate with specific routines/iterations as implemented by the ePWM peripheral.

13.2 Hardware Details

Effective PWM peripherals aim to generate complex pulse width waveforms with minimal CPU employment or interrupts. The ePWM modules reported here in the following address these requirements by allocating all needed timing and control resources. Cross-coupling or resource sharing are avoided. This peripheral is built up from smaller single-channel modules with separate resources that can operate together, as shown in Figure 13.3.

Each ePWM module mounted on the F28069M LaunchPad™ board has two PWM outputs, i.e., ePWMxA and ePWMxB, where the x is the module number. The corresponding output pins are selected through the GPIO multiplexer. Additionally, the ePWM module can generate an interrupt to the PIE block as well as a SOC signal to the ADC. Inputs to the ePWM module include signals from the input cross-bar, from the ePWM cross-bar and selected trip zone signals.

All ePWM modules installed onboard C2000™-based devices are identical. They can operate stand-alone, but they can be chained as well to work together as a single system via clock synchronization scheme. Note that this operation can be extended to other peripherals.

Here in the following, a recap list of the features supported by each ePWM module is reported:

- Dedicated 16 bit time-base counter with both period and frequency control.
- Two PWM outputs (EPWMxA and EPWMxB) that can be used in the following configurations:
  - Two independent PWM outputs with single-edge operation;
  - Two independent PWM outputs with dual-edge symmetric operation;
  - One independent PWM output with dual-edge asymmetric operation.

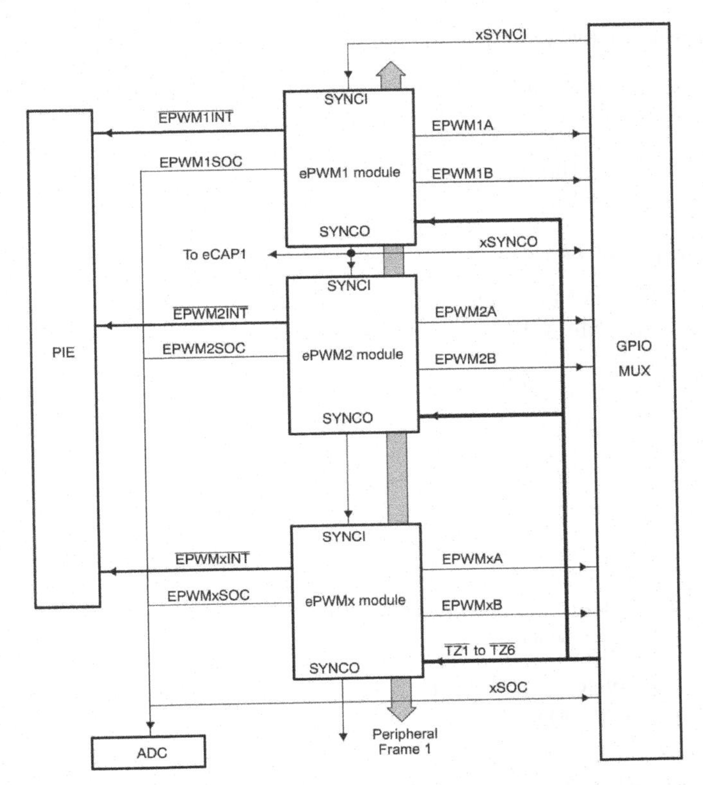

Figure 13.3 Scheme of the ePWM peripheral mounted on the LAUNCHXL F28069M board [21].

- Asynchronous over-ride control of PWM signals through software.
- Programmable phase-control support for lag or lead operation with other ePWM modules.
- Hardware-locked (i.e., synchronized) phase relationship with other modules on a cycle-by-cycle basis.
- Dead-band generation with independent rising and falling edge delay control.
- Programmable trip zone allocation of both cycle-by-cycle and one-shot trip in fault conditions.
- A trip condition which can force either high, low, or high-impedance state logic levels at PWM outputs.
- All events of the module can trigger both CPU interrupts and ADC SOC signals.

- Programmable event prescaling for minimizing CPU overhead or interrupts.
- PWM chopping by high-frequency carrier signal, useful for pulse transformer gate drives.

Moreover, a list of the main input and output signals for each ePWM module is reported here below:

- **PWM output (EPWMxA and EPWMxB)**

  They are the PWM output signals which are made available external to the device through the GPIO peripheral (see Chapter 11).

- **Trip-zone** ($\overline{\mathrm{TZ1}}$ – $\overline{\mathrm{TZ6}}$)

  These input signals alert the ePWM module of fault conditions external to the ePWM module. Each of them can be configured to either use or ignore any of the trip-zone signals. The $\overline{\mathrm{TZ1}}$ – $\overline{\mathrm{TZ3}}$ trip-zone signals can be set as asynchronous inputs through the GPIO peripheral. $\overline{\mathrm{TZ4}}$ is connected to an inverted EQEP1 error signal (EQEP1ERR) from the EQEP1 module. $\overline{\mathrm{TZ5}}$ is connected to the system clock fail logic, whereas $\overline{\mathrm{TZ6}}$ is connected to the $\overline{\mathrm{EMUSTOP}}$ output from the CPU. This allows the configuration of a trip action when the clock fails or the CPU halts.

- **Time-base synchronization input (EPWMxSYNCI) and output (EPWMxSYNCO)**

  They are the synchronization signals which are able to chain the ePWM modules together. Each module can be configured to either use or ignore its synchronization input. The clock synchronization input and output signals are brought out to pins only for ePWM1 (i.e., ePWM module 1). The synchronization output for ePWM1 (EPWM1SYNCO) is also connected to the SYNCI of the first enhanced capture module (eCAP1).

- **ADC start-of-conversion (EPWMxSOCA and EPWMxSOCB)**

  Any ePWM module can trigger a start of conversion. Indeed, each of them is associated with two ADC start of conversion signals. Whichever event triggers the SOC should be configured in the Event-Trigger sub-module of the ePWM peripheral.

- **Comparator output (COMPxOUT)**

  These are the output signals from the comparator module. In correspondence with the trip zone signals they can generate digital compare events.

- **Peripheral Bus**

  The peripheral bus is 32 bit wide and it allows both 16 bit and 32 bit writing operations in the ePWM register file.

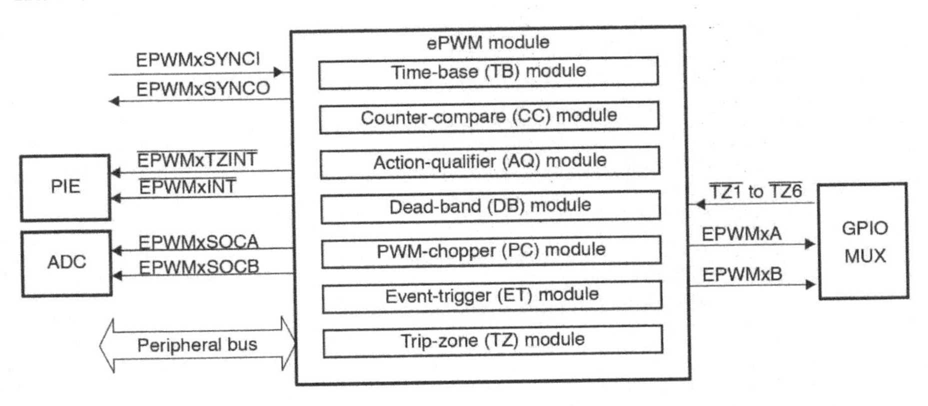

Figure 13.4 ePWM sub-modules [21].

13.2.1 ePWM sub-modules

Each ePWM module can communicate with other on-board peripherals through dedicated routines which are summarized in eight sub-modules, as shown in Figure 13.4. Namely, they are:

- Time-base;
- Counter compare;
- Action qualifier;
- Dead band;
- PWM chopper;
- Trip zone;
- Digital compare;
- Event trigger;

Almost all the working options of each sub-module can be programmed in Simulink® through the ePWM Block Parameter window. The features of these sub-modules are explained in details in the following:

- The **time-base sub-module** consists in a dedicated 16 bit counter along with built-in synchronization logic to allow multiple ePWM modules to work together as a single system. A clock prescaler is placed between the system clock and the counter. A period register is used to control both frequency and period of the generated waveform. The period register has a shadow register, which acts like a buffer to allow the updates to be synchronized with the counter. This avoids corruption or spurious operation due to the asynchronous modification of the register by the software. In addition, all the shadow registers support simultaneous global load, which avoids partial loads from the shadow registers to the active ones. The time-base counter operates in three modes: up count, down count, and up/down count (see Figure 13.5).

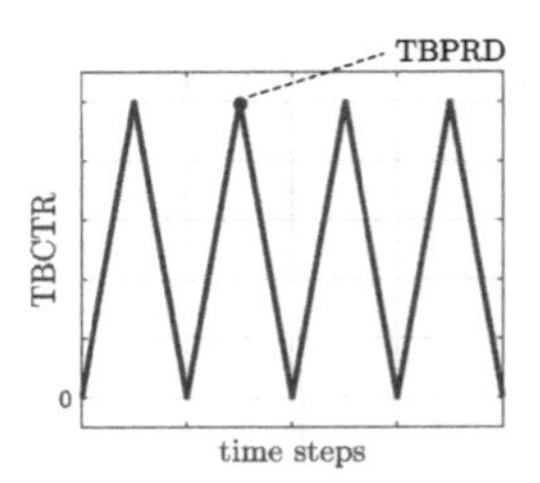

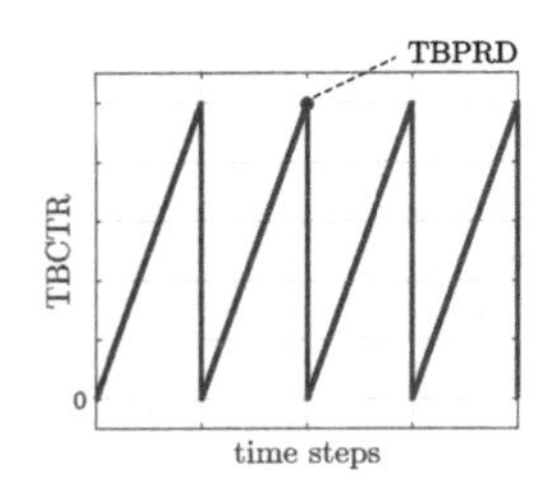

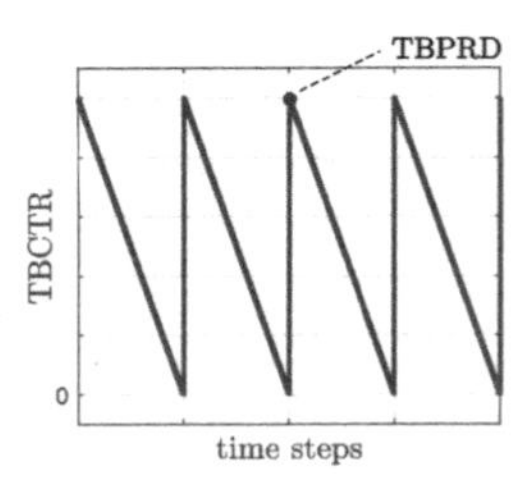

Figure 13.5 Counting modes available in the adopted MCU board. From left to right they are: up/down, up and down count.

- The **counter compare sub-module** continuously compares the time-base count value to four counter compare registers: Compare A, B, C, and D. They generate four independent compare events. Every time the time-base counter equals a compare register value, the compare event is fed by the action qualifier and event trigger sub-modules. Typically, compare A and B are used to control the duty cycle of the generated PWM waveform on a single leg.
- The **action qualifier sub-module** is a key element for ePWM peripherals because it is responsible to generate the switched PWM waveforms. It exploits events coming from the time-base and counter compare sub-modules to perform actions on the ePWMxA and ePWMxB output pins. These actions can be chosen among the following allowed solutions:

  - Do nothing;
  - Clear the pin low;
  - Set the pin high;
  - Toggle the pin based independently on count-up and count-down time-base match events.

  Instead, the triggering events can be generated by one of these occurrences:

  - The time based counter equals zero;
  - The time-base counter equals Compare A;
  - The time-base counter equals Compare B;
  - The time-base counter equals the period register value;
  - A trigger event T1 and T2 based on a comparator, trip, or synchronization signal.

  It is important to note that zero and period actions are fixed in time, whereas Compare A and B actions are movable in time by programming their corresponding registers. These actions are configured independently for each output using shadowed registers.

- The **dead band sub-module** allows to delay the switching action of a transistor. Since power switching devices turn on faster than they turn off, a delay is needed to prevent any short-circuit path from the DC source to ground. Further explanation is available in Section 13.3.
- The **PWM chopper sub-module** modulates a high frequency carrier signal with the PWM waveform that is generated by the action qualifier and dead band sub-modules. Its purpose is to check the power switching elements used with pulse transformer-based gate drivers.
- The **trip zone sub-module** provides a mechanism to protect the output pins from abnormalities, such as over-voltage, over-current, and excessive temperature rise. The trip zone signals can be generated externally from any GPIO pin, internally from an eQEP error signal, system clock failure, or emulation stop output from the CPU. Additionally, several trip zone source signals can be generated from the digital compare sub-module.
- The **digital compare sub-module** increases the capabilities of the trip zone sub-module by comparing signals external to the ePWM module to directly generate PWM events or actions which are then used by the time-base, event-trigger, and trip zone sub-modules.
- The **event-trigger sub-module** manages the events generated by the time-base, counter compare, and digital compare sub-modules for generating an interrupt to the CPU and/or a start of conversion pulse to the ADC when selected events occur (EPWMxSOCA/EPWMxSOCB). These event triggers can occur when the time-base counter equals:
  - Zero;
  - Period;
  - Zero or period;
  - The up- or down-count match of Compare A, B, C, or D registers.

13.3 Generation of PWM signals

The ePWM peripheral includes a complex structure capable of implementing several features. In order to efficiently explain how to deal with all the required settings, a short overview on how the PWM logic is implemented on the F28069M LaunchPad™ board is reported in this Section. Particular attention is paid on how the peripheral manages the comparison between modulation and carrier signals by using specific counters as well as on how complementary behaviors should be enabled.

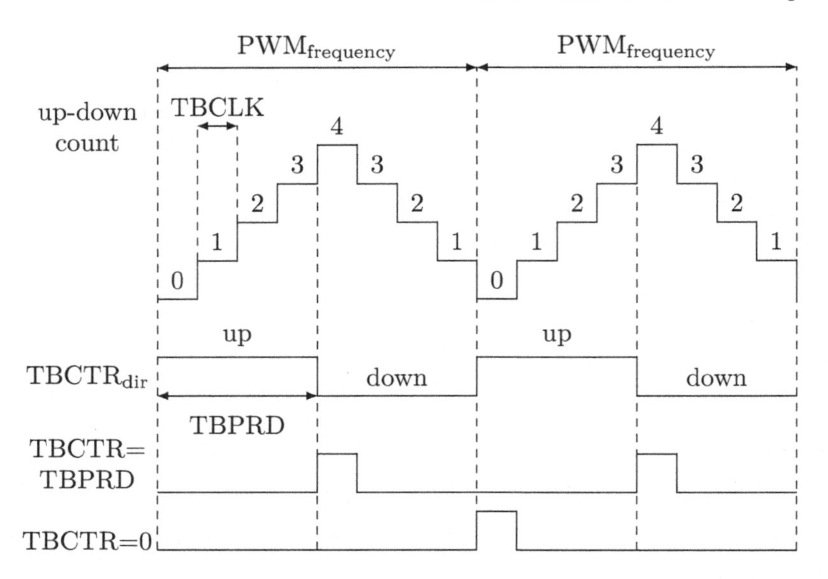

Figure 13.6 Example of up-down counting mode considering TBPRD = 4. The time instants (rising edge) when TBCTR = 0 and TBCTR = TBPRD occur are highlighted.

13.3.1 Counting modes

The ePWM modules generate PWM signals by processing discretized quantities (either modulating and carrier signals). The carrier waveform is defined in amplitude by a discrete counter, called Time-Base Counter (TBCTR), and in shape by the counting mode parameter. The TBCTR increases at every CPU clock cycle TBCLK defined by the frequency f_{ck} (also called $\mathrm{CPU}_{\mathrm{frequency}}$ or SYSCLKOUT) as $\mathrm{TBCLK} = 1/f_{\mathrm{ck}} = 1/\mathrm{CPU}_{\mathrm{frequency}}$. All the internal signals are defined with respect to this base time. Regarding the LAUNCHXL F28069M board, $f_{\mathrm{ck}} = \mathrm{CPU}_{\mathrm{frequency}} = 90\,\mathrm{MHz}$, thus, $\mathrm{TBCLK} = 11.1\,\mathrm{ns}$.

The peak value of the digital carrier is defined by the time-based period (TBPRD), which can be also called $\mathrm{PWM}_{\mathrm{counter\_period}}$. Based on that, every time the peak value is reached $\mathrm{TBCTR} = \mathrm{TBPRD} = \mathrm{PWM}_{\mathrm{counter\_period}}$. The triangular carrier was defined in a $[0, 1]$ range in Section 13.1 to the aim of providing a general introduction. In fact, at the embedded level, TBCTR varies in the interval $[0, \mathrm{PWM}_{\mathrm{counter\ period}}]$. Consequently, the modulation signal must be processed to span the same interval of values. Regarding the carrier shape, two counting modes are available for this peripheral:

- **Symmetrical carrier (up-down count)**, i.e., a *triangular carrier* waveform. In up-down-count mode, as long as the time base counter direction signal ($\mathrm{TBCTR}_{\mathrm{dir}}$) is high (1), the TBCTR counting increases till reaching $\mathrm{PWM}_{\mathrm{counter\ period}}$, while as long as $\mathrm{TBCTR}_{\mathrm{dir}}$ is low (0), the TBCTR

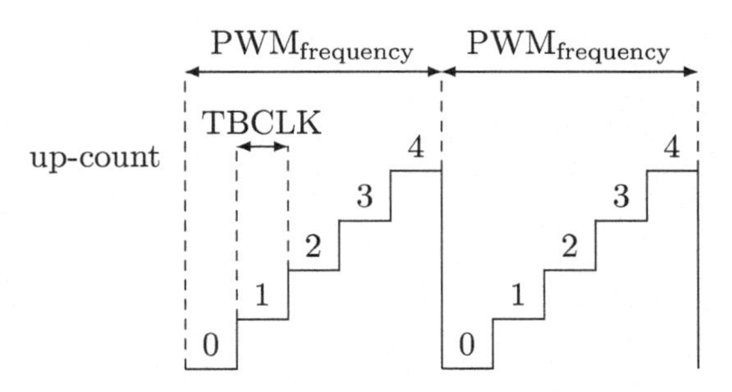

Figure 13.7 Example of up-count counting mode. $\mathrm{PWM_{frequency}}$ and TBCLK can be defined analogously for the down-count mode.

counting decreases down to 0. The value of $\mathrm{TBCTR_{dir}}$ is related to two events. When the event TBCTR = TBPRD occurs, $\mathrm{TBCTR_{dir}}$ switches from 1 to 0, while the opposite happens every time TBCTR = 0. For a symmetrical carrier, $\mathrm{TBCTR_{dir}}$ has to change value every time TBPRD is reached, which has to be defined as an integer multiple of TBCLK. These relationships are highlighted in Figure 13.6. In particular, the counter period for up-down counting modes is defined as:

$$\mathrm{TBPRD} = \mathrm{PWM_{counter\_period}} = \frac{\mathrm{CPU_{frequency}}}{2 \cdot \mathrm{PWM_{frequency}}} \tag{13.1}$$

where $f_c = f_{sw} = \mathrm{PWM_{frequency}}$ is the frequency of the carrier. For example, to implement a $f_c = 20\,\mathrm{kHz}$ carrier at into the F28069M LaunchPad™ board, the counter period should be set equal to:

$$\mathrm{PWM_{counter\_period}} = \frac{90\,\mathrm{MHz}}{2 \cdot 20\,\mathrm{kHz}} = 2250$$

- **Asymetrical carrier (up-count** or **down–count)**, i.e., *sawtooth carrier* waveform. In up-count mode, TBCTR starts from zero and it increases until TBCTR=TBPRD is reached. There, the time-base counter resets to zero and, then, it increases again from the next time step. This kind of carrier is shown in Figure 13.7. The down-count mode behaves similarly, with the exception that the counter starts from TBPRD and it decrements until it reaches 0. Therefore, the counter period for up or down countersign modes is defined as:

$$\mathrm{TBPRD} = \mathrm{PWM_{counter\_period}} = \frac{\mathrm{CPU_{frequency}}}{\mathrm{PWM_{frequency}}} \tag{13.2}$$

Considering again a $f_c = 20\,\mathrm{kHz}$ carrier signal, the counter period should be set equal to:

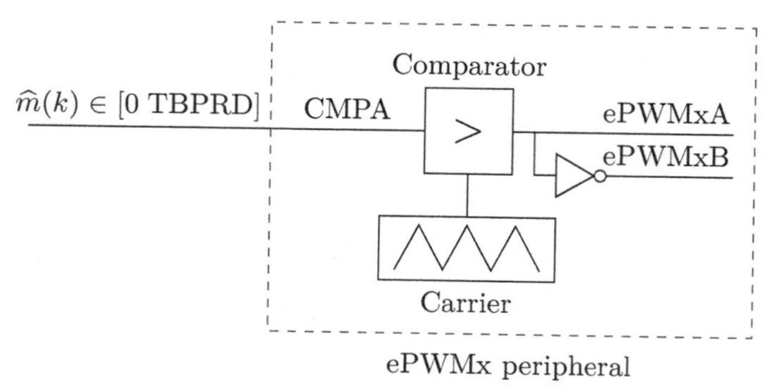

Figure 13.8 Representation of the inner structure of an ePWM module aimed to actuate a single-phase 2L-VSC.

$$\mathrm{PWM}_{\mathrm{counter\_period}} = \frac{90\,\mathrm{MHz}}{20\,\mathrm{kHz}} = 4500$$

Note that the value of $\mathrm{TBCTR}_{\mathrm{dir}}$ follows from the reasoning reported for the symmetrical carrier.

13.3.2 ePWMxA and ePWMxB sub-modules

In terms of implementation, the comparison between the modulating signal and the carrier is performed by setting the action qualifier (AQ) event related to comparator CMPA and CMPB. Recalling the example reported in Section 13.1 involving the 2L-VSC, it is assumed that the upper switch is now labeled with A while the lower one is called B. Since a single-leg is considered (i.e., first leg x = 1), the gate signals corresponding to the switches are driven by ePWM1A and ePWM1B, as shown in Figure 13.8.

The amplitude of the modulating signal is defined by the (instantaneous) value stored into the counter-compare A (CMPA) register. This value is continuously compared to TBCTR and when TBCTR = CMPA an appropriate event is generated according to the AQ settings. In this particular case, the AQ would be specified considering the output ePWM1A, which is the master, while ePWM1B is computed as its complement. This is the reason why one comparison between TBCTR and CMPA is performed only. Int is important to underline that $m(k)$ defined in Section 13.1 now becomes $\widehat{m}(k)$, since it is varying within the range $[0, \mathrm{TBPRD}]$. As an example, assuming an up-down counting mode, every time TBCTR = CMPA and the carrier slope is positive, **clear** is defined as AQ event on ePWM1A. Thus, ePWM1A = 0 and ePWM1B = 1. For consistency, when TBCTR = CMPA and the carrier slope is negative, **set** is defined as AQ event on ePWM1A leading to ePWM1A = 1 and ePWM1B = 0. This behavior is shown in Figure 13.9. Alternatively, the AQ event on ePWM1A can be set on **do nothing** to keep it at same level

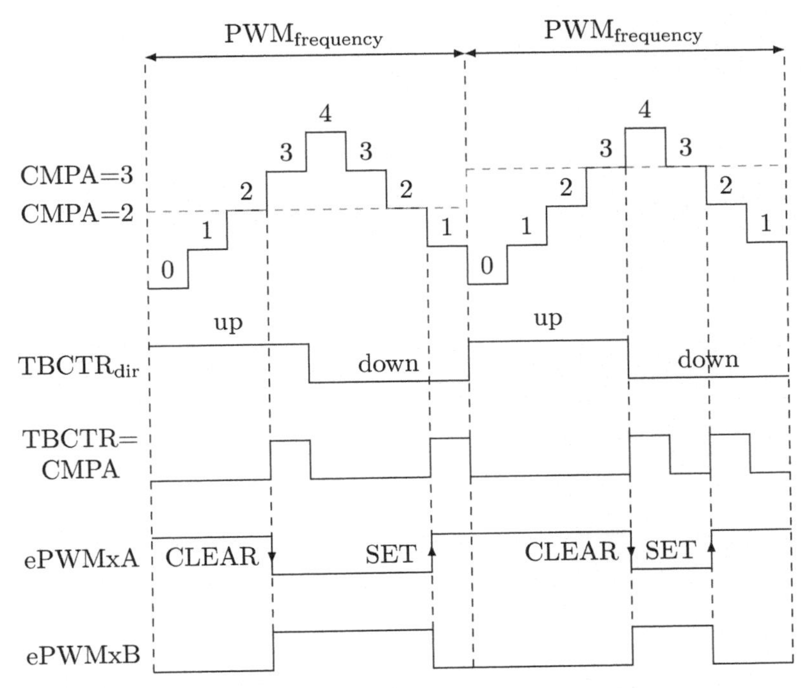

Figure 13.9 Counter compare operation considering CMPA = 2, 3 and with ePWMxA behaving as master.

as currently set. Instead, it is not possible to define AQ events related to CMPB, which is not used in this case. Note that the CMPA value is updated at the beginning of every switching period in Figure 13.9 (i.e., at a rate of $\mathrm{PWM}_{\mathrm{frequency}}$), which correspond to the time instant in which TBCTR = 0. This is not always the case and this behavior can be changed by modifying the *compare value reload conditions.*

Even if the aforementioned switching strategy (i.e., complementary ePWMxA/B) is the most common one for power electronic-based applications, ePWMxA and B can also operate independently. This is achieved using CMPA and CMPB with specific AQ events. Since many configurations are possible, ePWMxA and ePWMxB can be set separately. More details on the main options available in Simulink® are reported in the following sections.

13.3.3 Setting dead bands

The ePWM1A and B switching patterns are physically applied to the switches through specific gate driving circuits. These latter have to be designed according to the switch technology and the application requirements. It is important to underline that the design of gate driving circuits is out of the scope of this

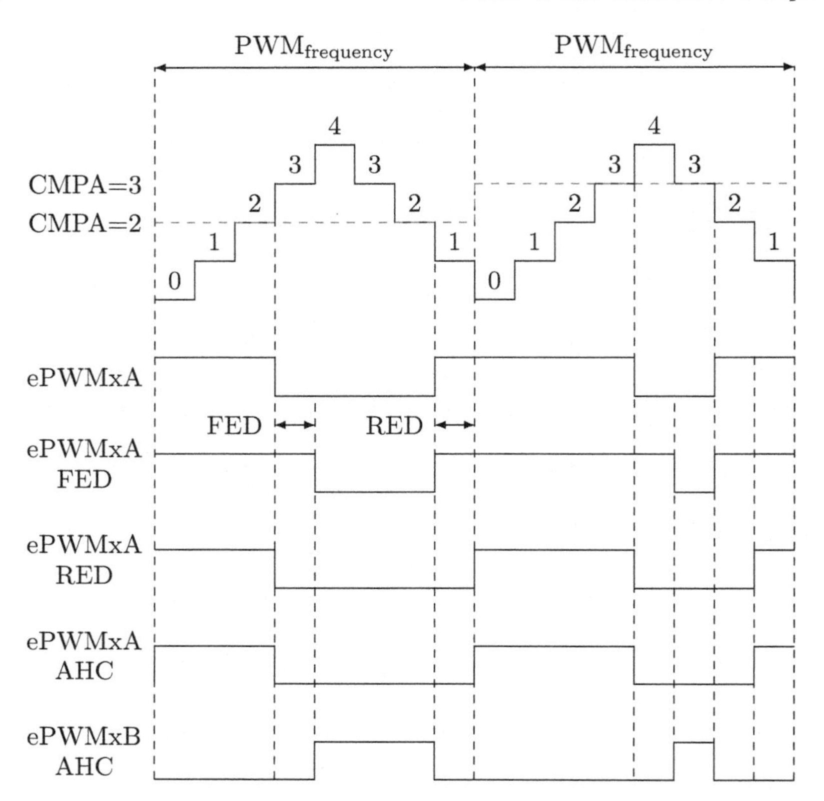

Figure 13.10 Counter compare operation for ePWMxA and ePWMxB with dead bands in the correspondence of both RED and FED.

book. Nevertheless, it is important to remind that such circuits may cause delay and/or particular actuation processing (e.g., bootstrap circuits) which could lead to simultaneous application of closing command to the switches and, thus, leg short circuit. To prevent such faulty condition, an interlock time (also called dead-time) is set for both ePWMxA and ePWMxB. According to the hardware realization and the application requirements, this time interval are in a range from 10 ns (e.g., low-power applications) up to 20 µs (e.g., high-power applications).[2] The ePWM peripheral settings of the F28069M LaunchPad™ board allow to add programmable delays that shift the switching patterns. Note that, the definition of interlock time or dead time refers to a quantity expressed in second. However, for embedded

[2]The gate driver circuits might add further dead-time aimed to increase the protection of the system, i.e., a delay is always present even if dead-time equals zero at the firmware level.

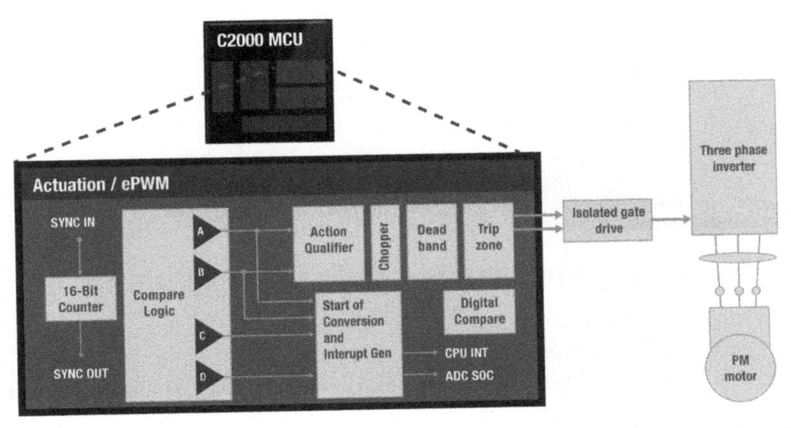

Figure 13.11 Summary of the features together with a possible application of the ePWM peripheral [8].

implementation, the dead time should be defined in clock cycles (i.e., as multiples of TBCLK). In this case, it is easier to refer to *dead band* which is used to define the time interval that separates the transition edges of two signals: output and complemented output.

In particular, dead times can be added in the correspondence of

1. Rising edges—defining a rising edge delay (RED);
2. Falling edges—defining a falling edge delay (FED);
3. Both rising (RED) and falling edges (FED).

By default, ePWMxA is used as signal source both for RED and FED. These latter identifies dead bands since they are defined in clock cycles. Furthermore, the choice of the dead time, thus, the length of the RED and FED dead bands should consider the value assumed by $PWM_{counter\_period}$. For instance, assuming $PWM_{counter\ period} = 2250$, the sum RED+FED must allow switching operations inside the switching period. Namely, each dead time is limited in the range [0 1023], so that $1023 + 1023 = 2046 < 2250$.

Referring to the 2L-VSC case, since ePWM1A is operated as master, the dead band polarity can be configured as Active High Complementary (AHC) to compute the ePW1B starting from ePW1A. Then, the resulting switching pattern provided to the gate driver circuits are ePW1A AHC and ePW1B AHC. By choosing equal values for RED and FED, the pattern keeps a centered symmetry. This is shown in Figure 13.10. However, this is not always the case in power electronics applications.

The aforementioned features and working principles of this peripheral can be summarized as shown in Figure 13.11. The reader should refer to the Section reported here in the following for the details on the implementation of these functions in Simulink®.

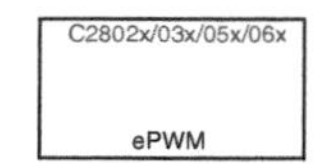

Figure 13.12 C2806x ePWM block.

13.4 Firmware Environment: ePWM Peripheral

The Simulink® block that allows to program the ePWM peripheral is the **ePWM** one (see Figure 13.12); it can be found in the *Embedded Coder Support Package for Texas Instruments C2000 Processors* library (C2806x subset).

13.4.1 C2806x ePWM

The ePWM block (see Figure 13.12) allows to set several settings and parameters which are split within ten different tabs inside the Block Parameters window. A detailed view of the available tabs is reported here below.

In the following, given the target of this book, only the main settings useful for power electronic-based applications are reported.

Block Parameters - General tab General

This menu allows to configure the carrier signal through the settings of the counters and registers. In particular, a key element is the TBCTR register from which the carrier shape is defined. The editable parameters are reported here in the following:

- **Module**

  There are 12 different ePWM modules available: ePWM1-12. Each of them has two outputs: ePWMxA and ePWMxB.

- **Timer period units**

  This setting specifies whether the unit of TBPRD is specified in seconds or directly in clock cycles (default choice). Note that, even if the timer period unit is seconds, the software converts this value to an equivalent quantity in clock cycles during the deploy to hardware. Therefore, it is suggested to specify TBPRD in a clock cycles units (TBCLK) to reduce computations and rounding errors. Regarding the F28069 LaunchPad™ board, it is important to remind that $f_{\text{ck}} = \text{CPU}_{\text{frequency}} = \text{SYSCLKOUT} = 90\,\text{MHz}$, that leads to $\text{TBCLK} = 11.1\,\text{ns}$.

- **Specify timer period via**

  This drop down menu allows to specify how the timer period value should be set. If **Specify via dialog** is selected, the following label becomes Timer period. If **Input port** is selected, the following label switches to Timer initial period and a timer period input port T is created on the block.

- **Timer period/Timer initial period (TBPRD)**

  This value sets the maximum count reachable by the TBPRD counter and it is stored in the corresponding register. This value must be compliant with the choice specified in **Time period units** and the selected counting mode, i.e., the type of the carrier waveform. Section 13.3 explains how to compute this value according to the counting mode. It is reminded that, in this book, TBPRD is also called $\text{PWM}_{\text{counter\_period}}$ and it represents the amplitude of the triangular carrier.

- **Reload for time-base period register (PRDLD)**

  This setting specifies the time at which the counter period is reset:

  - **Counter equals to zero**: the counter period refreshes when the value of the counter is 0. Namely, the shadow register contents are transferred to the active register TBCTR = 0;
  - **Immediate without using shadow**: the counter period refreshes immediately.

- **Counting mode**

  Three different counting modes are feasible: Up, Down, and Up-Down; this directly relates to the definition of TBPRD. More details are reported in Section 13.3.

- **Synchronization action**

  Every time different ePWM modules should work together, the source of a phase offset can be specified to act on the time-base synchronization input signal (EPWMxSYNCI), that is, a synchronization input signal from a previous ePWM module through the SYNC input port. Possible choices are:

  - **Set counter to phase value specified via dialog**, which creates the Time-Based Phase (TBPHS) *offset* value parameter;
  - **Set counter to phase value specified via input port**, which creates a phase (PHS) input port on the block;
  - **Disable** (default entry), which prevents the application of any phase offset to the ePWM module.

- **Counting direction after phase synchronization**

  This parameter appears when the **Counting mode** is set on **Up-Down**

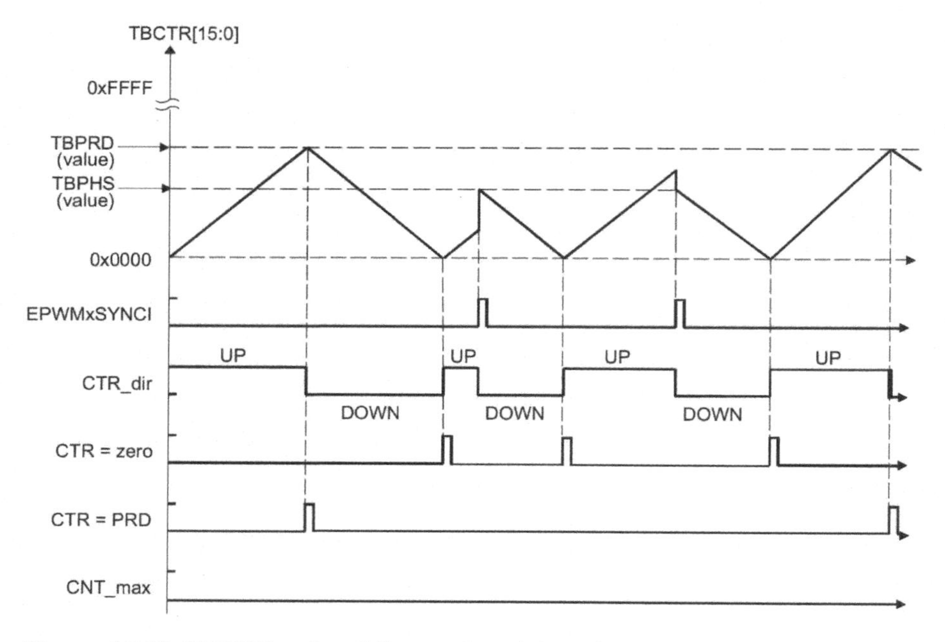

Figure 13.13 TBPHS role while synchronizing the counter: count down on synchronization event [21].

and the **Synchronization action** is **Set counter to phase value specified via dialog** or **Input port**. The timer can be configured to count up or down after reaching the value specified in **TBPHS** (see next option). This parameter corresponds to the PHSDIR field of the Time-Base Control (TBCTL) Register.

- **Phase offset value (TBPHS)**

  This field appears when **Set counter to phase value specified via dialog** is selected in **Synchronization action**. The offset value is stored in the TBPHS register and loaded in TBCTR at every Synchronization event. Each ePWM module can be configured to use or to ignore the synchronization input. If this parameter is set, then, the TBCTR is automatically loaded with the content of TBPHS (see Figure 13.13) when one of these two operations are configured:

  - **EPWMxSYNCI**, that is, **Synchronization Input Pulse**: the value of the TBPHS register is loaded into the TBCTR one when an input synchronization pulse is detected. Each synchronization attempt occurs at every TBCLK edge.
  - **Software Forced Synchronization Pulse**: every time 1 is specified in TBCTL control bit, a software forced synchronization is called.

This pulse acts as a synchronization input signal, and, therefore, it is equivalent to a pulse acting in EPWMxSYNCI.

- **Synchronization output (SYNCO)**

  This parameter specifies the event that generates a time-base synchronization output signal EPWMxSYNCO from the time-base sub-module and it corresponds to the SYNCOSEL field in the TBCTL register. This signal can be sent to next ePWM modules for synchronizing it. Indeed, the available choices are:

  - **Pass through (EPWMxSYNCI or SWFSYNC)**: a synchronization input pulse or software forced synchronization pulse are passed from an ePWM module to the following chained one. This option is used to achieve precise synchronization across multiple ePWM modules by daisy-chaining their time-base sub-modules.

    Remark: ePWM modules are chained in ascending order. Namely, ePWM1 sends synchronization signal to ePWM2, which, in turn, sends it to ePWM3, and so on.
  - **Counter equals to zero (CTR=Zero)**: the output signal is generated every time the time-base counter is equal to zero TBCTR = 0;
  - **Counter equals to compare B (CTR=CMPB)**: the synchronization signal is generated every time the time-base counter is equal to counter-compare B TBCTR = CMPB;
  - **Disable**: EPWMxSYNCO output is disabled (default choice).

- **Time base clock (TBCLK) and High speed time base clock (HSPCLKDIV) prescaler dividers**

  The default prescaler value is 1. Nevertheless, a different value can be set (i.e., additional prescaler) to reduce the size of the peripheral counters in order to not exceed the range of values allowed by `uint16` data type. This especially happens for low carrier frequencies. As an example, considering an up counting mode at $f_c = 1\,\text{Hz}$ and $\text{CPU}_{\text{frequency}} = 90\,\text{MHz}$, it follows that:

  $$\text{TBPRD} = \text{PWM}_{\text{counter\_period}} = \frac{\text{CPU}_{\text{frequency}}}{\text{PWM}_{\text{frequency}}} = 1 \cdot 10^6 \tag{13.3}$$

  which cannot be realized with 16 bit data. Hence, a frequency divider/prescaler is needed to adapt the representation range. The clock cycle used as base time can be recomputed as:

  $$\text{TBCLK} = \frac{1}{\text{PWM}_{\text{frequency}} \cdot 65535}$$

  Now, it can be used to cast $\text{PWM}_{\text{counter\_period}}$ up to the maximum value allowed by `uint16` data type.

- **Enable swap module A and B**

  Every time this checkbox is ticked, the options set in the following two tabs (ePWMA and ePWMB) are swapped.

Block Parameters—ePWMA tab ePWMA

The settings that can be edited in this tab are used to configure the comparison between the carrier and the modulating signals, the switching pattern behavior and the reload conditions, e.g., for the upper switch of a 2L-VSC leg (ePWMxA).

- **Enable ePWMxA**

  This checkbox is needed to allow any change in all the options for controlling the switch that corresponds to ePWMxA. By default, Enable ePWMxA is ticked (whereas Enable ePWMxB is not).

- **Action when counter = [· · ·]**

  The comparison between the input signal and the triangular carrier is defined by the AQ settings. In particular, there are four compare registers: CMPA, CMPB, CMPC, and CMPD, which are compared to TBCTR value to generate appropriate events. The values of interest of the counter are: ZERO, period (PRD), CMPA on up-count (CAU), CMPA on down-count (CAD), CMPB ojn up-count (CBU) and CMPB on down-count (CBD). In case of a 2L-VSC, CMPA is of particular interest. The reader is referred to Section 13.3 for further details.

- **Compare value reload condition**

  This drop-down menu determines if and when to reload the AQ operations. The available choices are:

  - Load on counter equals to zero (CTR=Zero);
  - Load on counter equals to period (CTR=PRD);
  - Load on either;
  - Freeze.

- **Add continuous software force input port**

  Every time this checkbox is enabled, an input port (software forced action-SFA) is created, which can be used to select the software force logic. Three different integer values can be sent to SFA port:

  - 0, that is, forcing disable: Do nothing (default option);
  - 1, that is, forcing low: Clear low;
  - 2, that is, forcing high: Set high.

- **Continuous software force logic**

  This drop-down menu appears in this tab only if the previous checkbox is disabled. It allows to select which type of software force logic to apply. The available choices are:

  – Forcing disable: Do nothing (default setting);

  – Forcing low: Clear low;

  – Forcing high: Set high.

- **Reload condition for software force**

  This drop-down menu allows to select when the reload operations occurs:

  – Zero (default setting);

  – Period;

  – Either period or zero;

  – Immediate.

Block Parameters - ePWMB tab ePWMB

The settings that can be edited in this tab are used to configure the comparison between the carrier and the modulation signals, the switching pattern behavior and the reload conditions, e.g., for the lower switch of 2L-VSC leg (ePWMxB). In this case, if a complementary switching is desired, ePWMxB has to be the opposite of ePWMxA. This is achieved by selecting the **inverted version of ePWMxA** checkbox, without requiring other settings. All the other parameters that can be found in this tab are the same as those described previously.

Block Parameters - Counter Compare tab Counter Compare

This tab allows to set specific parameters of the counter compare sub-module. The same options are repeated for the high and low switch. Namely, the available settings are:

- **CMPA units**

  This drop-down menu allows to specify the units used by the register which is dedicated for the comparison. The available choice are:

  – Clock cycles (default option);

  – Percentages.

- **Specify CMPA via**

  This second drop-down menu is used to specify in which way the control signal is read by the ePWM module. The available choices are:

  - **Dialog** (default selection). Namely, the triangular carrier wave is compared to the constant value specified in the **CMPA value** label;
  - **Input Port**. Namely, a new input port labeled with WA appears on the ePWM block.

- **CMPA value**

  This field appears when **Specify CMPA via** is set on **dialog**. A constant control signal is specified in this option in CMPA units.

- **CMPA initial value**

  This field appears when **Specify CMPA via** is set on **input port**. This label is intended to assign the initial value to CMPA, which is considered by the ePWM block every time it starts counting. Then, the input signal through WA port drives the modulation of the high switch.

- The same settings can be provided for **CMPB** such as **CMPB units**, **Specify CMPB via Input port** (which creates a new port WB), **CMPB value** and **CMPB initial value**.

Block Parameters - Deadband unit tab Deadband unit

This tab allows to program dead bands on the PWM signals generated by ePWMxA and ePWMxB modules to avoid any overlapping between the generated signals and, thus, any short-circuit in power electronic applications. The same settings are repeated for both ePWMxA and ePWMxB.

- **Use deadband for ePWMxA/ePWMxB**

  These two check-boxes enable/disable the use of dead bands. By default, these two boxes are left unchecked. By ticking one or both of them, some parameters appear.

- **Deaband polarity**

  During the dead band time, ePWMxA and/or ePWMxB outputs have to be settled to an inactive state. To this aim, this field configures the dead band polarity, that is, the logic state of the inactive state. This polarity depends on the specified settings. If only one of the two aforementioned check-boxes are ticked, the available choices are: **Positive** (default setting) and **Negative**. Instead, every time both check-boxes are selected, the available choices are four, that is: **Active high (AH, default setting)**, **Active low (AL)**, **Active high complementary (AHC)** or **Active low complementary (ALC)**. Active high/low means that the system is active when the ePWM output signal is set to a high/low logic state, respectively. Instead, active high/low complementary is used every time ePWMxB is the inverse of ePWMxA.

- **Signal source for raising edge (RED)**

  This entry selects the signal source on which RED has to be applied. This field is available only if **Use deadband for ePWMA** is enabled. By default ePWMxA signal is selected.

- **Signal source for falling edge (FED)**

  This entry selects the signal source on which FED has to be applied. This field is available only if **Use deadband for ePWMB** is enabled. By default ePWMxA signal is selected.

- **Deadband period source**

  This option specifies whether the RED and/or FED dead band period is specified through input port or through dialog. **Specify via dialog** is the default selection.

- **Deadband Rising edge (RED) period (0∼1023)**

  This label appears only if the **Deadband period source** is set on **Specify via dialog** and **Use deadband for ePWMxA** is enabled. This field specifies the rising edge dead band as multiple of TBCLK. For instance, considering $CPU_{frequency} = 90\,MHz$, $RED = 15$ means $15 \cdot TBCLK \approx 167\,ns$.

- **Deadband Falling edge (FED) period (0∼1023)**

  This label appears only if the **Deadband period source** is set on **Specify via dialog** and **Use deadband for ePWMxB** is enabled.

Block Parameters - Event trigger tab Event Trigger

This tab allows to configure ADC Start of Conversion (SOC) signals through one or both ePWMxA and ePWMxB outputs. Typically, the ePWMxA is kept as the master module and it is used to define the ADC Start of Conversion Event (ePWMSOCxA). PWM interrupt options are not investigated in this book.

- **Enable ADC start of conversion for module A/B**

  These two check-boxes enable to edit the parameters for the generation of SOC signals in the correspondence of a specified event.

- **Number of event for start of conversion A (SOCA) to be generated**

  If **Enable ADC start of conversion for module A** is selected, this drop down menu appears and it sets the number of the event that triggers ADC Start of Conversion for Module A (SOCA). Namely, three choices are available:

  - **First event**: SOC is triggered with every event (default option);
  - **Second event**: SOC is triggered with every second event;

  - **Third event**: SOC is triggered with every third event.

- **Start of conversion for module A event selection**

  If **Enable ADC start of conversion for module A** is selected, this field specifies the event (that is, a counter match), that triggers an ADC start of conversion event. There are several choices. Among all the possibilities, it is reported:

  - **Counter equals to period (CTR=PRD)**: a SOC signal is generated every time the ePWM counter reaches the period value TBPRD, i.e. the maximum carrier peak;
  - **Counter equals to zero or period (CTR=0 or CTR=PRD)**: an SOC signal is generated both at the the maximum or minimum carrier peaks.

  These options are particularly important for what concerns current measurement. Indeed, by synchronizing the SOC of an ADC reading related to a current sensor with CTR=0 or CTR=PRD (or both), the sensed data are coincident with the average value of the current, thus, avoiding to sense/introduce high frequency contribution (ripple) in the control loop. This concept is exploited in the examples reported in the following Chapters.

13.5 Example with ePWM block

Build a firmware aimed to realize a PWM logic with triangular carrier at 3 kHz and complementary output by using an ePWM peripheral

Referring to the implementation of the PWM logic proposed in Section 12.4, this example still uses one of the external potentiometer available on the extPot3 to vary the modulating signal. This time, the PWM logic is embedded into the ePWM block. The digitalized measurement read from ADCA3 must be scaled between 0 and 1 first and, after that, between 0 and TBPRD to match with the range of variation of the carrier implemented by the ePWM peripheral. This exercise is carried out aiming to realize the actuation of a 2L-VSC (single) leg using a triangular carrier. Reminding that for the LAUNCHXL F28069M board $f_{\text{ck}} = \text{CPU}_{\text{frequency}} = 90\,\text{MHz}$, for an up-down counting mode it follows:

$$\text{TBPRD} = \text{PWM}_{\text{counter\_period}} = \frac{\text{CPU}_{\text{frequency}}}{2 \cdot \text{PWM}_{\text{frequency}}} = 15000 \tag{13.4}$$

Firmware Environment

c28069_PWMbasics_F.slx (solver: fixed step-ODE4, step size: $T_s = 1/3\,\text{kHz}$)

Open a new blank Simulink® project and configure the environment as shown in Chapter 9. Then:

- Open the **Model Configuration Parameters** window and use the same settings adopted for **Example 1** in Section 10.5.
- Insert an **ePWM** block, double-click on it and set the following parameters:
  - **General** (Tab)
    * select **module** ePWM1;
    * set **Timer period units** to Clock cycles (TBCLK);
    * set **Timer period** equal to $\text{PWM}_{\text{counter\_period}}$;
    * verify that **Counting mode** is set on Up-Down;
    * leave the other default settings.
  - **ePWMA** (Tab)
    * tick **Enable ePWM1A** check-box;
    * **Action when counter=ZERO**: do nothing;
    * **Action when counter=period (PRD)**: do nothing;
    * **Action when counter=CMPA on up-count (CAU)**: clear;
    * **Action when counter=CMPA on down-count (CAD)**: set;
    * **Action when counter=CMPB on up-count (CBU)**: do nothing;
    * **Action when counter=CMPB on down-count (CBD)**: do nothing;
    * set in **Compare value reload condition** as Load on counter equals to zero (CTR=zero);
    * leave the other default settings.
  - **ePWMB** (Tab)
    * tick **Inverted version of ePWMxB** check-box. This option enables the implementation of the complement of ePWM1A;
    * leave the other default settings.
  - **Counter Compare** (Tab)
    * select clock cycles as **CMPA units**;
    * select input port in **Specify CMPA via**;
    * set **CMPA initial value** on 0;
    * select Counter equals to zero in **Reload for compare A register (SHDWAMODE)**;

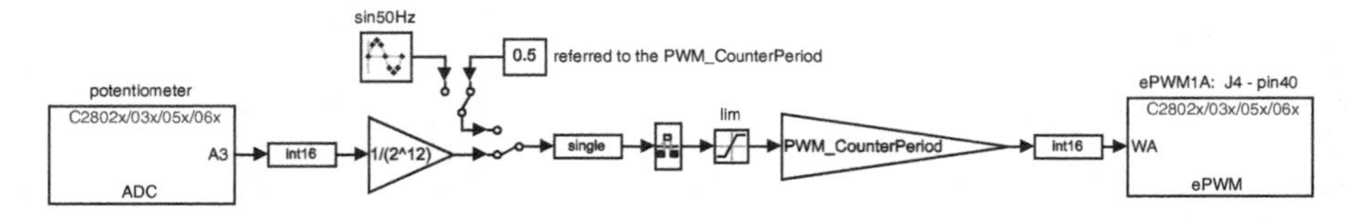

Figure 13.14 Simulink® scheme included in c28069_PWMbasics_F.slx.

  * select clock cycles in **CMPB units**;
  * select Specify via dialog in **Speicfy CMPB via**;
  * set **CMPB initial value** 0;
  * select Counter equals to zero in **Reload for compare B register (SHDWBMODE)**.
- **Deadband unit** (Tab)
  * tick the **Use deadband for ePWM1A** check-box;
  * tick the **Use deadband for ePWM1B** check-box;
  * select Active high complementary (AHC) as **Deadband polarity**;
  * select ePWMxA as **Signal source for raising edge (RED)**;
  * select ePWMxA as **Signal source for falling edge (FED)**;
  * select Specify via dialog in **Deadband period source**;
  * set **Deadband Raising edge (RED) period (0~1023)**: 15;
  * set **Deadband Falling edge (RED) period (0~1023)**: 15.
- **Event Trigger** (Tab)
  * select **Enable ADC start of conversion for module A**;
  * set **Number of event for SOCA to be generated** to First event;
  * select Counter equals to period (CTR=PRD) as **Start of conversion for module A event selection**.
- Leave the other tabs set on their default options.

Figure 13.14 shows the complete scheme and its connections. In particular, the ADC conversion chain from the ADCA3 reading up to the saturation block is the same as that one reported in Section 12.4. Note that the rate transition is used to reduce the discretization of the modulating signal, with the value of $m(k)$ being updated every T_s. It is common practice to set the sampling time related to the controller, i.e., T_s, according to the chosen switching period $T_s = 1/\text{PWM frequency}$ or half of it $T_s = 1/(2 \cdot \text{PWM frequency})$. If $T_s = 1/\text{PWM frequency}$ is chosen, this imply a constant modulating signal over $T_c = 1/f_c = 2 \cdot \text{TBPRD}$. Therefore, in the **ePWMA** tab, the **Compare value reload condition** has to be set as *Load on counter equals to zero* because a new CMPA value is evaluated at CTR=ZERO.

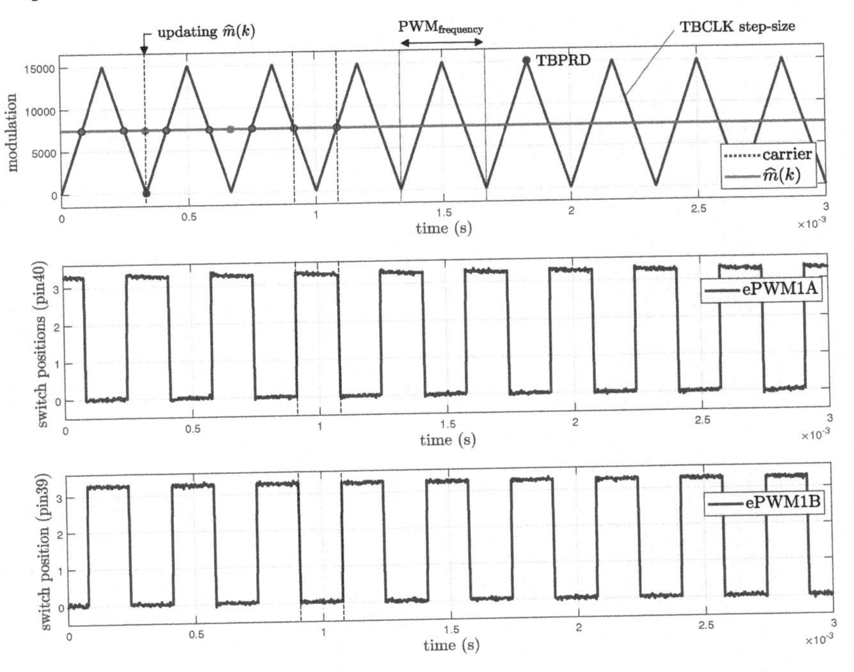

Figure 13.15 Measured signals at the pins 39 and 40 when the modulating signal is a constant $m(k) = 0.5$.

All the system parameters have to be defined and initialized before the firmware deploy through a separate MATLAB® `.m` file or included in *Model Properties/Main/InitFcn*. An example of such script is reported here in the following.

```
%% F28069M clock frequency
CPU_frequency        = 90e6;
%% carrier frequency definition
PWM_frequency        = 3e3;
Tsw                  = 1/PWM_frequency
%% TBPRD (counting mode) definition
PWM_counterperiod    = CPU_frequency/(2*PWM_frequency)
%% sampling time definition
Ts                   = Tsw;   or Ts = Tsw/2
```

The modulating signal must be scaled in the range $[0, \text{PWM}_{\text{counter\_period}}]$. Therefore, a gain block is used to compute $\widehat{m}(k) = m(k) \cdot \text{PWM}_{\text{counter\_period}}$, where $m(k) \in [0, 1]$. The result is converted to an `int16` data type through a Data Type Conversion block to be compatible with the ePWM peripheral registers. The firmware allows to choose (by double-clicking on two different

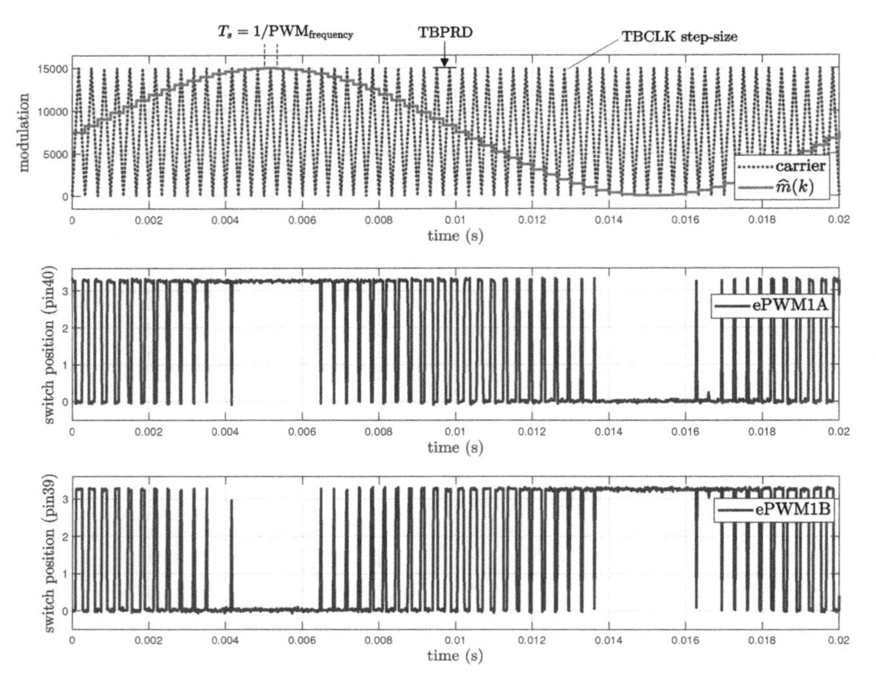

Figure 13.16 Measured signals at pin 39 and 40 when the modulating signal is a 50 Hz sine wave.

manual switches) between three operating modes: a constant modulating signal equal to 0.5 $\rightarrow$ 1125; a sample-based sine wave at $f_1 = 50\,\text{Hz}$; a variable modulating signal provided by the potentiometer. In any case, ePWM1 is used. Therefore, ePWM1A is sending its output on pin 40, whereas ePWM1B on pin 39. The output signals connected to these pins can be visualized by using an oscilloscope, with ground (GND) connection to, e.g., pin 20 of J2. In case $m(k) = 0.5$, it is possible to highlight the behavior of the counter as well as the sample updates for $\widehat{m}(k)$, as shown in Figure 13.15. Regarding the sample-based sine-wave block, the reader is referred to the settings reported in Section 10.5, with the exception of the sample time. This latter has to be set based on the *Compare value reload condition*. Thus, the sine wave is sampled at T_s leading to a pulse number equal to $n_s = 1/(f_1 \cdot T_s) = 60$. The resulting waveforms are shown in Figure 13.16. In particular, Figure 13.15, 13.16 and 13.17 show the voltage level at the output pins (TTL logic) related to both ePWM1A and B. The triangular carrier is discretized every TBCLK even if it is not visible ($TBCLK \approx 11\,\text{ns}$). It is important to note that the cases $m(k) = 0$ (left) and $m(k) = 1$ (right) do not match exactly with a duty cycle equal to 0 or 1, as shown in Figure 13.17. This is due to the potentiometer, which does not provide an internally generated signal (like the constant

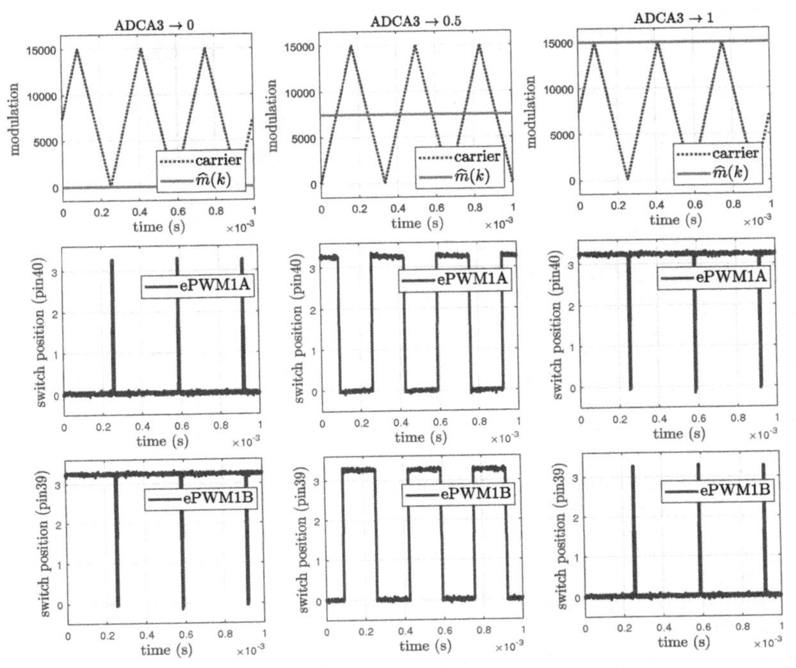

Figure 13.17 Measured signals at pin 39 and 40 when the modulating signal comes from a potentiometer.

or sine-wave blocks) and, therefore, is never exactly equal to 0 or 1.[3] The ePWM peripheral settings enable the ADC SOC for module A which can be used to synchronize the ADC reading. This is not particularity useful for this kind of exercise, but it will be fundamental in case of ADC reading related to measurements, e.g., phase current.

13.6 DAC Peripheral—Filtered PWM

The F28069M LaunchPad™ board has two ePWM modules which are dedicated to provide special Digital to Analog Converters (DACs) features and they are simply called DACx, where x is the number of the module. The difference between standard ePWM and DAC peripherals is that these latter are equipped with RC low pass filters right before their outputs, as shown in Figure 13.18 (a). If the spectrum of a switching pattern generated by a PWM logic (e.g., a chopped voltage waveform as that one reported in Figure 13.16) is considered, the aim of filtered PWM signals is to try

[3]Sometimes, it may happens that minimum RED and/or FED are introduced by default for safety reason, i.e., to avoid short-circuits, but this is not the case in this example.

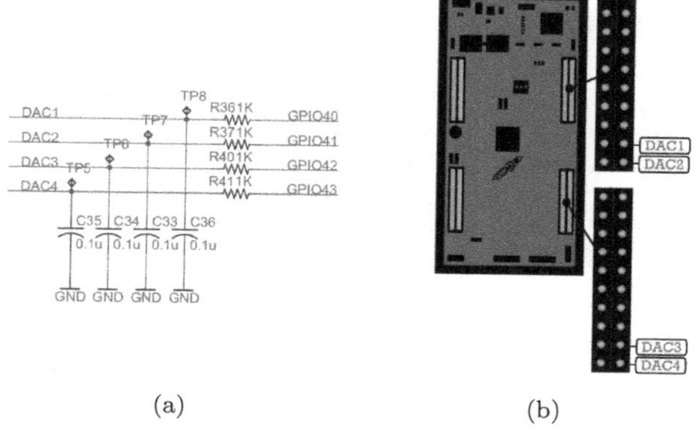

Figure 13.18 Hardware configuration of the ePWM peripherals equipped with the DAC: (a) schematic of the installed output filters on DAC channels and (b) pins on the F28069M LaunchPad® board connected to those channels [16].

approximate modulating signal in a more accurate way. The cut off frequency of the RC filter is constant because it is related to the physical components mounted on the F28069M LaunchPad™ board. Therefore, the quality of the filtered signal strongly depends on the choice of the switching frequency at which the PWM logic operates. The LaunchXL F28069M board provides 4 DAC channels, which are connected to ePWM7A, ePWM7B, ePWM8A, and ePWM8B. Their corresponding output voltages are connected to GPIO40, GPIO41, GPIO42 and GPIO43, as shown in Figure 13.18 (b).

There are no dedicated Simulink® blocks for DAC peripheral since, at the firmware level, the same ePWM settings still hold. Nevertheless, the correct ePWM module must be selected.

13.7 Examples with DAC Peripherals

Example 1

Build a firmware aimed to internally generate a 10 Hz sinusoidal waveform and visualize it through a DAC channel

Knowing how to operate with an ePWM peripheral, a DAC can be used to visualize a an internally generated signal to debug a processing chain, e.g., for debugging purpose.

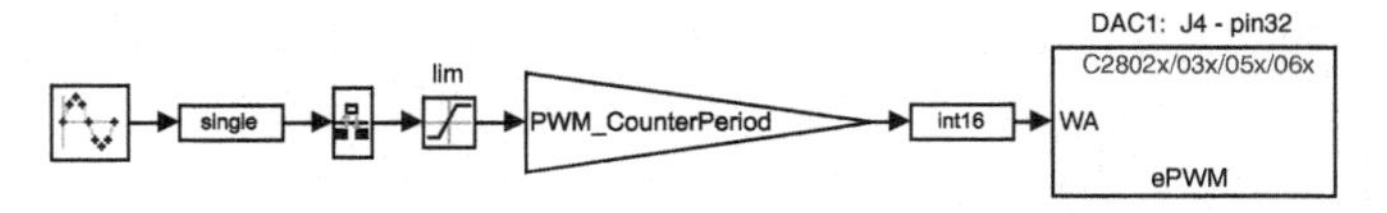

Figure 13.19 Simulink® scheme included in c28069_sinDAC_F.slx.

Firmware Environment

c28069_sinDAC_F.slx (solver: fixed step - ODE4, step size: $T_s = 100\,\mu s$)

Open a new blank Simulink® project and configure the environment as shown in Chapter 9. Then:

- Open the **Model Configuration Parameters** window and use the same settings adopted in Section 10.5, **Example 1**.
- Insert a **Sine Wave** block and double click on it:
  - Set sample based as **Sine type**;
  - Select use simulation time in **Time (t)**;
  - Set an **Amplitude** equal to 0.5;
  - Set a **Bias** equal to 0.5. i.e. sine-wave ranges between 0 and 1;
  - Set the number of **Samples per period** equal to $n_s = 100$;
  - Set the **number of offset samples** equal to 0;
  - Set the **sample time** $T_{sig} = 0.001$ s. which implies

$$f_1 = \frac{1}{n_s \cdot T_{sig}} = 10\,\text{Hz}$$

- A **gain block** is placed in the scheme. Its value is set equal to $\text{PWM}_{\text{counter\_period}}$. This multiplication allows to re-scale the modulation signal, making it comparable with the carrier.
- The **Saturation**, **Data type conversion** (`int16`) and **Rate trasition** blocks have the same meaning (i.e., protection and scaling) as the those used in previous examples (see Section 13.5).
- An ePWM module is added to the scheme. Module ePWM7 is used in this example. Its setting for the General and ePWMA tabs are the same as those reported in Section 13.5. Sub-module ePWM7B is not used (disabled). All the other settings are left in default conditions.

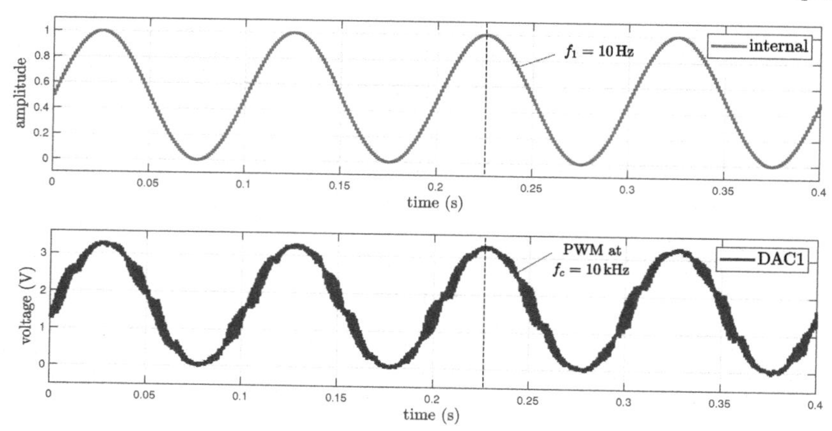

Figure 13.20 Modulating signal and measured signal at pin 32, referred to the ground of the board.

Connect the previously mentioned blocks as shown in Figure 13.19.
In order to have a good approximation of the 10 Hz sine wave, the carrier frequency is set equal to $\text{PWM}_{\text{frequency}} = 10\,\text{kHz}$. Considering an up-down counting mode, it follows that $\text{TBPRD} = \text{PWM}_{\text{counter\_period}} = \text{CPU}_{\text{frequency}}/2 \cdot \text{PWM}_{\text{frequency}} = 4500$. The sampling time is dependent on the modulating signal updates, i.e., $T_s = 1/\text{PWM frequency} = 100\,\mu\text{s}$. The settings of the ePWM block are the same as reported in Section 13.5. The parameters are initialized by a script similar to the one reported there. The resulting voltage signal can be checked with an oscilloscope by connecting a probe to pin 32 of J4, which corresponds to DAC1 (the reader can use pin 20 of J2 to refer the measurement to ground). The result is shown in Figure 13.20.

Example 2

Build a firmware aimed to internally generate a 10 Hz sinusoidal waveform, visualize it through a DAC channel, read it via an ADC and visualize the delayed results through another (different) DAC channel

This exercise is an extension of the previous one (Example 1) to underline how the processing delay affects both measurements and computations. The internal signal is again sent to DAC1 which is now physically connected to ADC via cable. The signal is processed and sent to a different DAC channel (e.g., DAC3) for debugging purpose. The results are visualized through an oscilloscope.

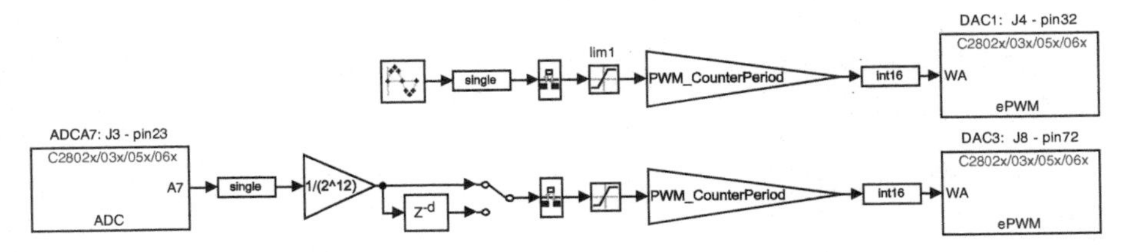

Figure 13.21 Simulink® scheme included in c28069_sinDACinout_F.slx.

Firmware Environment

c28069_sinDACinout_F.slx (solver: fixed step - ODE4, step size: $T_s = 100\,\mu s$)

Make a copy of the Simulink® file from the previous exercise (Example 1) keeping the same settings. Then:

- Copy and paste the previous ePWM block. Double-click on the new copy and select module ePWM8. Leave all the other settings unchanged, thus, only ePWM8A is enabled.
- Add an **ADC** block and use the same configurations as those adopted in Section 12.4. Note that ADCA7 has to be selected, with sample time equal to T_s. Use the same approach to process the data in terms of scaling (to obtain a $[0, 1]$ range) and data type (`single`).
- Keep the same data processing in terms of saturation, $\text{PWM}_{\text{counter\_period}}$ scaling, rate transition (T_s), and data type (`int16`).
- Insert a manual switch and a **Delay** block; set **Sample time** as T_s and **Delay length** equal to 100.

Arrange the aforementioned blocks as shown in Figure 13.21. Channel ePWM8A corresponds to DAC3, which is connected to pin 72 from J8. Since DAC1 (ePWM7A) corresponds to pin 32 from J4, ADCA7 is now selected because it is located in a handy position, i.e., pin 23 from J3. A short cable is then connecting DAC1 to ADCA7. Therefore, the sine wave approximated by DAC1 is read, re-processed/manipulated by the LaunchXL F28069M board, and, then, sent to DAC3. Both ePWM7A and ePWM8A operate with a $f_c = 10\,\text{kHz}$ carrier frequency, with $T_s = 1/\text{PWM frequency} = 100\,\mu s$. The oscilloscope measurements can be referred to GND connecting pin 20 from J2.

Since the sine-wave is subject to a double processing (first generated through a DAC and, then, measured by an ADC), such workflow may be affect the result introducing both attenuation and computational delay.

Regarding the former, Figure 13.22 shows DAC1 almost aligned with the internally generated sine wave, while DAC3 presents a slightly different amplitude. This might be due to the ADC reading which processes a noisy waveform

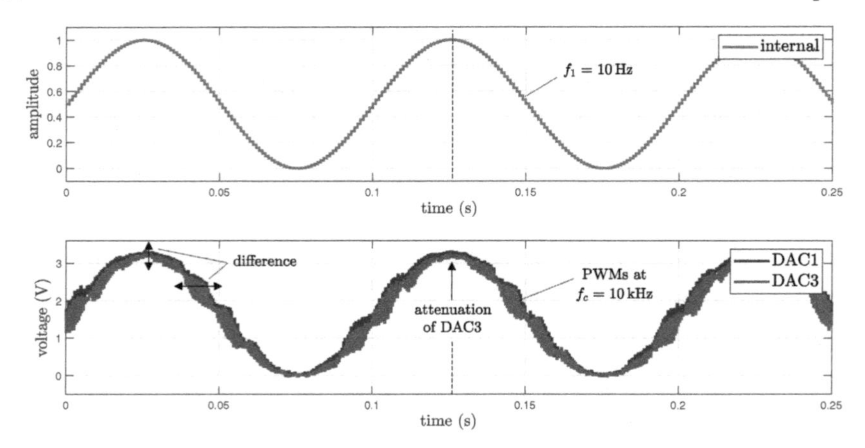

Figure 13.22 Modulating signal and measured signals at pin 32 and 72, both referred to the ground of the board.

(from DAC1), thus, returning rounded data. Moreover, also data type conversion blocks are included in the processing. The cable connection from DAC1 to ADCA7 might introduce additional noise as well.

Regarding the computational delay, DAC3 is lagging DAC1, even if this time delay is not fully recognizable from Figure 13.22. This can be somehow expected since the data read/actuation is scheduled within TBCLK, while the processing/computations occurs every T_s. Thus, the new data provided by the ADC reading (i.e., board data in) is delayed with respect to the output values (i.e., board data out). The processing effect can be emphasized by manually adding a larger delay in the ADC conversion chain. The manual switch inserted in the scheme (see Figure 13.21) allows to include a delay block which is configured to add $T_d = 100T_s = 0.01\,\text{s}$ to the computational time at each step. This is evident in Figure 13.23, where DAC1 is still aligned with the internally generated signal, while DAC3 is now delayed by T_d (i.e., it is shifted to the right).

To make a comparison, this approach is also suitable to describe the effects observed in Example 2 reported in Section 10.5, where the overall delay was mainly due to serial communication sampling.

Remark

If the delay is too large, the workflow of the firmware might be corrupted in terms of synchronization between input/output. As an example, given that the internal sine-wave frequency is set equal to $f_1 = 10\,\text{Hz}$, if $T_d > 0.05\,\text{s}$ (i.e., greater than $1/2f_1$) the MCU starts to adapt the frequency of the sine-wave approximated by DAC1 (which still operates at carrier frequency, $f_c = 10\,\text{kHz}$), resulting in lower values than f_1. As an example, running the previous firmware with $T_d = 600T_s = 0.06\,\text{s}$, DAC1 returns a 5 Hz sine-wave. Hence, in

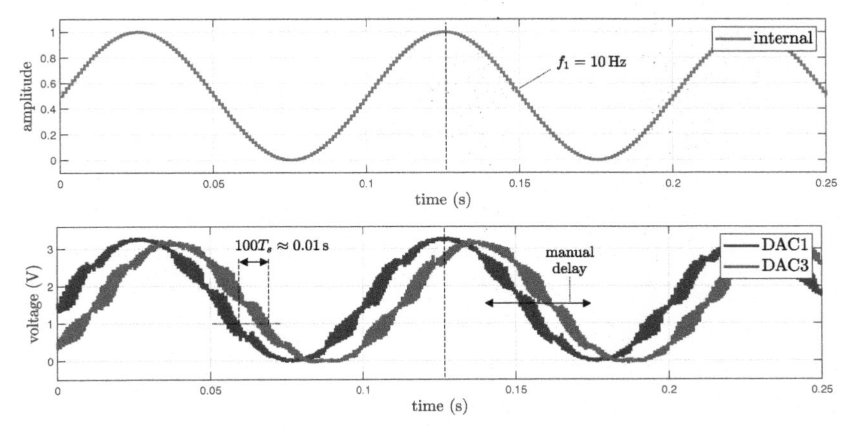

Figure 13.23 Modulating signal and measured signals at pin 32 and 72, both referred to the ground of the board in presence of large latency $T_d = 0.01\,\text{s}$.

case of large delay, the PWM carrier frequency should be increased to obtain more accurate outputs. Note that, no SOC synchronization between ADC and ePWM/DAC has been introduced yet.

13.8 Synchronization between Multiple ePWM Modules

Synchronization events can be settled between peripherals of different kind or between modules of the same family. In the previous Sections, all discussions described the operation of a single ePWM module. Nevertheless, multiple ePWM modules may need to work together in the same firmware. To facilitate the understanding of this scenario, the ePWM module can be represented by the simplified block diagram shown in Figure 13.24 . This picture shows the key resources needed to explain how multiple ePWM modules may working together by synchronizing their actions. Namely signals *SyncIn*, *SyncOut*, and

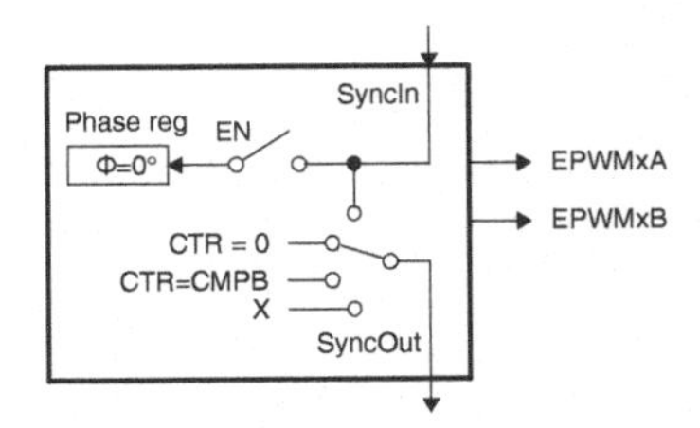

Figure 13.24 Simplified representation of the features of an ePWM module for synchronizing peripherals [21].

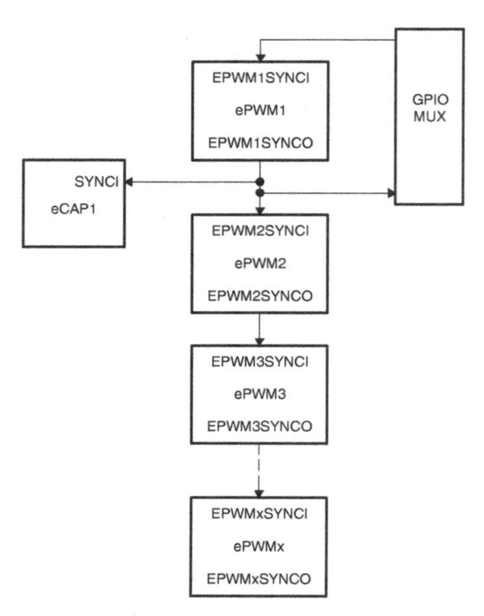

Figure 13.25 Time-Base Counter Synchronization Scheme [21].

Phase reg are explained to this aim. A time-base synchronization scheme connects all the ePWM modules installed on a board. Each ePWM module has a synchronization input (EPWMxSYNCI), also called *SyncIn*, and a synchronization output (EPWMxSYNCO), also called *SyncOut*. The synchronization input for the first instance (ePWM1) comes from an external pin. The synchronization scheme for the remaining ePWM modules is shown in Figure 13.25. Each ePWM module can be configured to use or ignore the synchronization input/output. In the Simulink® environment, such feature can be configured by acting on:

- **Synchronization action** (*input*)

  This option allows to specify any possible delay between counters by setting the reset value of ePWM modules. Actually, this quantity may be specified through an input port with **Set counter to phase value specified via input port** or via dialog with **Set counter to phase value specified via dialog** (see Section 13.4.1). The latter setting generates two new parameter fields: **Counting direction after phase synchronization**, which can be set equal to **Count up after sync** or **Count dwon after sync**, and **Phase offset value (TBPHS)** which has to be defined. This feature enables to configure the selected direction of the time-base counter immediately after a synchronization event, such as TBCTR=TBPHS.

- **Synchronization output (SYNCO)**

  Figure 13.25 shows that the synchronization input of the second ePWM module (ePWM2) can be fed by the synchronization output of ePWM1. In this case, the carrier waveform internally generated in ePWM2 is synchronized with a suitable triggering signal generated inside ePWM1. Moreover, all the ePWMx modules are chained with increasing x. Therefore, a specific**Synchronization output (SYNCO)** mode should be selected for each module. Indeed, this feature define if the selected ePWM module generates its own synchronization output to be passed to the next one, i.e., by setting **Counter equals to zero (CTR=Zero)** or **Counter equals to compare B (CTR=CMPB)**, or if it just passes the SYNCO signal generated by the previous module, that is, setting the option **pass trough (EPWMxSYNCI or SWFSYNC)**, as explained in Section 13.4.1.

Combining these two features, an ePWM module can be synchronized with another one. Lead or lag phase control can be also added to the waveforms generated by different ePWMs.

Example

Build a firmware aimed to synchronize three ePWM modules in a master-slave chain configuration

This exercise is an extension of the Example reported in Section 13.5. In this case, three consecutive ePWM modules, i.e., ePWM1, ePWM2, and ePWM3, are considered. The main settings to create a master-slave chain configuration between ePWM modules in Simulink® is presented here in the following.

Firmware Environment

c28069_PWMsync_F.slx (solver: fixed step - ODE4, step size: $T_s = 1/3\,\text{kHz}$)

Modify the Simulink® file adopted for the Example proposed in Section 13.5 as follows:

- Make two copies of the **ePWM** block in order to have three of them in the scheme;

ePWM1 settings:

- **General** (Tab)
  - Set **Synchronization action:** to **Disable**;
  - Set **Synchronization output (SYNCO):** to **Counter equals to zero (CTR=Zero)**;
  - Leave the other parameters unchanged.

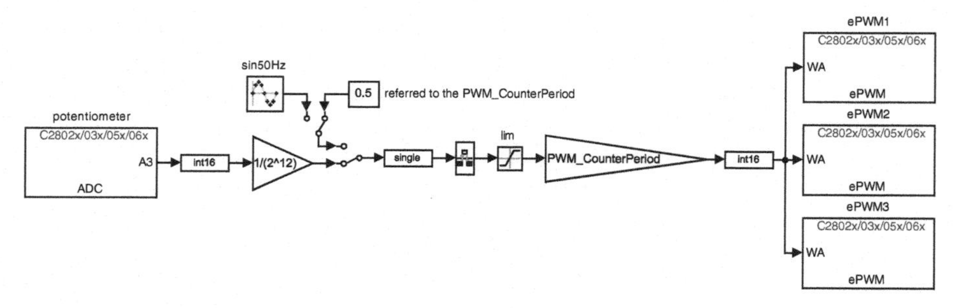

Figure 13.26 Simulink® scheme included in c28069_PWMsync_F.slx.

ePWM2 settings:

- **General** (Tab)
  - Update the **Module** to **ePWM2**
  - Set **Synchronization action:** to **Set counter to phase value specified via dialog**;
  - Set **Counter direction after phase synchronization:** to **Count up after sync**;
  - Set **Phase offset value (TBPHS):** equal to zero;
  - Set **Synchronization output (SYNCO):** to **Pass through (EPWMxSYNCI or SWFSYNC)**;
  - Leave the other parameters unchanged.
- Keep the ePWM2B as the complement of ePWM2A.

ePWM3 settings:

- **General** (Tab)
  - Update the **Module** to **ePWM3**
  - Set **Synchronization action:** to **Set counter to phase value specified via dialog**;
  - Set **Counter direction after phase synchronization:** to **Count up after sync**;
  - Set **Phase offset value (TBPHS):** equal to zero;
  - Set **Syncronization output (SYNCO):** to **Disable**;
  - Leave the other parameters unchanged.
- Keep the ePWM3B as the complement of ePWM3A.

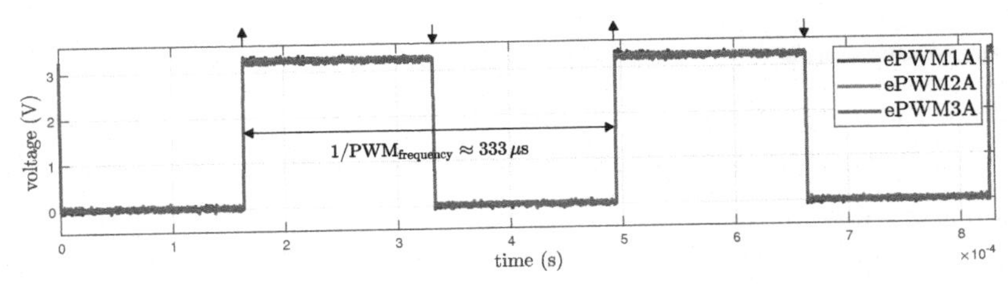

Figure 13.27 Voltage waveforms measured on the pins of the LaunchXL F28069M board corresponding to ePWM1A, 2A and 3A.

- Leave the settings in the **Model Configuration Parameters** windows and those for the remaining blocks (i.e., ADC reading process) of the scheme unchanged.

The updated Simulink® scheme is reported in Figure 13.26. These settings make ePWM1 providing a triggering source EPWMxSYNCI to synchronize both ePWM2 and ePWM3. In particular, the EPWMxSYNCI signal goes from ePWM1 to ePWM3 through ePWM2 without any phase offset (TBPHS=0). For ePWM2 and ePWM3, the **Counting direction after phase synchronization** is set on *Count up after sync* to synchronize the carriers both in frequency (counter period length) and shape in all three modules. The **SYNCO** signal is disabled in the ePWM3 module since there are no further modules to be triggered (i.e., **Synchronization output (SYNCO):** *Disable*). The resulting outputs of the three modules can be measured through an oscilloscope. The corresponding pins to be visualized are pin 40 for ePWM1A, pin 39 for ePWM1B, pin 38 for ePWM2A, pin 37 for ePWM2B, pin 36 for ePWM3A and pin 35 for ePWM3B. Figure 13.27 reports the voltage waveform corresponding to ePWM1A, 2A and 3A.

To provide a better understanding of the adopted master-slave configuration, Figure 13.28 reports a simplified block diagram.

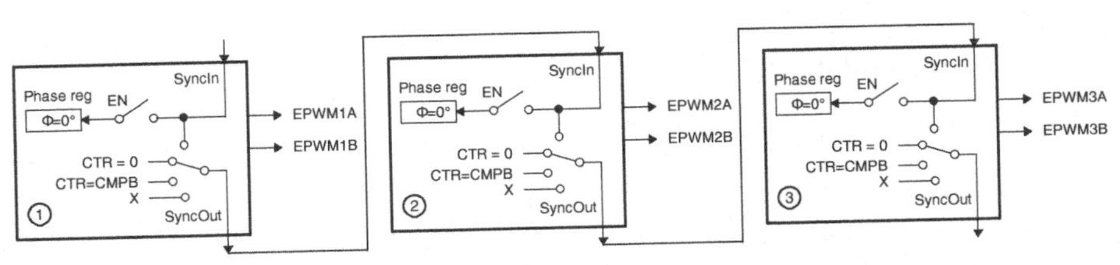

Figure 13.28 Simplified block diagram explaining the master-slave synchronization between three ePWM modules [21].

13.9 Synchronization between ADC and ePWM Modules: *Average* Measurements

The synchronization between an ADC and an ePWM block is an important subject in many applications involving switching devices. Particularly, it is fundamental every time average values of measurements must be processed into closed-loop control schemes. An ePWM module is able to send a synchronization signal that triggers the SOC of an ADC module (or multiple modules) when the carrier signal (i.e., TBCTR) equals a specific value (i.e., TBPRD). To this aim, the following settings are summarized here below:

- **Event Trigger** (Tab)
  - Select **Enable ADC start of conversion for module A**;
  - Set **Number of event for SOCA to be generated** to First event;
  - Select Counter equals to period (CTR=PRD) as **Start of conversion for module A event selection.**

When the carrier signal reaches the maximum value TBPRD, a SOC is sent to the ADC. Nevertheless, this latter has to be configured as well:

- Select ADCINA0 as **Conversion channel** in the **Input Channels** tab. Then, go back in the **SOC Trigger** tab;
- **Sampling mode**: Single sampling mode;
- **SOC trigger number**: SOC0;
- **SOCx acqusition window**: 7;
- **SOCx trigger source**: ePWM1_ADCSOCA;
- **ADC will trigger SOCx**: No ADCINT;
- **Sample time**: T_s;
- **Data type**: uint16;
- Flag **Post interrupt at EOC trigger** (optional);
- **Interrupt selection**: ADCINT1 (optional);
- Flag **ADCINT1 continous mode** (optional).

The key parameter in this list of instruction is the **SOCx trigger source**, which is set to follow a signal coming from the ePWM module.

This features is really useful when a current measurements (typically made of a fundamental and several high frequency components) has to be processed

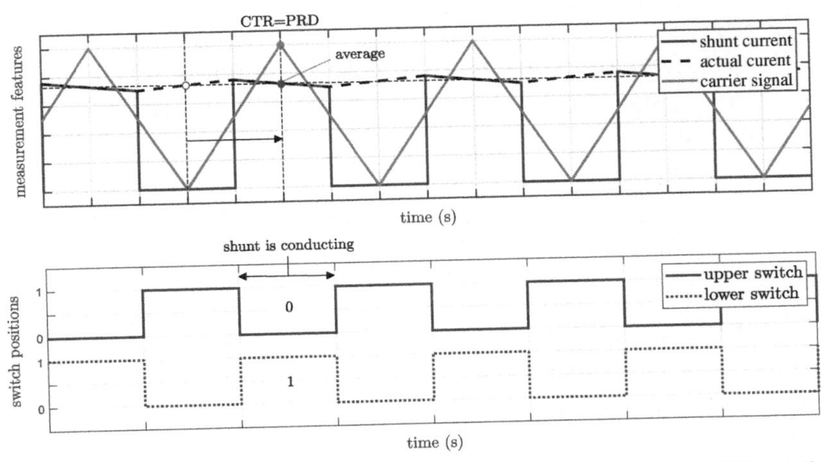

Figure 13.29 Example of synchronization between ePWM and ADC modules: measured current is compared to the actual current flowing through the low-side shunt. Moreover, switching patterns of the considered leg is reported as well in the bottom plot. The converter installed on the DRV8301 expansion board is operated as a DC/DC step-down converter.

into a closed-loop control based on linear controllers. Indeed, these latter may work poorly in presence of high ripple. Thus, averaging may be necessary.

As an example, a setup which comprise a control platform (LaunchXL F28069M) and a converter (BOOSTXL-DRV8301) equipped with current sensors is considered. In this particular case, phase current measurements are provided by some low-side shunts (see Section 3.2). By using the previous synchronization settings, ADCA0 starts the acquisition window when ePWM1 module equals CTR=PRD (an up-down counting mode is adopted). This SOC event is chosen since it occurs in the correspondence of the theoretical mean value of the current, avoiding to process any ripple inside the MCU. This result is shown in Figure 13.29. Note that in presence of an highly distorted current, this approach still limits the amount of noise processed in the closed-loop scheme. Therefore, when an average measurements is required, the firmware should be settled to follow this proposed approach (i.e., ePWM and ADC synchronization).

Additional Consideration on Example 2

Since the sine-wave approximated by DAC1 has to be read by ADCA7 channel, the latter has to by synchronized with ePWM7A module. To this aim, the **Event Trigger** tab settings are no more an optional and they must be correctly configured. In particular:

- **Event Trigger** (Tab)

  - Select **Enable ADC start of conversion for module A**;
  - Set **Number of event for SOCA to be generated** to First event;
  - Select Counter equals to period (CTR=PRD) as **Start of conversion for module A event selection.**

Instead, for ePWM8A module the settings can be the same as those used in Example 1. Both ePWM7A and ePWM8A still operate with a 10 kHz carrier. Once those module are set, the **ADC** block can configured for ADCA7 channel as follows:

- Select ADCINA7 as **Conversion channel** in the **Input Channels** tab. Then, go back to the **SOC Trigger** tab:
- **Sampling mode**: Single sampling mode;
- **SOC trigger number**: SOC0;
- **SOCx acqusition window**: 7;
- **SOCx trigger source**: ePWM7_ADCSOCA;
- **ADC will trigger SOCx**: No ADCINT;
- **Sample time**: T_s;
- **Data type**: uint16;
- Flag **Post interrupt at EOC trigger**;
- **Interrupt selection**: ADCINT1;
- Flag **ADCINT1 continous mode.**

The rest of the scheme is left unchanged.

13.10 Events Execution within Sample Time

The synchronization between ADC and ePWM modules to obtain *average* measurements is an example of how important is to identify what is the goal of a closed-loop scheme (and, therefore, which kind of measurements/actuation is going to be used/adopted) and how to choose the specific settings to avoid issues with related to peripheral and task scheduling early during the design stage. Many challenges are associated to **integrate MCU scheduling and peripherals in power electronics-based applications**. Typical examples are:

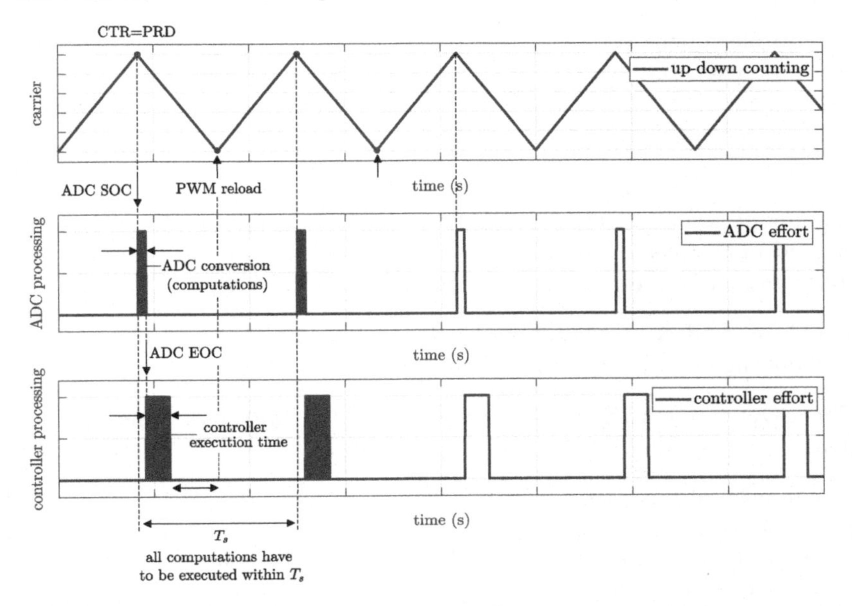

Figure 13.30 Example of possible operations (event sequence, interrupt trigger, and control algorithm execution time) to be performed on a target hardware within the sample time T_s.

- ADC-ePWM synchronization to achieve current sensing in the correspondence of its theoretical average value (see previous Section);
- Compensate sensor delays to achieve the desired controller response for a closed loop system;
- Studying different modulation strategies (or ePWM settings) while designing special algorithms.

The importance of understanding the hardware details of each peripheral (including the identification of the effect of its parameters/registers/operating modes) becomes more and more crucial as the complexity of a firmware increases. As an example, on the F28069M LaunchPad™ Piccolo™ board the ADC hardware contains a sample and hold circuit to sense analog inputs (see Chapter 12). To ensure complete ADC measurement, the minimum acquisition time must be adjusted to account for the combined effects of input circuit and the capacitor installed in the sample and hold circuit. This means, in practice, to update the **SOCx acquisition window** accordingly, i.e., choosing values grater than 7. Another important aspect is the insight of the execution of events. To this purpose, a firmware involving one ADC channel, one ePWM module and a controller which use the ADC reading to compute the modulating signal for the ePWM is considered. An example of possible operations that can occur during each sample time is reported in Figure 13.30. Indeed, the controller operation or other processing/computation routines are scheduled to be executed within $T_s = T_{sw} = 1/f_{sw}$. Therefore, the controller is synchronized with the ADC interrupt. The event sequence which is scheduled on the target hardware foresees that:

- At the beginning of T_s, when the PWM counter value equals the PWM counter period (TBCR=TBPRD), the ePWM peripheral (which is center-aligned in this case, that is, up-down counting mode) triggers the start-of-conversion (SOC) event for the ADC module;
- The ADC module converts the sampled analog signal into digital counts and triggers the end-of-conversion (EOC) event;
- The EOC triggers the ADC interrupt for the controller, which reads the phase current value;
- Assuming that any other reference value is already provided, the controller starts performing the required operations necessary to the computation of the modulating signal for the ePWM module. The number as well as the complexity of these operations define the length of this execution window;
- When the controller execution is done, the resulting modulating signal is sent to the ePWM input and hold on until the end of the T_s.

14

Encoder Peripheral

The enhanced quadrature encoder pulse (eQEP) modules are used for direct interface with linear or rotary encoders to get position, direction and speed information from the rotating parts of a system. Typically, they are used for measuring the speed or the position of rotors in electrical machines driven by power electronics. Thus, this peripheral is used for closing speed or position control loops. This Chapter provides some insight in the working principles of this peripheral and of optical rotary encoders.

14.1 Operating Principle of Incremental Encoders

A rotary incremental encoder is a disk with a track of slots placed along its periphery, as shown in Figure 14.1. These slots create an alternating pattern of dark and light lines which can be detected, for example, by optical sensors. A disk count is defined as the number of dark and light line pairs that occur to complete a revolution (lines per revolution). A second track is added on the disk to generate a signal that occurs only once per revolution. This is typically known as index signal. Different encoder manufacturers may call this signal using several other terms with similar meaning such as marker, home position or zero reference. Commonly, to derive direction information, the first track on the disk is read by two different photo-elements/optical sensors that provide output (digital) pulses whenever a dark/light line pairs is measured. The two sensors are placed in such a way to look at the disk pattern with a mechanical shift of 1/4 of the pitch running between two consecutive lines. Hence, as the

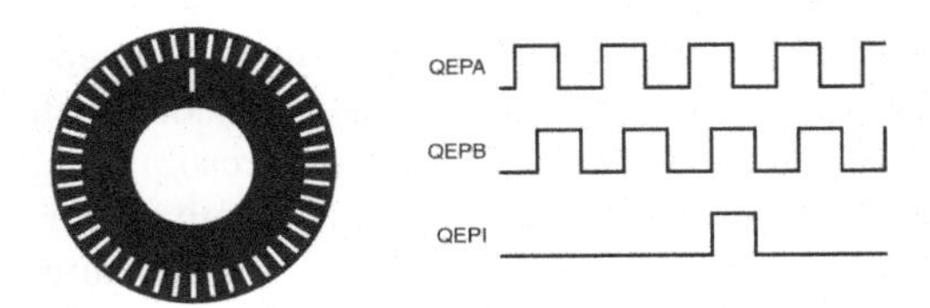

Figure 14.1 Generic structure of an optical rotary encoder disk and corresponding channel signals [20].

DOI: 10.1201/9781003196938-14

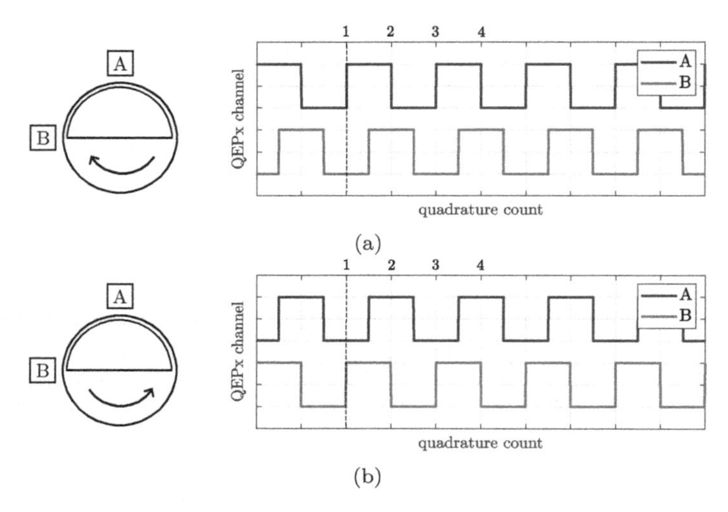

Figure 14.2 Working principle of a quadrature encoder: clockwise (a) and counterclockwise (b) rotation.

disk rotates, the two optical sensors generate square waves signals that are shifted by 90° out of phase from each other.

For the F28069M LaunchPad™ Piccolo™ board, these waveforms are called QEPA (or simply A) and QEPB (or simply B) signals, while the index signal QEPI (or simply I). QEPA and QEPB channels in quadrature (i.e., **quadrature encoders**) means that when the incremental encoder is moving at a constant speed, the duty cycle of each pulse is 50% (i.e., the waveform is a square wave) realizing the 90° phase difference between A and B. Typically, the clockwise direction is defined as the QEPA channel going positive before the QEPB channel, as shown in Figure 14.2. Unlike absolute encoders, an incremental encoder is not able to indicate absolute position. Indeed, it only reports variations in position and the direction of the motion for every angular change.

In general, direct coupled encoders (i.e., mounted on the rotor shaft before a gearbox) make one revolution for each complete rotation of the motor. Therefore, the frequency of the digital signal coming from the QEPA and QEPB outputs varies proportionally with the motor speed; high frequencies indicate high speeds, whereas low frequencies mean slow speeds. The **resolution** of an incremental encoder refers to the precision of the produced position information. In the case of rotary encoders, resolution is specified as the number of pulses per revolution (ppr) or cycles/counts per revolution (cpr). The *ppr* determines the number of pulses that can be counted on each signal (QEPA and/or QEPB) in each complete revolution. Nevertheless, quadrature encoders are able to increase their effective resolution by combining the

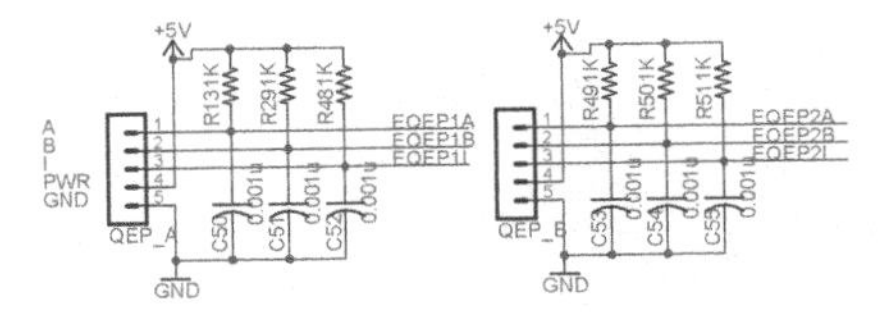

Figure 14.3 Correspondence between eQEP channels and pin of the LaunchPad™ F28069M board [16].

information carried by QEPA and QEPB to define the *cpr*.

$$\text{cpr} = 4 \cdot \text{ppr} \tag{14.1}$$

For instance, if a 1 ppr quadrature encoder is considered, its effective resolution is increased up to 4 cpr (see the numbers reported in Figure 14.2).

Therefore, the **measurement resolution of the encoder** is defined as the smallest position change that the encoder can detect. Every signal edge on QEPA or QEPB indicates a detected position change. Since each square-wave cycle on A (or B) encompasses four signal edges (rising A, rising B, falling A and falling B), the measurement resolution of the encoder equals one-fourth of the variation represented by a full QEPA (or QEPB) output cycle.

14.2 Hardware Details

The LaunchXL F28069M can be connected up to two linear and/or rotary encoders, thanks to the presence of two eQEP channels. The interface between board and encoder is reported in Figure 14.3. As already discussed, A, B and index signals are called QEPxA, QEPxB and QEPxI, respectively, where x denotes the considered module. The signals A and B are the output of the optical sensors and they can be combined to increase the resolution of the encoder, other than providing information on the direction of rotation. The two channels use the index signal to assign the reference point from which the position is encoded. It is possible to program the peripheral in such a way that this signal resets the position counter at each revolution. In addition, it can be used to initialize or latching the position counter every time a desired event occurs on the index pin.

These modules are able to deal with strobe inputs eQEPxS as well. This kind of signal can initialize or latch the position counter when a desired event occurs on the strobe pin. This signal is typically connected to a sensor or to a limit switch to notify that the motor reaches a certain position. Actually, the default settings of the board LaunchPad™ F28069M do not foresee any physical pin to be connected to this signal. Indeed, it is possible to program the correspondence between hardware pins and eQEP signals in the

Figure 14.4 LPD3806-600BM-G5-24C incremental rotary encoder.

Configuration Parameters menu, Hardware Implementation, Target hardware resources, eQEP.

14.3 Optical Rotary Encoder LPD3806

In this book, the optical rotary encoder LPD3806-600BM-G5-24C is used. This piece of equipment is reported in Figure 14.4. This latter provides A/B channel quadrature-encoded with 600 ppr (thus, 2400 cpr). The index signal is not available. The supply voltage can range from 5 up to 24 V. The maximum measurable mechanical speed is 5000 rpm, with a frequency response in the range $[0, 20]$ kHz. The body height is 38 mm, whereas the size of the shaft is 6×13 mm.

The encoder interface is an electronic circuit (LED/photo-transistor module-based) that processes the A/B signals, making the resulting information available to external circuitry. The encoder interface has open collector outputs with 20 Ω resistors for protection. The circuit draws around 30 mA. Since it can not source current, the open-collector circuit must be connected to positive DC voltage through a pull-up resistor. The eQEP pins of the F28069M LaunchPad™ board have pull-up resistors[1] (i.e., they are supplied with the onboard 5 V voltage level), see Figure 14.3. Attention must be paid to limit the cable length connecting the encoder to the board, since it is preferable to have pull-up resistors in close proximity to the encoder interface to improve noise immunity. Note that the encoder interface also has a 5 V linear regulator onboard (LM7805), which power dissipation has to be kept within reasonable limits. The internal LM7805 regulator in D-PAK package relies on a big ground plane that acts as a heat-sink. Moreover, the encoder case is connected to the chassis ground and not to circuit ground. More hardware details are given in Appendix B.

[1] If the encoder is not connected to the MCU board, the A/B signals are not visible even with an oscilloscope; the open collector output has no voltage when there is no pull-up resistor.

Figure 14.5 Encoder connected to the F28069M LaunchPad™ board through the extPot3 custom. eQEP2 is considered. In this picture, the encoder is mechanically coupled to a DC motor.

In the examples reported in the following Chapters, extPot3 custom board is considered since it provides a direct connection of the encoder terminals with the MCU board, as shown in Figure 14.5. The wiring is:

- Red wire: voltage supply (VCC)
- Black wire: ground (GND)
- Green wire: A
- White wire: B

A or B wires must not be connected to VCC. Otherwise, the output stage is shot-circuited. The correct connection between encoder and the board is reported in Figure 14.5.

14.4 Speed Computation

Incremental encoder interfaces detect mechanical displacement without directly measuring the angular speed. Consequently, speed must be indirectly computed either by counting or by timing the encoder output pulses (edges). The resulting value is a frequency or a period, respectively, from which speed can be calculated. The speed is proportional to frequency, and inversely proportional to period.

In discrete time the signals are inherently quantized (e.g., interface circuits). The encoder interface samples A and B output signals frequently enough to detect every A/B state change before the following variation occurs. Upon detecting a state change, the position counts $x_p(k)$ is incremented or decremented based on whether A leads or lags B. This is typically done by storing a copy of the previous A/B state and, upon state change, compare the current and previous A/B states to determine movement direction.

Therefore, signals QEPA and QEPB are both processed for computing angular speeds. Nevertheless, regarding the implementation, estimating the speed from a digital position sensor is a cost-effective strategy in motor control. Hence, two different first order approximations for the speed computation may be written starting from the position counter $x_p(k)$, namely:

- **Standard approach:**

$$\omega(k) \approx \frac{x_p(k) - x_p(k-1)}{T_w} = \frac{\Delta x(k)}{T_w} \tag{14.2}$$

where $\omega(k)$ is the speed computed at time step k, $x_p(k)$ and $x_p(k-1)$ are the position counters at time step k and $k-1$ respectively, T_w is the acquisition time window, which is the inverse of speed calculation rate (and is known in advance). This approach is based on the fact that the encoder count (position) is read once at the beginning of each acquisition time period, and the width of each pulse is a function of the motor speed and the resolution of the sensor.

However, this method has an inherent accuracy limit directly related to the ratio between the resolution of the position sensor and the acquisition time period T_w that has to be chosen.

As an example, considering the LPD3806-600BM-G5-24C encoder, a 600 ppr quadrature encoder results in 2400 cpr which implies a resolution of

$$\text{res} = \underbrace{\frac{1}{\text{cpr}}}_{\frac{1}{\text{revolution}}} 360^\circ = 0.15^\circ \tag{14.3}$$

Considering a speed calculation rate of 400 Hz, thus, $T_w = 1/400 = 2.5\,\text{ms}$, it follows that:

$$\omega_{\text{min}} = \frac{\text{round}}{\text{min}} \text{ or } \frac{\text{revolution}}{\text{min}} = \underbrace{\frac{1}{\text{cpr} \cdot T_w}}_{\frac{\text{revolution}}{\text{sec}}} 60 = 10\,\text{rpm} \tag{14.4}$$

Assuming that this motor has to measure the angular frequency of a motor, which rated speed is 2000 rpm, ω_{min} represents the 10 rpm/2000 rpm $\approx$ 0.42% of this quantity. While this resolution may be satisfactory from moderate to high speeds, it would behave poorly at very slow rotations. Indeed, the estimated speed would erroneously be zero much of the time for pulsations lower than 10 rpm. Note that, by varying T_w, different minimum speed can be computed, e.g., for $T_w = 1/1\text{e}3 = 1\,\text{ms}$ it results $\omega_{\text{min}} = 25\,\text{rpm}$.

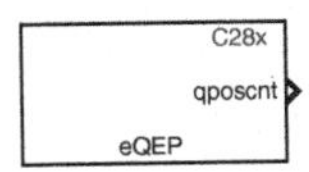

Figure 14.6 C2806x eQEP block.

- **Low speed approach:**

$$\omega(k) \approx \frac{X}{t_p(k) - t_p(k-1)} = \frac{X}{\Delta t_p} \tag{14.5}$$

  where $t_p(k)$ and $t_p(k-1)$ are the time instant at time step k and $k-1$ in which a channel A/B present a rising edge, while X is the position step, that is, one. Equation (14.5) provides a more accurate approach at low speed rather than (14.2). Such approach computes the speed by measuring the elapsed time between consecutive edges. An acquisition window is not present, while the encoder interface is required to work with a high sampling time in order to detect rising edges even at high speed. Indeed, this approach leads to better results at low speeds while suffering every time the motor spins at such speed for which the sensor makes the time interval Δt_p small, difficult to measure and greatly influenced by the timer resolution. Therefore, considerable speed estimation errors can be introduced.

For motor control applications characterized by large speed ranges, these two approaches can be combined by using (14.5) at low speed while switching to (14.2) every time the motor speed overcomes a specified threshold.

14.5 Firmware Environment: eQEP Peripheral

This section shows the main characteristics of the eQEP block (see 14.6) available in Simulink® inside the library *Embedded Coder Support Package for Texas Instrument C2000 Processors*, subset C2806x.

14.5.1 C2806x eQEP

This block (see Figure 14.6) can be included into a firmware to compute both the position and the angular speed of a rotating element. Therefore, these measurements can be used for closing both speed or position control loops. Its main settings are briefly explained here in the following.

Block Parameters—General tab

- **Module**

This drop down menu allows to select which module is in use. The adopted board has two modules only, i.e., eQEP1 and eQEP2.

- **Position counter mode**

  Depending on the structure of the encoder, there are four ways in which the device and the peripheral can be interfaced:

  - **Quadrature count**: this option allows to use the quadrature decoder, which exploits two 90° out-of-phase signals (QEPxA and QEPxB) to generate quadrature-clock and direction signals. Therefore, the resolution of the encoder is increased;
  - **Direction count**: direction and clock signals are provided directly from an external source. Some position encoders have this type of output instead of quadrature output. The QEPxA pin is used for the clock input XCLK, whereas the QEPxB pin for the direction input XDIR;
  - **Up-count**: the position counter is exploited for measuring the frequency of the QEPA signal. The direction of the counter is hard-wired for up-count;
  - **Down-count**: the position counter is exploited for measuring the frequency of the QEPA signal. The direction of the counter is hard-wired for down-count.

- **Positive rotation**

  This option allows to select the convention for positive speed, i.e., clockwise or anti-clockwise, depending on how the encoder is mechanically coupled with a motor.

- **Sample Time**

  This setting specifies how often the position counter is read. In case of the standard speed computation approach, Equation (14.2), this refer to T_w acquisition time period.

Block Parameters—Position Counter tab

In this tab, it is possible to enable (flag) the **output position counter**, to set the **Maximum position counter value** and to choose the **position counter reset mode**. Particular attention should be paid while setting the maximum position counter value since it should be set equal to the maximum feasible integer value depending on the chosen data type, which can be set in the **Signal data types tab.**

As an example, if `int32` is chosen in the **Signal data types tab**, the **Maximum position counter value** is a number that ranges between 0 and $2^{32} - 1$. By enabling the **software initialization**, it is possible to choose the **Software initialization source** as **Set to init value at startup** or via **Input port**. The **Initialization value** is a number between 0 and $2^{32} - 1$;

0 is set by defualt. There are different options for **Position counter reset mode**, e.g., **Reset on the maximum position**.

Block Parameters—Speed Calculation tab

The edit of the settings available in this tab is enabled through the checkbox **Enable eQEP capture**. Once ticked, different timer and values can be sent as outputs. Moreover, this tab allows to set prescaler applied on the system clock to simplify speed computations at low speeds.

Among all the check-boxes that are reported in this tab, two of them are briefly described here in the following:

- **Output capture timer**: this option enables the block to output the capture timer value in clock cycles stored inside the quadrature capture timer (QCTMR) register, in the correspondence of the last quadrature signal. Port qctmr is generated;
- **Output capture period timer**: this option enables the block to output the capture period timer value (in clock cycles) stored in the quadrature capture period (QCPRD) register, that is, the time elapsed between two consecutive quadrature signals (i.e., positions). This value can be used for speed computation purposes.

Block Parameters—Other tabs

- **Compare output tab**: enables synchronization signal for position comparison in case of multiple blocks or triggered output signals.
- **Watchdog unit tab**: enables a Watchdog timer to detect misoperations.
- **Interrupt tab**: there, several kind of Interrupts can be enabled. In particular, eQEP module can generate Interrupt signals but it cannot be triggered directly by any interrupts unlike the ADC and ePWM blocks.

14.6 Example with eQEP block

Build a firmware aimed to read encoder positions, compute angular speeds and visualize pulse changes through the red led

Angular speed must be indirectly computed by counting the encoder output pulses on QEPxA and QEPxB. When the encoder interface detect a state variation, it increments the counter $x_p(k)$. This information is available at the eQEPx output and used to compute a frequency or period from which speed can be calculated. This exercise is carried out by considering a LPD3806-600BM-G5-24C incremental rotary encoder, aiming to implement the standard computation approach by using Equation (14.2).

Firmware Environment

c28069_encoder_F.slx (solver: fixed step-ODE4, step size: $T_s = 100\,\mu s$)

Open a new blank Simulink® project and configure the environment as shown in Chapter 9. Edit an initialization script, e.g., in *Model Properites/Callbacks/InitFcn*. Then:

- Insert a **eQEP** block, double-click on it and set the following parameters:
  - **General** (Tab)
    - select **module** eQEP2;
    - set **Position counter mode** to Quadrature-count;
    - set **Positive rotation** equal to Clockwise;
    - set **Sample time** equal to T_w;
    - leave the other default settings.
  - **Position counter** (Tab)
    - flag **Output position counter**;
    - set **Maximum position counter** to $2^{32} - 1$;
    - flag **Enable software initialization** ;
    - set **Software initialization source** equal to Set to init value at start up;
    - set **Initialization value** equal to 0;
    - set **Position counter reset mode** equal to Reset on the maximum position;
    - Leave the other default settings.
  - **Signal data types** (Tab)
    - set **Position counter value data type** equal to `int32`;
  - Leave the other tabs with their default settings.
- Insert a **Delay** block with **Sample time** equal to T_w and **Delay length** equal to 1 (i.e., one step delay).
- Insert an **Add** block followed by a **Data type conversion** block set to `single`. This process the difference between $x_p(k)$ and $x_p(k-1)$ which is then multiplied by a **Gain** block set to $60/(\text{cpr} \cdot T_w)$. The result is in rpm, which can be translated in rad/s by multiplying this quantity by $\pi/30$.

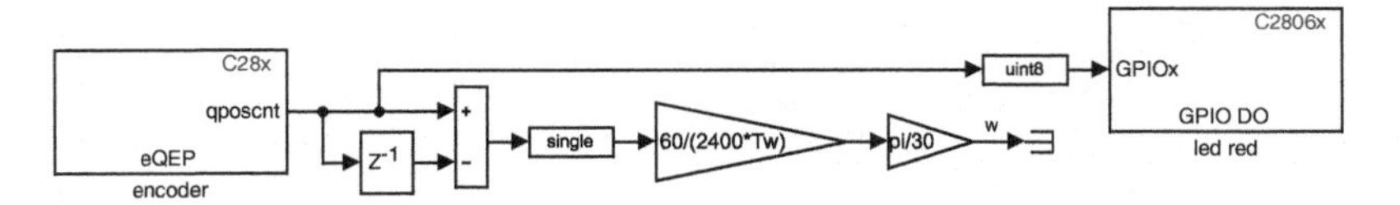

Figure 14.7 Simulink® scheme included in c28069_encoder_F.slx.

- Insert a **GPIO** block with the related data type as reported in Chapter 11 (red led is connected to GPIO34);

The whole Simulink® scheme is shown in Figure 14.7. Every time the position signals QEP2A e QEP2B are sampled (discrete time signals), the pulses (or pulse edges) are detected and counted by the encoder interface. The MCU has read access to the interface and the eQEP module returns directly the position counter through `qposcnt`. The latter is then used to compute the speed by taking the actual position count $x_p(k)$, i.e., `qposcnt`(k), and the previous one $x_p(k-1)$, i.e., `qposcnt`(k-1), within a specific acquisition period T_w. This is realized by using an add and delay block, with sample time T_w. The result is then multiplied by $60/(\text{cpr}\cdot T_w)$. Considering a speed calculation rate of 400 Hz, it follows $T_w = 1/400 = 2.5\,\text{ms}$. Note that, the sample time T_s is related to the switching period of the modulator, e.g. $T_s = T_{\text{sw}} = 1/f_{\text{sw}} = 1/10\text{e}3 = 100\,\mu\text{s}$. Hence, it follows $T_w = 25 \cdot T_s$, which underline that T_w is (quite) often a multiple of T_s to keep an hierarchical scheduling of firmware executions.

A coefficient equal to 60 is used since $1/(\text{cpr} \cdot T_w)$ is expressed as counts per unit time (e.g., counts per second), while, in practice, it may be necessary to translate the speed in units such as revolutions per minute (rpm). Such scaling factor takes into account the relationship between counts and desired distance units, as well as the ratio between the sampling period and desired time units. The other coefficient depends on the fact that LPD3806-600BM-G5-24C encoder produces 2400 counts per revolution (cpr). The speed is finally multiplied by another scaling factor, i.e., $\pi/30$, to move from rpm to rad/s. Given the encoder connection on eQEP2 as shown in Figure 14.5, the firmware can be tested by manually rotating the encoder shaft. The `qposcnt` value is connected to the red led (GPIO34) to visualize the pulse frequency changes

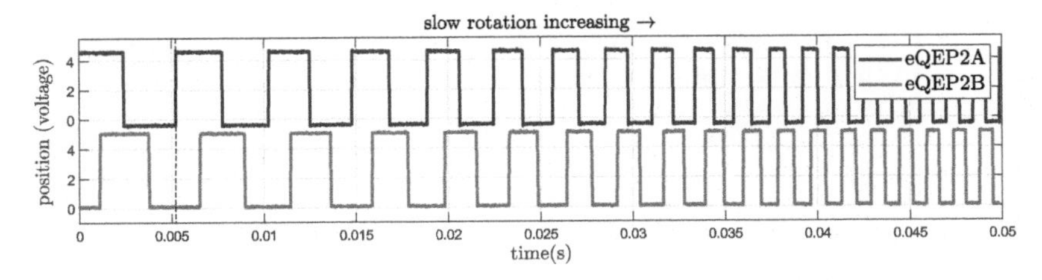

Figure 14.8 QEP2A (pin 54) and QEP2B (pin 55) waveforms, measured through an oscilloscope for slow rotational speeds.

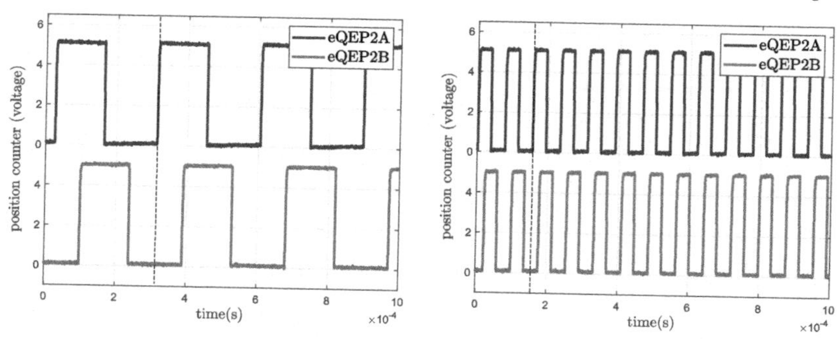

Figure 14.9 Detailed QEP2A (pin 54) and QEP2B (pin 55) waveforms in case of slow and fast rotation scenarios.

as led blinking. Nevertheless, the **qposcnt** behavior can be better understood by connecting two voltage probes to QEP2A (pin54) and QEP2B (pin55) through the available test points on the extPot3 to visualize the corresponding waveforms. By manually increasing the rotating speed of the encoder, the pulse frequency increase, thus, increasing **qposcnt**. This reflects on the behavior of QEP2A (pin 54) and QEP2B (pin 55), which is shown in Figure 14.8. The waveforms related to voltage signals varies in a $[0, 5]$ V range.[2] Two detailed views of these signals are reported in Figure 14.9, in which slow and fast rotations are compared, respectively.
The effect is that pulse frequency changes as fast as the shaft rotation, which can be increased further by (manually) creating really quick speed transition from slow to high rotational speed, as reported in Figure 14.10 (voltages measured at pin 54 and 55).

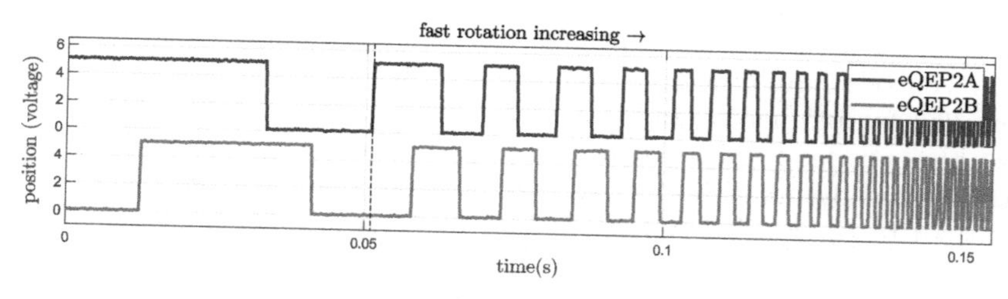

Figure 14.10 QEP2A (pin 54) and QEP2B (pin 55) waveforms in case a fast transition from slow to high rotational speed is considered.

[2]Both cables and encoder interface circuits (even considering the voltage regulator) impact on the output with a voltage drop.

Part IV

Real-Time Control in Power Electronics: Applications

15

Open Loop Control of a Permanent Magnet DC Motor

After a brief overview of the main peripherals and their main settings for the F28069M LaunchPad™ Piccolo™ board, the reader is guided in this part of the book where some experimental open- and close-loop implementations are described step-by-step. Starting from simple examples with DC motors and custom RL(C) loads, the reader will be shown how to solve all the proposed control problems starting from the preliminary Simulink® simulations, then going through the design of the PI controller and finally implementing the required control loops. In particular, this Chapter shows some examples of open-loop control algorithms for DC Motors using the Simulink® blocks introduced in the previous Sections coming from the *Embedded Coder Support Package for Texas Instruments C2000 Processors* library. The objective is to drive a two-pole permanent magnet DC (PMDC) motor through a BOOSTXL-DRV8301 converter connected to the LaunchXL F28069M board. The DRV8301 boosterpack presents three independent MOSFET-based switching legs. It is worth noting that the actuation of a PMDC motor does not require the use of all of them (i.e., due to its DC nature). However, the converter legs can be physically connected to create some specific topologies. In particular, two conversion stage configurations are investigated:

- Half-Bridge (also known as Two-Quadrants converter);
- Full-Bridge (also known as H-Bridge or Four-Quadrants converter).

Both stages aim to convert a fixed input DC voltage, V_{DC}, to a variable/controlled output DC voltage, which is provided at the PMDC terminals. Such DC-DC converters are also called *choppers*.

15.1 Required Hardware

- TI LaunchPad™ Piccolo™ F28069M and BOOSTXL-DRV8301 boards;
- Power supply;
- PMDC motor;

DOI: 10.1201/9781003196938-15

- extPot3 custom board (see Appendix B for technical details).

The reader is referred to Section 3.2 for all the details on BOOSTXL-DRV8301 board. Since the three legs of the converter are independent, it is possible to realize the two aforementioned configurations by changing the physical connections between the motor and the power converter. Then, the related ePWM modules must be enabled.

It is important to note that every time the converter is mounted on the microcontroller board, it is preferable to feed both of them through the power supply only. Indeed, jumpers JP1 and JP2 should be removed (further details are reported in Section 3.1.3) to isolate the USB port. In this case, the USB connection is used to deploy of the firmware and for the serial communication between the LaunchPad™ and the PC only. Moreover, it should be remembered that the BOOSTXL-DRV8301 converter must be supplied with a DC voltage which is $V_{\mathrm{DC}} > 6\,\mathrm{V}$, to allow the proper working of the boosterpack regardless of the state of the jumpers JP1 and JP2. The reader is referred to the rated voltage limits of the converter reported in Section 3.2.

Now, the LaunchPad™ F28069M board connected to the BOOSTXL-DRV8301 converter are programmed to drive the PMDC motor in open loop by manipulating the duty cycle $d(t)$. Both the possible converter configurations are considered. This is achieved by using one of linear potentiometers mounted on the extPot3 custom board. As shown in Section 12.4, the rotation of the potentiometer changes an internal resistance, i.e., varying the ratio of a voltage divider. The resulting voltage value is read by the ADC peripheral that converts it into a digital value. By using a proper scaling (like those presented in Chapter 13), this latter can be used as a modulating signal for the ePWM peripheral. Depending on the chosen topology, one or two ePMW modules are operated in up-down counting mode at $f_{\mathrm{sw}} = \mathrm{PWM}_{\mathrm{frequency}} = 20\,\mathrm{kHz}$. Reminding that for the F28069M board $f_{\mathrm{ck}} = \mathrm{CPU}_{\mathrm{frequency}} = 90\,\mathrm{MHz}$, it follows:

$$\mathrm{TBPRD} = \mathrm{PWM}_{\mathrm{counter-period}} = \frac{\mathrm{CPU}_{\mathrm{frequency}}}{2 \cdot \mathrm{PWM}_{\mathrm{frequency}}} = 2250 \tag{15.1}$$

where the sampling time is $T_s = T_{\mathrm{sw}} = 1/\mathrm{PWM}_{\mathrm{frequency}}$.

15.2 Linear Model of a PMDC Motor

A PMDC motor has a stationary set of magnets in the stator and an armature with one or more windings of insulated wire wrapped around a soft iron core that concentrates the magnetic field. The windings usually have multiple wire turns around the core, with their ends connected to a commutator. The latter allows each armature coil to be energized in turn and connects the rotating

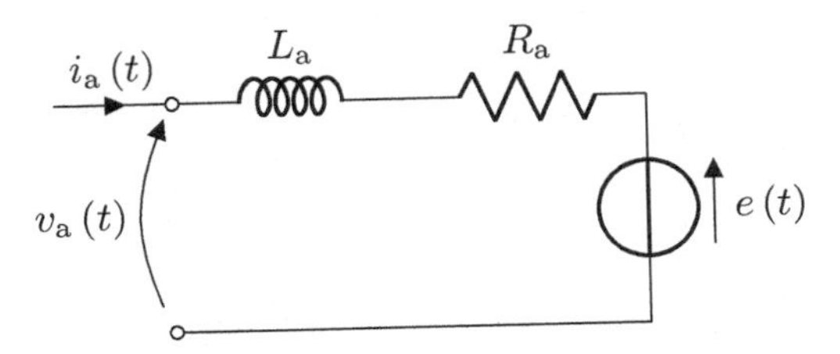

Figure 15.1 Equivalent circuit of a PMDC motor.

coils with the external power supply through brushes. A coil of wire with a current running through it generates an electromagnetic field aligned with the center of the coil. Direction and magnitude of the magnetic field produced by the coil can be changed with the direction and magnitude of the current flowing through it.

Focusing on the system to control, the equivalent circuit of the PMDC motor is shown in Figure 15.1 and its electrical and mechanical equations are:

$$L_a \frac{di_a(t)}{dt} = v_a(t) - R_a i_a(t) - e(t) \tag{15.2}$$

$$J \frac{d\Omega(t)}{dt} = m_e(t) - m_l(t) - \beta\Omega(t) \tag{15.3}$$

where L_a is the armature inductance, R_a is the armature resistance, $i_a(t)$ is the armature current (i.e., the current flowing through the winding of the motor), $v_a(t)$ is the armature voltage (i.e., the voltage applied to the terminals of the machine), e is the back electromotive force (back-emf) induced on the armature winding, J is the equivalent moment of inertia of the machine, β is the rolling friction coefficient, $\Omega(t)$ is the mechanical speed of the rotor, $m_e(t)$ is the torque generated by the motor and $m_l(t)$ is load torque that may be applied to the electrical machine. In DC machines, both electromagnetic torque $m_e(t)$ and back-emf $e(t)$ depend on the magnetic flux linked with the armature windings ψ_{ae} and the related constants k_T and k_e, respectively. As the name says, the main feature of PMDC machines is the presence of permanent magnets (PM) on the stator to provide the magnetic field against which the rotor field interacts to produce torque.[1] The use of PMs is convenient in small motors to eliminate the power consumption of the field windings, thus, increasing their efficiency. Moreover, large PMs are costly and difficult to assemble; this favors wound fields for large machines. To minimize overall weight and size, small PMDC motors may use high energy magnets made by peculiar alloys, e.g., with rare earth materials such as *neodymium-iron-boron* or *samarium-cobalt.*

The PMs create a constant flux density B (if no field weakening technique is applied) and, accordingly, a constant excitation flux which is defined

[1]Therefore, the PMs replace the excitation circuit installed in separately excited DC motor, which is not considered here.

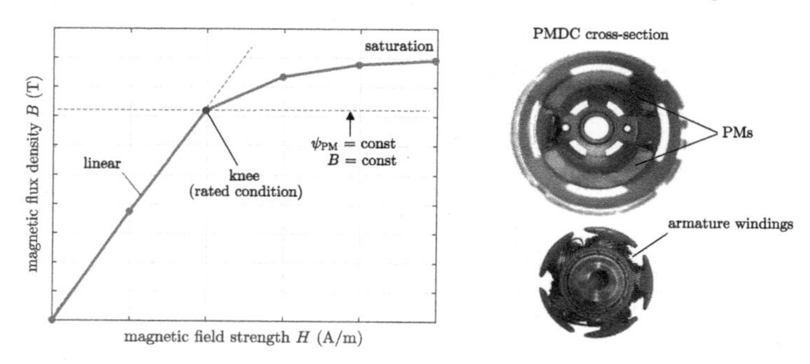

Figure 15.2 First magnetization curve of two PMs (no hysteresis is represented here) and internal structure of the PMDC motor used in this exercise.

ψ_{PM} ($\psi_{\mathrm{PM}} = \psi_{ae}(t)$), see Figure 15.2. This is the case for all the examples reported in this book which use PMDC motor. Moreover, k_{T} and k_{e} can be approximately considered equal, assuming that the energy conversion is almost perfect. Thus, a new constant K can be defined as $K = \psi_{\mathrm{PM}}k_{\mathrm{T}} = \psi_{\mathrm{PM}}k_{\mathrm{e}}$. Therefore, the electrical torque and the back emf can be computed as:

$$m_e(t) = Ki_{\mathrm{a}}(t) \tag{15.4}$$

$$e(t) = K\Omega(t) \tag{15.5}$$

with $m_e(t)$ which is function of $i_{\mathrm{a}}(t)$ only. The voltage $v_a(t)$ is a controllable variable, i.e., $u(t)$, which allows to increase/decrease the current flowing into the PMDC motor (according to the system parameters). Assuming that $m_l(t)$ is a controllable disturbance related to the characteristic of the mechanical load, the vector of the inputs is $\boldsymbol{u}(t) = [v_a(t)\ m_l(t)]^T$. As shown in Chapter 5, by considering[2] $\boldsymbol{x}(t) = [\Omega(t)\ i_{\mathrm{a}}(t)]^T$ and $\boldsymbol{y}(t) = \boldsymbol{x}(t)$, the equations (15.2)-(15.5) can be summarized by the following state-space representation:

$$\frac{\mathrm{d}\boldsymbol{x}(t)}{\mathrm{d}t} = \boldsymbol{A}\boldsymbol{x}(t) + \boldsymbol{B}\boldsymbol{u}(t) \tag{15.6}$$

$$\boldsymbol{y}(t) = \boldsymbol{C}\boldsymbol{x}(t) \tag{15.7}$$

where the matrices are defined as:

$$\boldsymbol{A} = \begin{bmatrix} -\beta/J & K/J \\ -K/L_{\mathrm{a}} & -R_{\mathrm{a}}/L_{\mathrm{a}} \end{bmatrix} \quad \boldsymbol{B} = \begin{bmatrix} -1/J & 0 \\ 0 & -1/L_{\mathrm{a}} \end{bmatrix} \quad \boldsymbol{C} = \boldsymbol{I}_2 \tag{15.8}$$

where $\boldsymbol{I}_2$ is a two-dimensional identity matrix. To simplify the exercise, zero load torque is assumed, i.e., $m_l(t) = 0$, which implies $\boldsymbol{u}(t) = [v_a(t)\ 0]^T$. Therefore, the input term can be simply interpreted as $u(t) = v_{\mathrm{a}}(t)$. It is important to note that, under this assumption, even matrix $\boldsymbol{B}$ has to be

[2]Bold symbols refer to both vectors of elements or matrices.

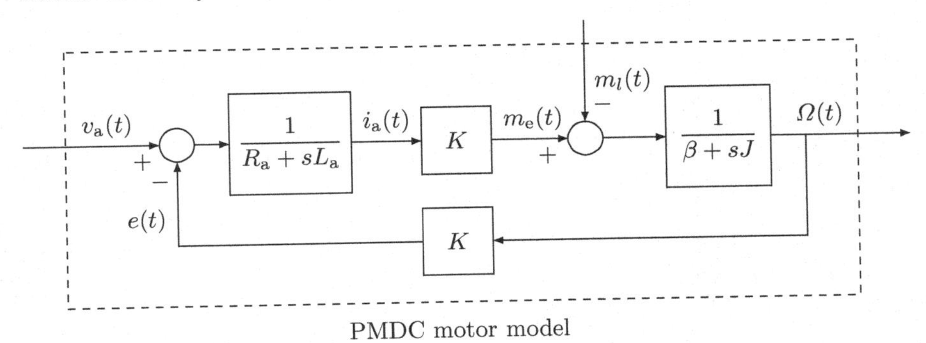

Figure 15.3 Block scheme representation of a PMDC motor.

updated. Indeed, the state-space representation in matrix notation becomes:

$$\frac{\mathrm{d}}{\mathrm{dt}}\begin{bmatrix}\Omega(t)\\ i_a(t)\end{bmatrix} = \begin{bmatrix}-\beta/J & K/J\\ -K/L_a & -R_a/L_a\end{bmatrix}\begin{bmatrix}\Omega(t)\\ i_a(t)\end{bmatrix} + \begin{bmatrix}-1/J\\ 0\end{bmatrix} v_a(t) \quad (15.9)$$

$$\begin{bmatrix}\Omega(t)\\ i_a(t)\end{bmatrix} = \begin{bmatrix}1 & 0\\ 0 & 1\end{bmatrix}\begin{bmatrix}\Omega(t)\\ i_a(t)\end{bmatrix} \quad (15.10)$$

The equations reported in (15.2)–(15.5) can be also visualized by means of a block scheme (Figure 15.3), in which the Laplace operator substitutes the derivative one $s \leftarrow \mathrm{d/dt}$. Therefore, based on Figure 15.3, the the continuous-time linear system is constituted by two transfer functions defined as follows:

$$G_i(s) = \frac{i_a(t)}{v_a(t)} = \frac{1}{R_a + sL_a} \qquad G_\Omega(s) = \frac{\Omega(t)}{m_e(t)} = \frac{1}{\beta + sJ} \quad (15.11)$$

where $G_i(s)$ and $G_\Omega(s)$ refer to the electrical and mechanical contributions respectively. These latter are linked through the back-emf $e\,(t)$ (acting as a physical feedback) and $m_e(t)$, i.e., denoting an electromechanical system. Due to the discrete nature of the MCU, both the state-space representation and transfer functions should be mapped into discrete-time domain with k and z notation, that is, in terms of $G_i(z)$ and $G_\Omega(z)$. The reader is referred to Chapters 5 and 6 for further details on this mapping. Apart from the modeling description, it is important to understand the operating principle of the system. Both using the Half- or Full-Bridge configuration for the converter stage, a modulator is necessary anyway (see Figure 15.4). The modulation signal can be related to the rotations of a potentiometer, like in this Chapter. However, for the considered application, the modulating signal is a (variable) dc quantity, not sinusoidal, to be compared with a triangular carrier. Therefore, constant modulating signal means constant duty cycle value. Hence, by manually manipulating $d(t) \rightarrow d(k)$ it is possible to increase the average voltage at the PMDC motor terminals $v_a(t)$, thus, the current flow $i_a(k)$. This latter directly impacts on the torque value $m_e(k)$ and on the speed $\Omega(k)$ of the shaft as well.

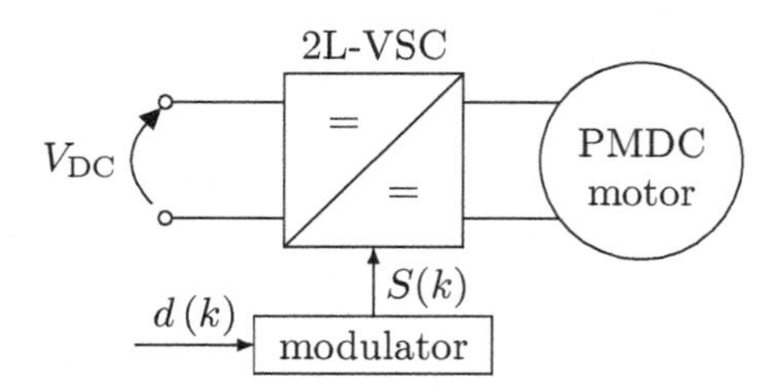

Figure 15.4 Operating principle of the open-loop control scheme in discrete-time domain ($d(k)$) including the converter stage.

15.3 System Simulations

In this book, system simulations are used to analyze and design the considered control systems. The simulation of the open-loop control of the PMDC motor is carried out with MATLAB® and Simulink®. This is done for both converter configurations. To this purpose, the latter and the PMDC motor are modeled through Simscape™ Electrical elements, in particular referring to Specialized Power Systems™ library (*Simscape/Electrical/Specialized Power Systems/*) as shown in Section 6.1.

The **Mosfet** blocks (available in the subset */Fundamental Blocks/Power Electronics*) are used to represent the switches mounted into the BOOSTXL-DRV8301 board. Each Mosfet block uses a macro model[3] made by an ideal switch controlled through an input port (i.e., a logical gate signal $g(t)$) and a diode connected in parallel so that it turns on when the MOSFET is reverse biased ($v_{ds}(t) < 0$). The MOSFET behavior follows the logic reported here below:

- $g(t) > 0 \rightarrow$ ON, with the drain-source voltage $v_{ds}(t)$ positive or negative;
- $g(t) = 0 \rightarrow$ OFF.

Referring to Chapter 13, $g(t)$ is equivalent to $S(t)$. In addition, the Mosfet block allows to include more physical parameters such as the internal resistance and inductance for the switch and the diode (R_{on}, L_{on} and R_d, L_d, respectively) as well as R_s and C_s forming the snubber circuit. The internal diode forward voltage V_f can be inserted too. These values can be found and computed starting from [14].

The two-level half-bridge and full-bridge configurations are obtained by arranging these blocks. The switches are driven by a PWM modulator having a carrier frequency of $f_c = f_{sw} = 20\,\text{kHz}$ built from standard blocks already presented (e.g., Relational operator, Repeating Sequence, and so on).

[3]This model does not take into account neither the geometry of the device or its complex physics.

The PMDC motor is modeled through the **DC Machine** block which can implement both a wound-field and permanent magnet DC machine by selecting the motor model in the *Field type* frame. The armature circuit (A+, A-) consists of an inductor L_a and resistor R_a connected in series with the back-emf. In the permanent magnet DC machine, there is no field current creating the excitation flux, which is established by the magnets. Thus, k_e and k_T are constants as expected. It is important to underline that *Torque constant* has to be selected in the *Specify* frame. The DC Machine block already include a lumped (first-order) mechanical model based on the inertia J, the viscous friction coefficient β (or B_m), and, as a further option, the Coulomb friction torque (which is not used in this book, i.e., = 0). The load torque $m_l(t)$ can be provided as an external input by selecting *Torque TL* in the *Mechanical input* frame. By selecting −1 as sample time, this block uses the largest integration step-size allowed by the **Powergui** block. The latter is essential to manage the numerical integration of the Simscape™ elements, as explained in Section 6.1.2.

15.4 Half-Bridge Configuration

A Half-Bridge configuration foresees the use of just a single-leg of the adopted converter, for which it is enough to use the module ePWM1 only. Namely, the switch 1_A is the master and 1_B the follower (dual). The terminals of the PMDC motor are connected between the central point of leg A and ground. The *average* armature voltage is function of the duty cycle $d(k)$ (or $d_A(k)$) is therefore:

$$V_a(k) = v_{a,avg}(k) = d(k)V_{DC} \tag{15.12}$$

where the variables are already considered within discrete-time steps k, i.e., in view of the implementation. The voltage V_{DC} is set by the power supply (i.e., the voltage on the DC bus) while $d(k) \in [0,\ 1]$. As a consequence, it is possible to feed the electrical machine with positive voltages only, since $v_a(k) \in [0,\ V_{DC}]$, which is reflected in $V_a(k)$. However, the motor can spin in clockwise or counter-clockwise direction according to the physical connection of its terminals + and −, e.g., V_{DC} connected to + and GND to −, or vice-versa. So, its rotation can not be reversed once the connection is set. The BOOSTXL-DRV8301 board adopt a two-level topology for each leg, thus, Figure 15.5 shows the equivalent circuit for the Half-Bridge topology connected to a PMDC motor. The PMDC motor is characterized by $R_a = 0.5290\,\Omega$, $L_a = 0.8651\,\text{mH}$, $K = 0.0232\,\text{Nm/A}$, $\beta = 0.1803\,\text{mN m s}$ and $J = 3.559\,\mu\text{g m}^2$. The supply voltage is $V_{DC} \approx 10\,\text{V}$ and the switching frequency is set equal to $f_{sw} = 20\,\text{kHz}$. These parameters can be initialized in *Model Properites/Callbacks/InitFcn* or in a separate `m`-file as follows:

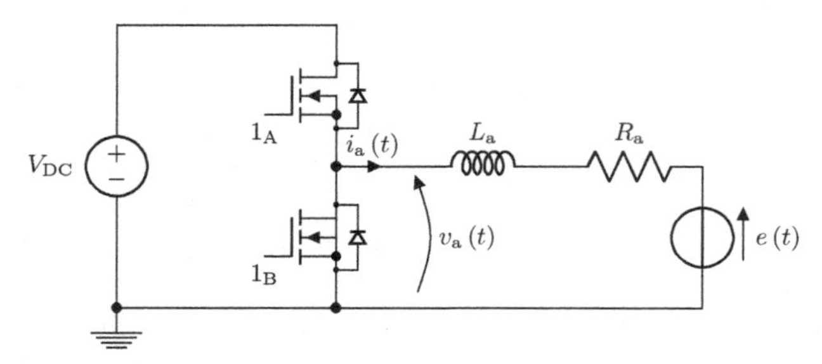

Figure 15.5 Equivalent circuit of a PMDC motor driven by a Half-Bridge topology converter.

```
%% power supply
Vdc     = 10;
%% carrier frequency definition
fsw     = 20e3;
Tsw     = 1/fsw;
%% sampling time definition
Ts      = 50e-6;
%% step-size definition
Tsim    = Tsw/400;
%% motor parameters
Ra      = 0.5290;
La      = 8.6507e-4;
K       = 0.0232;
B       = 1.8026e-4;
J       = 3.559e-6;
```

The equivalent circuit shown in Figure 15.5 is then realized in Simulink® as reported in Figure 15.6. This scheme is included in the file:

HBsimDC.slx (solver: fixed step - ODE4, step size: T_{sim})
(powergui: simulation type *Continuous*)

The simulation time step T_{sim} is small enough to approximate the switching behavior and high frequency effects that occurs in the scheme well, i.e., getting the simulation closer to the practice. The reference subsystem includes several kind of inputs to operate the system differently. Focusing on the PWM stage, the duty cycle $d(k)$ is compared to a triangular carrier at $f_c = 20\,\text{kHz}$, generating a switching pattern $S(k)$. This is done in the MCU by using the ePWM modules. The gate signals of the 2L-VSC topology are $S(k) = S_1(k)$ for 1_A and $S_2(k) = \text{not}(S_1(k))$ for 1_B, which are applied to the switch through the gate drivers. Therefore, the operation of these switches together with the given switching pattern creates a pulsed output voltage waveform, i.e., the

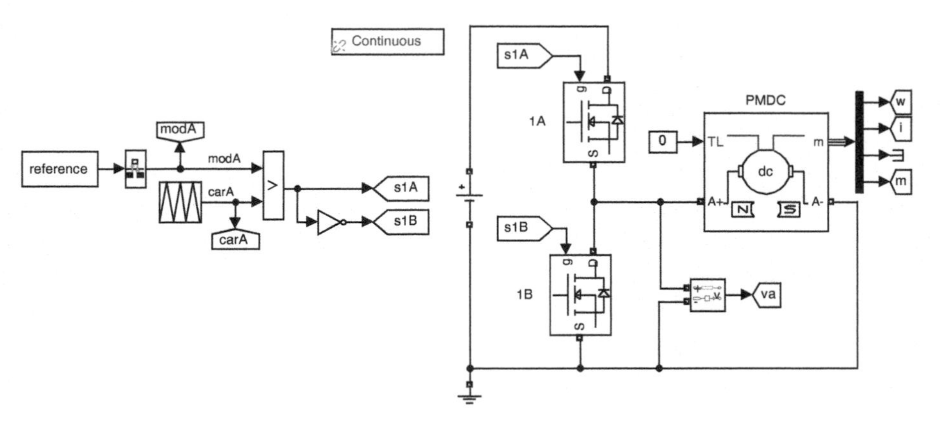

Figure 15.6 Simulink® scheme to simulate the open-loop dynamics of the PMDC motor driven in half-bridge configuration included in HBsimDC.slx.

armature voltage $v_{\mathrm{a}}(k)$. This is the reason why the average value of this latter $v_{\mathrm{a,avg}}(k)$ is considered in Equation (15.12). Indeed the actual voltage $v_{\mathrm{a}}(k)$ is computed as

$$v_{\mathrm{a}}(k) = S(k)V_{\mathrm{DC}} \tag{15.13}$$

which is discontinuous and bounded within two voltage levels, i.e., $[0, V_{\mathrm{DC}}]$. Considering the ramp waveform ranging from 0 to 1 as reference signal, $d(k)$ reaches 1 in 0.3 ms after which such value is hold. The effects of the reference increase are reflected into the switching pattern of 1_{A} and 1_{B}, which is aimed to make the average armature voltage $V_{\mathrm{a}}(k)$ increasing by enlarging the pulse width of $v_{\mathrm{a}}(k)$. This is evident in Figure 15.7. This latter also shows that the speed $\Omega(k)$ is increasing almost quadratically. In can be noted that the same occurs for the current $i_{\mathrm{a}}(k)$ and the torque $m_e(k)$, but the switching effects are more evident (see the dashed line interval reported in Figure 15.7). Indeed, the inertia J limits these abrupt variations on $\Omega(k)$.

A longer simulation with trapezoidal duty cycle $d(k)$ is then considered, as shown in Figure 15.8. The carrier waveform is dropped out to improve the plot readability. A longer simulation allows to analyze the steady-state interval for each of the considered quantities. In particular, Figure 15.8 shows how fast is the current $i_{\mathrm{a}}(k)$ reaction to the voltage increase. Its dynamic is related to L_{a}. When $v_{\mathrm{a,avg}}(k)$ reaches the first steady-state point, the current $i_{\mathrm{a}}(k)$ is at its peak value and, after that, it starts decreasing due to the presence of $e(k)$ (the motor is spinning) and R_{a}. The speed $\Omega(k)$ requires, as expected, a longer time to reach the steady-state due to the mechanical inertia J. It is fundamental to remind that the considered system is running in open-loop. Thus, no forcing action to adjust the dynamics of the motor is applied.

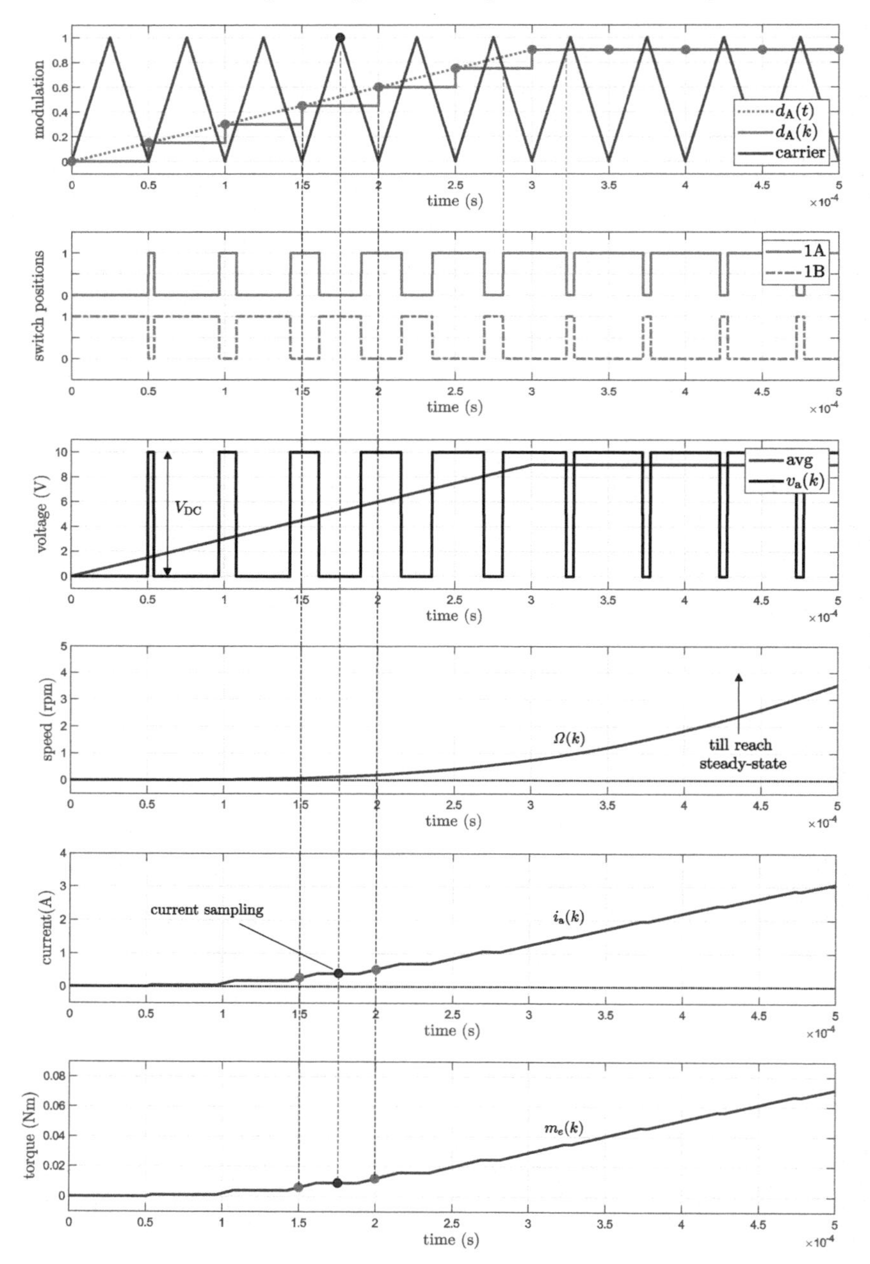

Figure 15.7 Open-loop dynamics of the PMDC motor with a ramp reference signal and Half-Bridge converter.

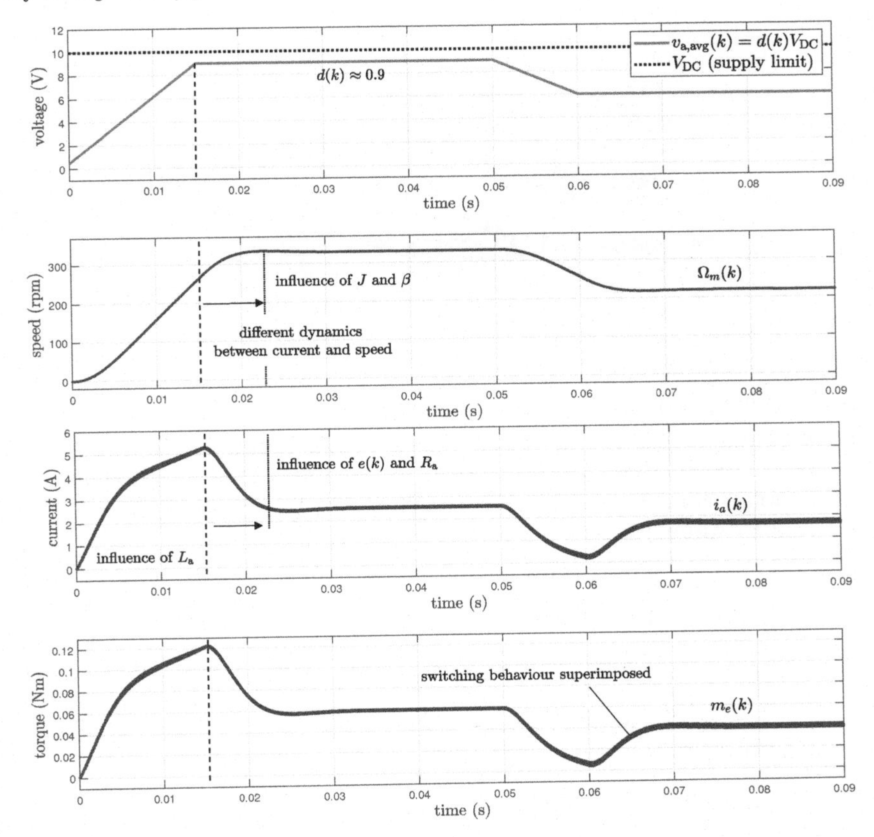

Figure 15.8 Analysis of steady-state and transient intervals for the open-loop dynamics of the PMDC motor.

15.4.1 Control implementation

The open-loop dynamics are tested through a test-bench which includes a TI BOOSTXL-DRV8301 converter mounted on the TI LaunchXL F28069M Piccolo™ and connected to a PMDC motor, as shown in Figure 15.9. By managing the cable connections from the BOOSTXL-DRV8301 to the motor terminals, both Half-Bridge and Full-Bridge configurations can be realized. Regarding the Half-Bridge configuration, only leg (phase) A is used, as shown in the connection scheme reported in Figure 15.10. A step-by-step procedure for programming the microcontroller for this task is reported here in the following.

Figure 15.9 Test-bench including a TI BOOSTXL-DRV8301 converter mounted on a LaunchXL F28069M Piccolo™ board and a PMDC motor (see the possible connections in Figure 15.10 and 15.23). This setup includes the extPot3 board connected to the LaunchXL F28069M Piccolo™. The reader is referenced to Appendix B for further details.

Firmware Environment

c28069_openDC_hbF.slx (solver: fixed step - ODE4, step size: $T_s = 50\,\mu s$)

Open a new blank Simulink® project and configure the environment as shown in Chapter 9. Then:

- Open the **Model Configuration Parameters** window and use the same settings adopted for the **Example 1** in Section 10.5.
- Add a **GPIO** block. Remember that the BOOSTXL-DRV8301 is enabled by settling **GPIO50** high (see Section 3.2). Namely, the GPIO peripheral is driven by a **Constant** block with value 1 and sample time T_s.
- Insert an **ADC** block to read the potentiometer mounted on the extPot3 custom board; double click on it:
  - Select ADCINB5 as **Conversion channel** in the **Input Channels** tab. The, go back to the **SOC Trigger** tab:
  - **Sampling mode**: Single sampling mode;
  - **SOC trigger number**: SOC0;
  - **SOCx acqusition window**: 7;
  - **SOCx trigger source**: CPU0_TINT0n *or* Software;
  - **ADC will trigger SOCx**: No ADCINT;
  - **Sample time:** $T_{\text{ADC}} = 0.001\,\text{s}$;
  - **Data type:** uint16;

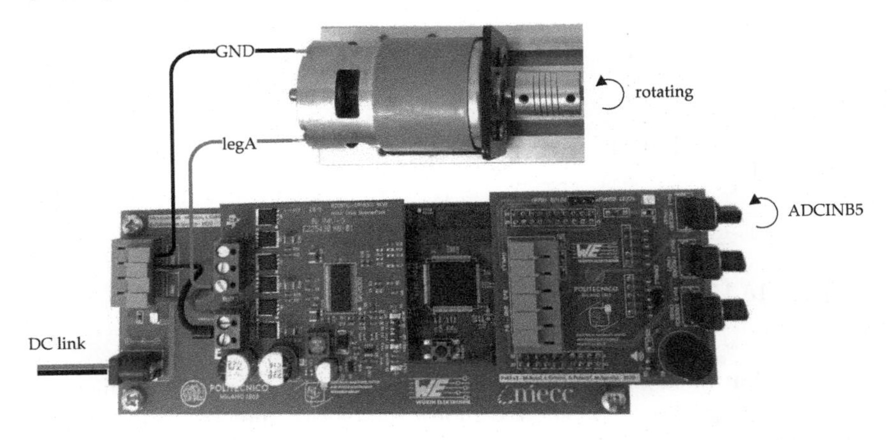

Figure 15.10 Connections of the PMDC motor with the BOOSTXL-DRV8301 converter realizing a Half-Bridge configuration.

- Flag **Post interrupt at EOC trigger** (optional);
- **Interrupt selection**: ADCINT1 (optional);
- Flag **ADCINT1 continous mode** (optional).

- Add a **Data Type Conversion** block set to int16 (since positive signal only are used) or `single` to exploit 32 bit.

- Place a **Gain** block for scaling the digitalized signal within [0, 1]. This is fundamental because a 12-bit reading returns a value in a range from 0 (0 V) up to $2^{12} - 1 = 4095$ (3.3 V). Hence, set the gain as $1/(2^{12} - 1)$ or, to be more conservative, as $1/2^{12}$, even if it, in this last case, the representation range is not fully exploited.[4]

- Insert an **ePWM** block and use the same settings reported in Section 13.5:
  - Select **ePWM1** module with TBPRD related to $f_c = 20\,\text{kHz}$, i.e., $PWM_{counter-period} = 2250$;
  - Remember to unflag Enable ePWM1B and tick Inverted version of ePWMxA. This option allows to drive 1_B as the complement of 1_A;
  - Keep all the other settings as reported in Section 13.5.

- Add another **gain** block to normalize the modulation signal over the peak value of the carrier signal $PWM_{counter-period}$. Include a **data type conversion** set to `int16` to represent the duty cycle within a 16-bit base.

[4]Alternatively, this operation can be carried out by using an arithmetic shift of 12 bit, which is the preferable choice in microcontrollers. Indeed, this kind of operation is quicker than multiplications or divisions with this hardware. In particular, the right shift means division (see Section 7.3).

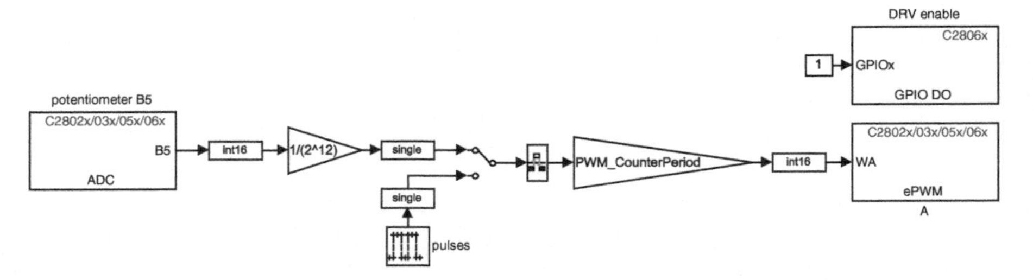

Figure 15.11 Simulink® scheme included in c28069_openDC_hbF.slx.

The aforementioned blocks should be connected as shown in Figure 15.11. **Power Settings:** Based on the DC motor datasheet, the power supply should be manually set to provide $V_{\mathrm{DC}} = 10\,\mathrm{V}$ and to saturate its output current to $2\,\mathrm{A}$.

Once the firmware is downloaded into the MCU, the resulting behavior is the motor spinning at a speed which can be adjusted through the potentiometer. As a matter of fact, the PWM nature of v_{a} allows to adjust the average voltage (within the PWM period) applied on the terminals of the motor by rotating the potentiometer. When $d(k) \approx 1$ the PMDC is fed with the full V_{DC}. Thus, it rotates at maximum speed. In principle, to further increase the motor speed, the voltage provided by the power supply can be, i.e. $V_{\mathrm{DC}} > 10\,\mathrm{V}$. However, since the motor is running in open-loop, no protection or limitation are included into the control platform, thus, potentially allowing to exceed the rated value of the electrical machine.

15.5 Full-Bridge Configuration

A Full-Bridge configuration (or H-bridge) refers to use two legs of the adopted boosterpack (legs A and B), for which two ePWM modules are required, e.g., ePWM1 and ePWM2. The terminals of the PMDC motor are connected to the central point of legs A and B. Hence, the switches 1_{A}, 1_{B} and 2_{A}, 2_{B} are considered. The equivalent circuit of this test case is reported in Figure 15.12. A benefit of a H-bridge configuration with respect to the Half-Bridge one is the chance to turn the motor in both directions (clockwise or counter-clockwise) without changing the electrical connections of the setup. The system is simulated considering the parameters reported in Section 15.4, which can be initialized in *Model Properites/Callbacks/InitFcn* or in a separate `m`-file. The equivalent circuit shown in Figure 15.12 is then realized in Simulink® as reported in Figure 15.13. This scheme is included in the file:

FBsimDC.slx (solver: fixed step - ODE4, step size: T_{sim})

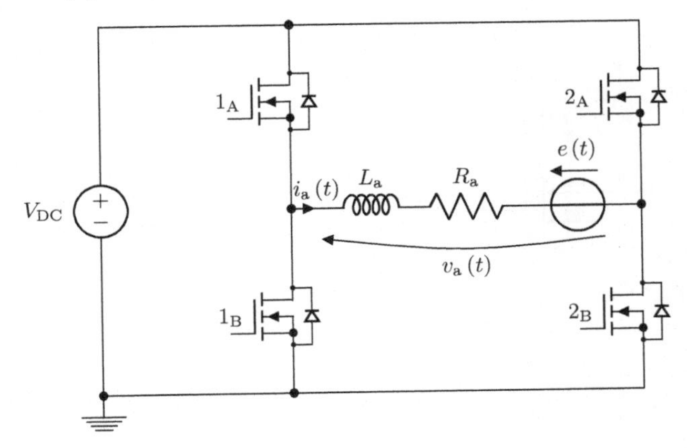

Figure 15.12 Equivalent circuit of a PMDC motor driven by a Full-Bridge topology converter.

(powergui: simulation type *Continuous*)

Similarly to Section 15.4, the simulation time step T_{sim} is set small. Moreover, the reference subsystem includes different kind of inputs to operate the system differently. A unipolar or bipolar voltage switching approaches can be selected by acting on the manual switch, which sets the carrier phase-shift (`carB`) processed by the modulator stage related to leg B.

Moreover, the actuation of the two legs (i.e., two duty cycles for two modulation stages) is done by manipulating one modulation signal $d(k)$ only, from which the gating signals of the four switches are generated. Such switching pattern derivations (that is, how to compute the duty cycles) can be done by using different strategies. Therefore, before going into the implementation details, the following sections provide the descriptions of the different converter operations.

15.5.1 Modulation strategies

The legs A and B are driven by two different duty cycles $d_A(k)$ and $d_B(k)$. The voltages measured from the central point of each leg and ground are:

$$v_A(k) = d_A(k)V_{DC} \tag{15.14}$$

$$v_B(k) = d_B(k)V_{DC} \tag{15.15}$$

The armature voltage $v_a(k)$ results from the following difference:

$$v_a(k) = v_A(k) - v_B(k) = (d_A(k) - d_B(k))\, V_{DC} \tag{15.16}$$

The two legs should be synchronously operated to control the current flow $i_A(k)$ of the motor and regulate its rotation as well. Thus, $d_A(k)$ and $d_B(k)$

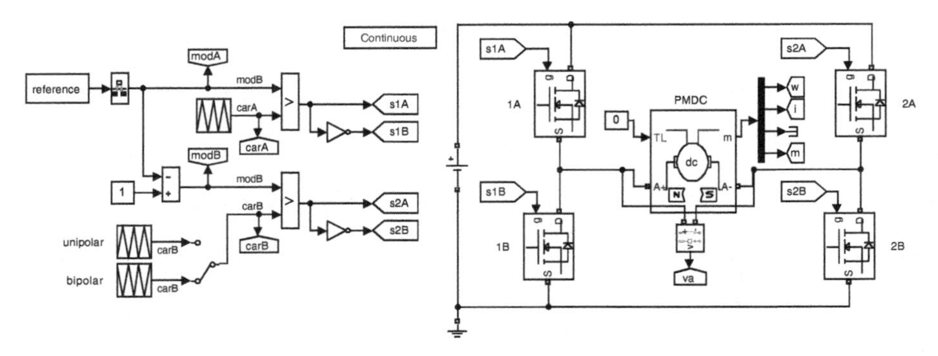

Figure 15.13 Simulink® scheme included in FBsimDC.slx.

have to be linked together even it their modulation stages may be separated. This need combined with the duty cycle manipulation, leads to define two possible modulation strategies which are called **bipolar** and **unipolar** voltage switching.

15.5.2 Unipolar voltage switching

In this case, $v_a(k)$ is continuously switched from $+V_{DC}$ down to $-V_{DC}$, and vice-versa. This modulation strategy is obtained by imposing the following condition on the duty cycles:

$$d_B(k) = 1 - d_A(k) \tag{15.17}$$

Thus, by defining $d(k) = d_A(k)$, the armature voltage becomes function of one duty-cycle only:

$$v_a(k) = (2d_A(k) - 1)\,V_{DC} \tag{15.18}$$

The unipolar voltage switching behavior is achieved by using two modulator stages, which keep the same frequency and phase shift for both the triangular carriers (i.e., same settings inside ePWM1 and ePWM2). Therefore, $S_1(k)$, $S_2(k)$ are considered related to 1_A, 1_B of leg A, respectively, whereas $S_3(k)$, $S_4(k)$ to 2_A, 2_B of leg B, respectively. The switching pattern results from the application of the PWM logic. It can be observed that when $S_1(k)$, $S_3(k)$ or $S_2(k)$, $S_4(k)$ are turned on, the armature voltage $v_a(k)$ is zero. The voltage level of this latter jumps between 0 and $-V_{DC}$ or from 0 and $+V_{DC}$. Given the equivalence between modulating signal and duty cycle, only the computation of $d_A(k)$ is required. Indeed:

- If $d_A(k) = 0$, then $d_B(k) = 1$
  which imply $v_A(k) = 0$, $v_B(k) = V_{DC}$ and $v_a(k) = -V_{DC}$;
- If $0 < d_A(k) < 0.5$, then $d_B(k) > d_A(k)$
  which imply $v_B(k) > v_A(k)$ and $v_a(k) < 0$;

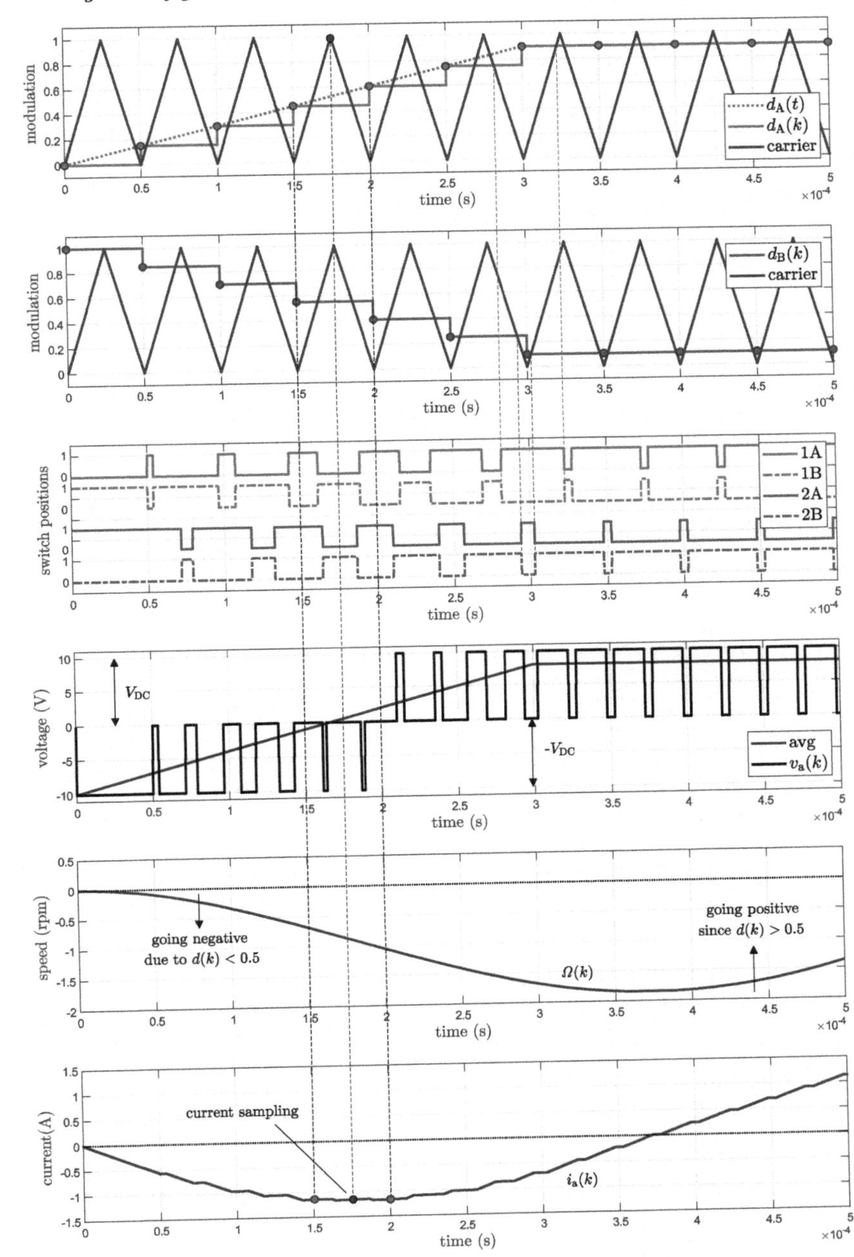

Figure 15.14 Example of unipolar voltage switching operation: associated waveforms and dynamics of the system.

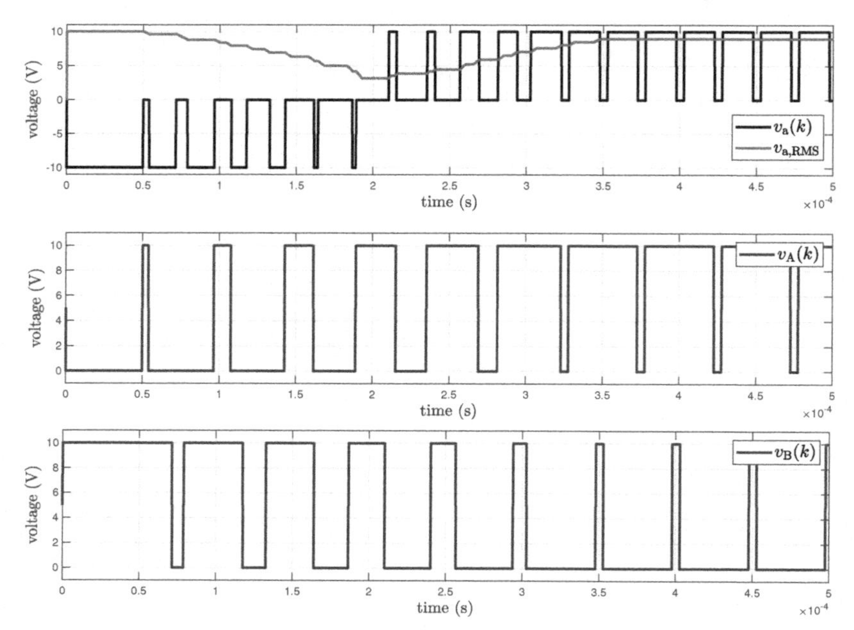

Figure 15.15 Leg voltages during a unipolar voltage switching operation.

- If $d_A(k) = 0.5$, then $d_A(k) = d_B(k) = 0.5$ which imply $v_A(k) = v_B(k) = V_{DC}/2$ and $v_a(k) = 0$;
- If $0.5 < d_A(k) < 1$, then $d_A(k) > d_B(k)$ which imply $v_A(k) > v_B(k)$ and $v_a(k) > 0$;
- If $d_A(k) = 1$, then $d_B(k) = 0$ which imply $v_A(k) = V_{DC}$, $v_B(k) = 0$ and $v_a = V_{DC}$.

Every time the potentiometer is in its centered position, it results $d_A(k) = 0.5$. Thus, the motor is standstill. The motor speed is assumed positive in clockwise direction. Therefore, depending on the terminal connections realized in the setup (see Figure 15.10), when $d_A(k) < 0.5$ the motor speed is negative (i.e., counter-clockwise rotations), while it is positive for $d_A(k) > 0.5$.

Figure 15.14 shows an example of unipolar voltage switching operations. In particular, a positive ramp profile is applied to $d(k) = d_A(k)$ which corresponds to a negative ramp profile for $d_B(k)$. At the beginning, $d(k) = 0 \rightarrow d_A(k) = 0$, $d_B(k) = 1$ leading to $v_A(k) = -V_{DC}$. Hence, the PMDC motor starts rotating in counter-clockwise direction ($\Omega(k) < 0$) until $d(k) > 0.5$. The current $i_a(k)$ shows a similar behavior but with a different (faster) dynamic. It can be noted that, even if $v_a(k)$ can be negative, the output voltages $v_A(k)$ and $v_B(k)$ are always positive on each leg as shown in Figure 15.15. Considering a longer simulation and a trapezoidal duty cycle $d(k)$ (similarly to what is reported in Section 15.4), the corresponding results are those shown

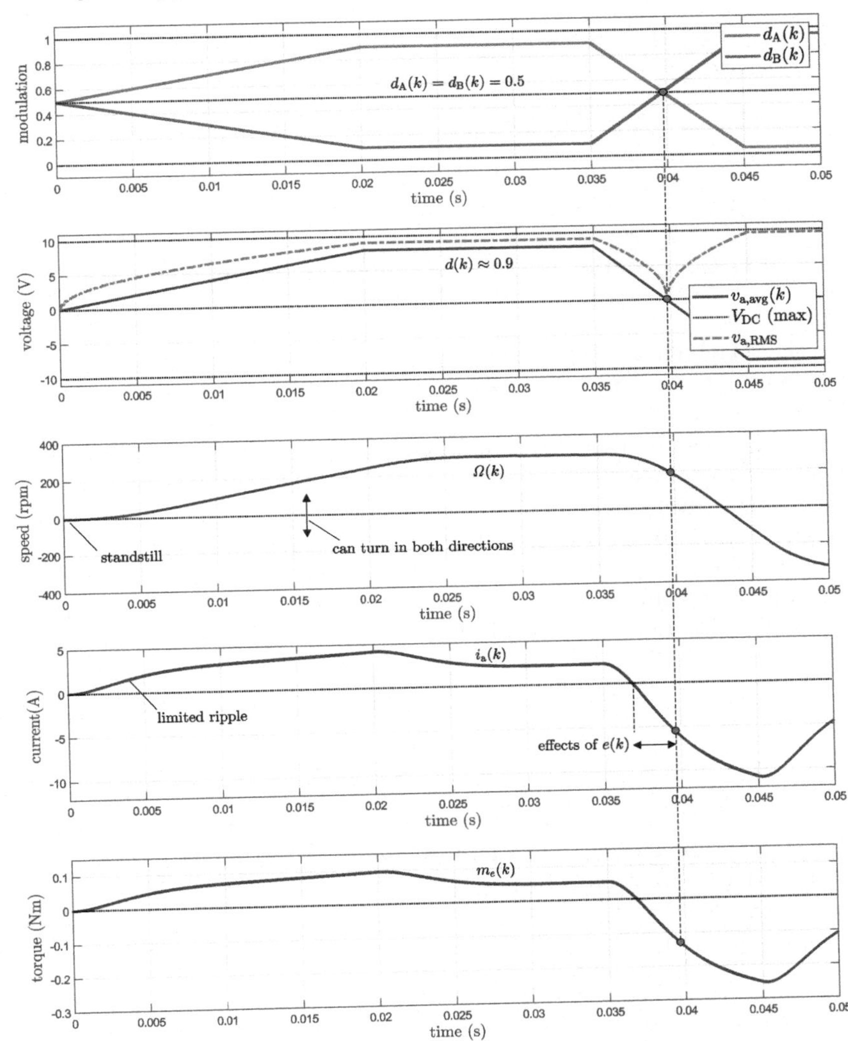

Figure 15.16 Analysis of a unipolar operation during a long simulation.

in Figure 15.16 (where the carrier waveforms are dropped out to improve the plot readability). In particular, the benefits in terms of lower current ripple in $i_a(k)$ is evident.

The unipolar voltage switching reduces the magnitude of the harmonic components of $v_a(k)$ and $i_a(k)$, which means lower ripple and harmonic distortion, while keeping the converter switching frequency f_{sw} constant. This goal is achieved by displacing $v_A(k)$ and $v_B(k)$, so that the *effective* switching frequency in $v_a(k)$ is $2f_{sw}$ and, consequently, in $i_a(k)$ as well. In standard

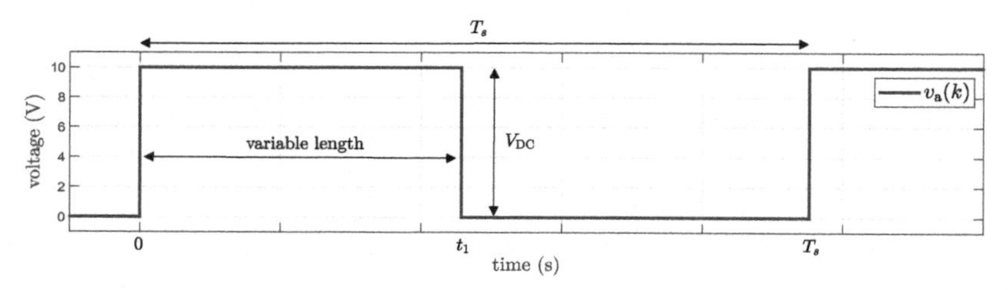

Figure 15.17 Computation of the RMS voltage during an unipolar operation.

cases (i.e., without a Full-Bridge configuration), such voltage/current ripple reduction might be achieved by actually operating the converter at twice the switching frequency. Therefore, this operation has the benefit of doubling the *apparent* f_{sw} seen by the PMDC motor (i.e., the load stage), without increasing the switching losses. This result is also a consequence of the introduction of a zero voltage level, which allows to have shorter jumps while reaching $+V_{\mathrm{DC}}$ or $-V_{\mathrm{DC}}$ (see Figure 15.15). Another important aspect regards the computation of the root mean square (RMS) value (also known as the quadratic mean) of the armature voltage. The RMS definition is the square root of an integral over the signal period, as shown in Figure 15.17. Namely, it is defined as:

$$v_{\mathrm{a,RMS}} = \sqrt{\frac{1}{T_s}\int_0^{T_s} v_{\mathrm{a}}^2(t)\ \mathrm{d}t} \tag{15.19}$$

Since the resulting $v_{\mathrm{a}}(k)$ is a pulse function, it can be equivalently represented in continuous-time domain by $v_{\mathrm{a}}(t)$. Inside one sampling period T_s (where the on time is defined as t_1, see Figure 15.17), the pulse function is constant and equal to V_{DC} for $0 \leq t < t_1$, while zero for $t_1 \leq t < T_s$. By replacing such values in (15.19), it follows:

$$v_{\mathrm{a,RMS}} = \sqrt{\frac{1}{T_s}\int_0^{t_1} V_{\mathrm{DC}}^2\ \mathrm{d}t} = V_{\mathrm{DC}}\sqrt{\frac{t_1}{T_s}} \tag{15.20}$$

Therefore, the RMS value of $v_{\mathrm{a}}(k)$ is varying in time, ranging between 0 and V_{DC}, as shown in Figures 15.15 and 15.16.
In conclusion, this solution has the following advantages and disadvantages:

Advantages

- Better harmonic content compared both to the Half-Bridge and Full-Bridge bipolar voltage switching configurations;
- One control signal is needed leading to three voltage levels ($-V_{\mathrm{DC}}$, 0, $+V_{\mathrm{DC}}$).

Disadvantages

- The RMS value of the armature voltage varies in time.

15.5.3 Bipolar voltage switching

The duty cycles relationship still holds

$$d_{\mathrm{B}}(k) = 1 - d_{\mathrm{A}}(k) \tag{15.21}$$

$$v_{\mathrm{a}}(k) = (2d_{\mathrm{A}}(k) - 1)\, V_{\mathrm{DC}} \tag{15.22}$$

Even in this case, one modulating signal is still enough to fully operate the system. Thus, the relationship between $d(k) = d_{\mathrm{A}}(k)$ (i.e., potentiometer rotation) and the motor speed is the same as for the unipolar voltage switching.

The main peculiarity of this solution, compared with the previous case, is that in a bipolar voltage switching approach the output of leg A, $v_{\mathrm{A}}(k)$, is equal and opposite to the output of leg B, $v_{\mathrm{B}}(k)$. This condition is carried out by complementing the behavior of ePWM1 and ePWM2 modules. Given the equivalent circuit reported in Figure 15.12, the diagonally opposite switches 1_{A}, 2_{B} and 1_{B}, 2_{A} are turned on or off at the same time. This logic leads to an armature voltage v_{a} switching between two levels: $-V_{\mathrm{DC}}$ and $+V_{\mathrm{DC}}$, which is the reason beyond the name *bipolar*.

In terms of implementation, only one ePWM module may be used to realize such approach theoretically speaking. For instance, if ePWM1 is settled to provide $S_1(k)$ and $S_2(k)$ (i.e., driving $1_{\mathrm{A}}, 1_{\mathrm{B}}$) at pins 40 and 39 respectively, it would be enough to physically connect $S_3(k)$ to pin 39 and $S_4(k)$ to pin 40 to control leg B.

On the other hand, the proposed implementation foresees to use two modulator stages (ePWM1/2). The bipolar voltage switching is achieved by using modulators operating with different triangular carriers. The latter would be in opposition (counter-phase) to each other. For instances, if the phase shift of the ePWM1 carrier is set equal to 0° (or 90°), then the one of ePWM2 should be set equal to 180° (or −90°), which is equivalent to have an inverted/mirrored carrier. Since the carrier is internally created by the ePWM blocks, the phase shift is introduced through a specific parameter setting.

Figure 15.18 shows an example of bipolar voltage switching operation. In particular, the same positive ramp profile adopted for the unipolar case is applied to $d(k) = d_{\mathrm{A}}(k)$ and complemented for $d_{\mathrm{B}}(k)$. At the beginning, $d(k) = 0 \rightarrow d_{\mathrm{A}}(k) = 0$, $d_{\mathrm{B}}(k) = 1$ leading to $v_{\mathrm{A}}(k) = -V_{\mathrm{DC}}$. Nevertheless, at the following time steps, Figure 15.18 shows how the zero voltage level is absent from $v_{\mathrm{a}}(k)$, leading to a switching between $-V_{\mathrm{DC}}$ and V_{DC} to regulate the average voltage. The resulting behavior of the output voltages $v_{\mathrm{A}}(k)$ and $v_{\mathrm{B}}(k)$ (which are always positive) are shown in Figure 15.19.

Comparing Figure 15.14 and 15.18, it is evident that the bipolar voltage switching returns an higher current and voltage harmonic distortion with respect to the unipolar one. The latter benefit of the cancellation of the all the

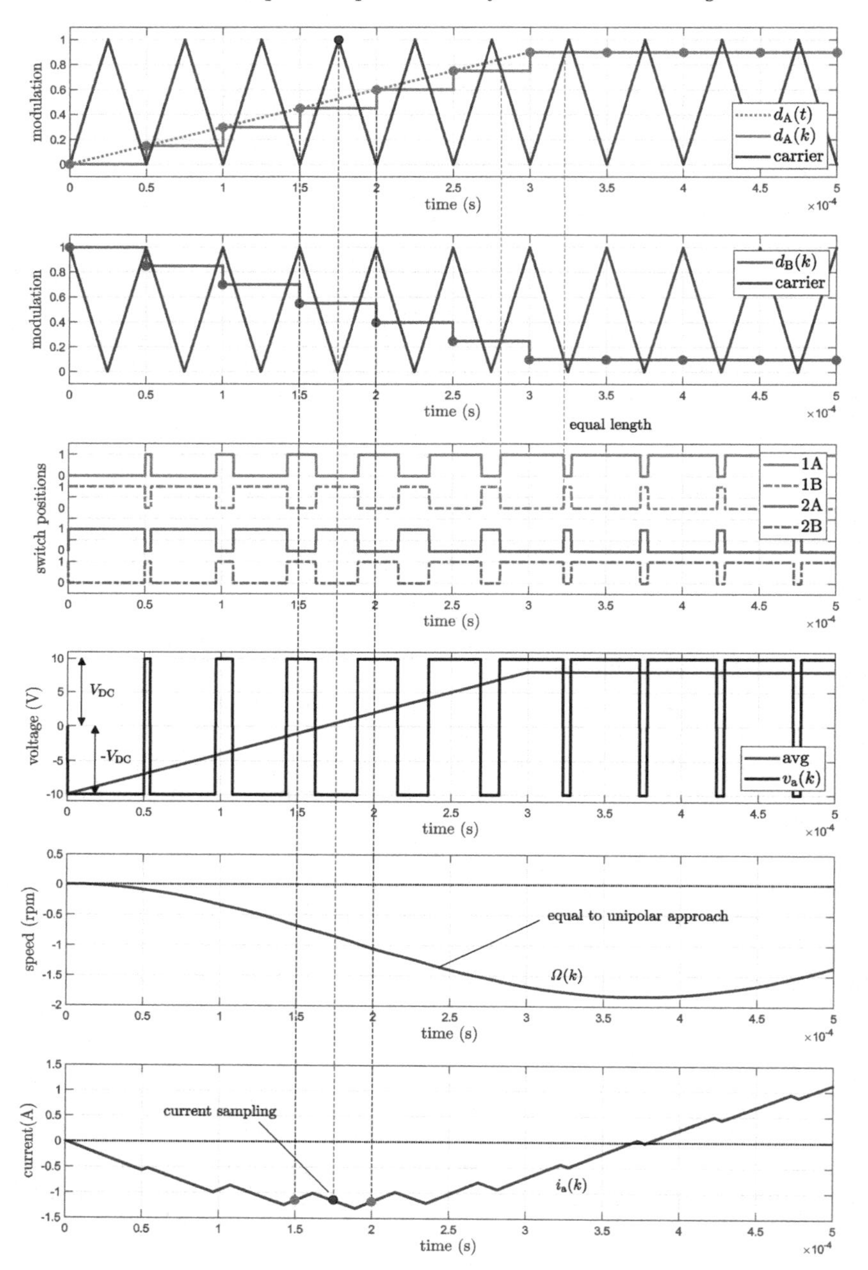

Figure 15.18 Example of bipolar voltage switching operation: associated waveforms and dynamics of the system.

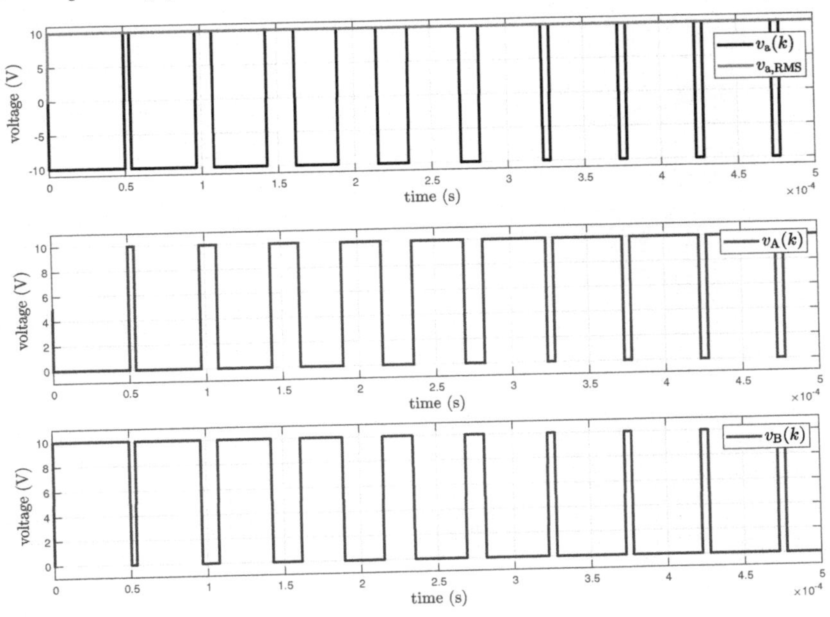

Figure 15.19 Leg voltages during a bipolar voltage switching operation.

odd harmonics of f_{sw} for $v_a(k)$ and, consequently, $i_a(k)$, which is not the case for the bipolar case. This can be proved and clarified by Figure 15.20, which compares the harmonic spectra of the armature voltages $v_a(k)$ when a fixed

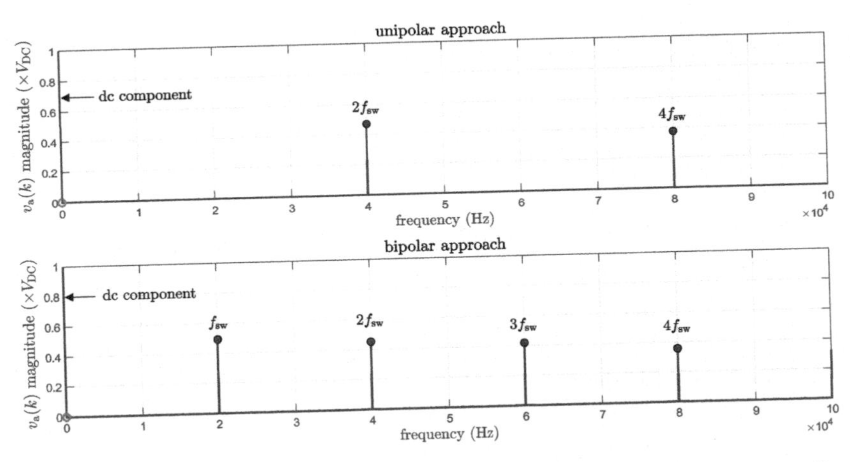

Figure 15.20 Spectra (truncated at the 5th harmonic) of the armature voltages obtained in unipolar and bipolar voltage switching operator. In both cases, the converter operates at the same f_{sw}.

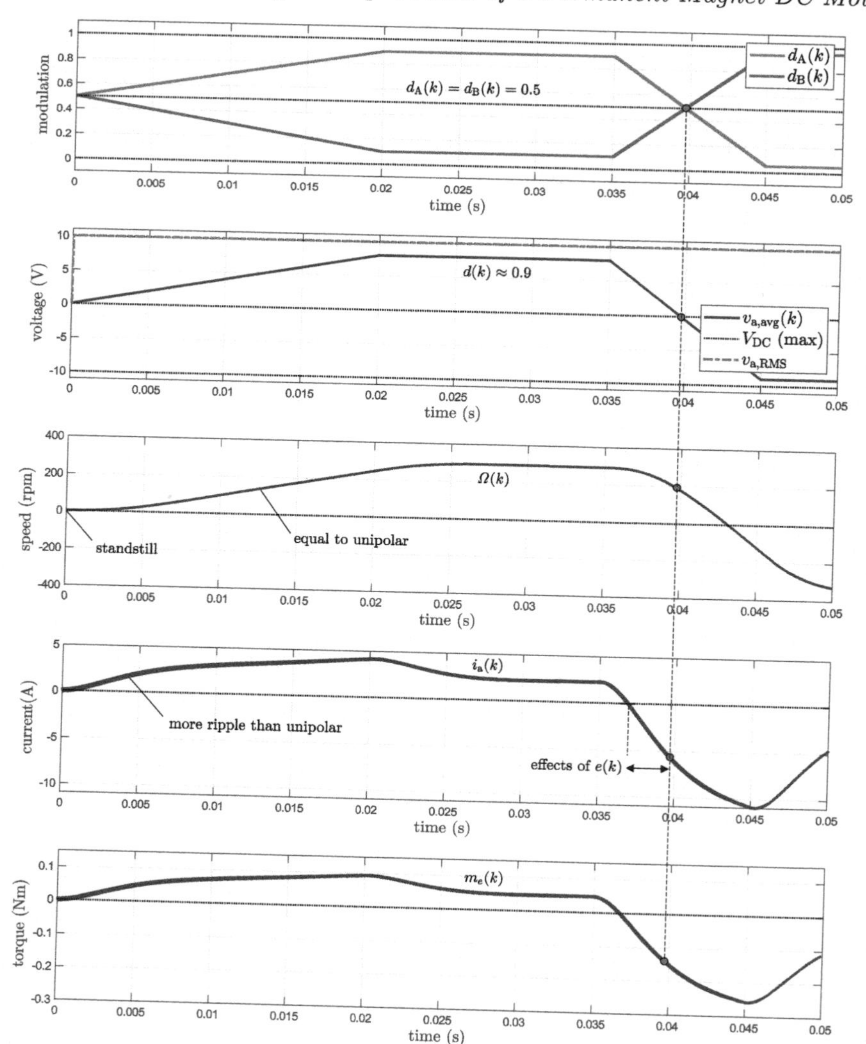

Figure 15.21 Analysis of a bipolar operation during a long simulation.

duty cycle $d(k) = 0.2$ (i.e., in steady-state conditions) is applied for the unipolar and bipolar operations, respectively. The spectra are computed through a discrete fast Fourier transform (DFFT) considering 800 periods. The converter legs are operated at the same f_{sw} in both cases. This comparison shows how the unipolar approach is able to operate the system at an effective switching frequency of $2f_{sw}$, leading to a lower current distortion.

Similarly to Section 15.5.2, a long simulation and a trapezoidal duty cycle $d(k)$ are considered for the bipolar operation. The results are shown in

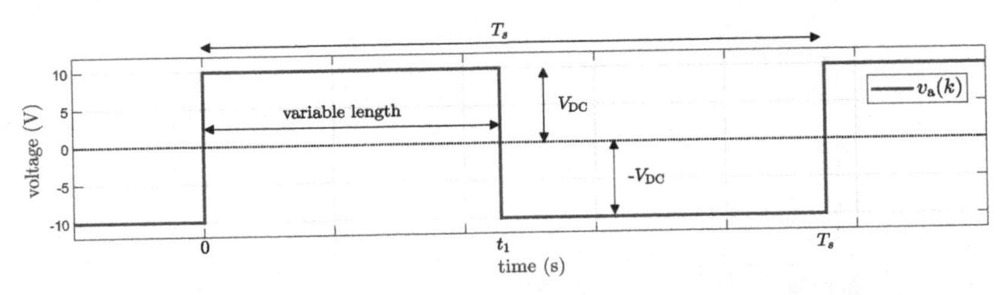

Figure 15.22 Computation of the RMS voltage during a bipolar operation.

Figure 15.21, where the carrier waveforms are dropped out to improve the plot readability. A larger ripple on $i_a(k)$ is particularly evident in comparison to Figure 15.16.

Regarding the RMS value of the armature voltage, $v_a(k)$ is now a bipolar pulse function. Inside one sampling period T_s (where the on time is defined as t_1, see Figure 15.22), $v_a(t) = V_{DC}$ for $0 \leq t < t_1$, while $v_a(t) = -V_{DC}$ for $t_1 \leq t < T_s$. By replacing such values in the general equation (15.19), it follows:

$$v_{a,RMS} = \sqrt{\frac{1}{T_s}\left(\int_0^{t_1} V_{DC}^2 + \int_{t_1}^{T_s}(-V_{DC})^2\right) dt} = V_{DC}\sqrt{\frac{t_1}{T_s}} + V_{DC}\sqrt{1 - \frac{t_1}{T_s}} = V_{DC} \tag{15.23}$$

Therefore, the RMS value of $v_a(k)$ in case of bipolar voltage switching is always constant and equal to V_{DC} independently from the duty cycle, as shown in Figure 15.19 and 15.16. As an example, it can be assumed to include a lighting device in the drive system. Light emissions are sensitive to the RMS value of the supply voltage (this is true for LEDs, incandescent and fluorescent lights). By connecting a lighting device in parallel to the PMDC motor terminal, the motor control is unaffected and this device is continuously emitting light. Car mirrors typically use this idea to regulate the position of the mirrors through small motors, while turning on/off the integrated light indicators by directly connecting, e.g., LEDs, to the motor terminals.[5] In conclusion, the bipolar solution shows the following advantages and disadvantages:

Advantages

- One modulation signal needed leading to two voltage levels ($+V_{DC}$, $-V_{DC}$);
- RMS value of the armature voltage is constant.

Disadvantages

- Higher voltage and current harmonic distortion compared to unipolar case.

[5] This approach is used to reduce the number of elements aimed to the electronic circuit.

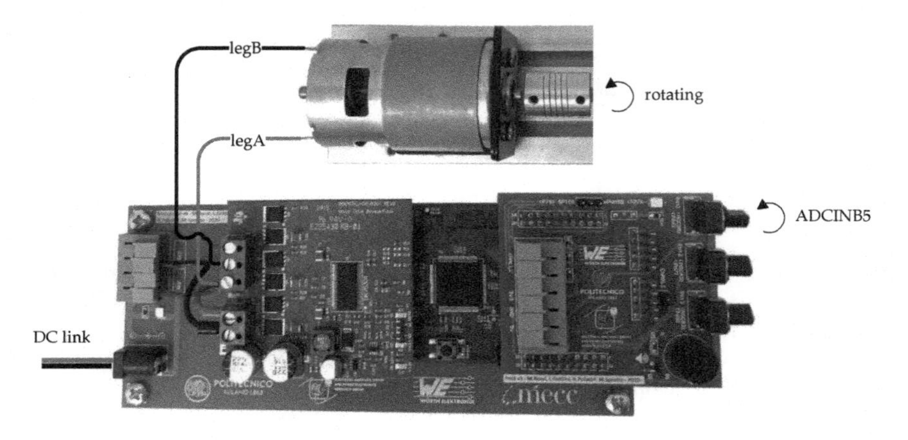

Figure 15.23 Connections of the PMDC motor with the BOOSTXL-DRV8301 converter realizing a Full-Bridge configuration.

15.5.4 Control implementation

The open-loop dynamics are tested through a test-bench which includes a TI BOOSTXL-DRV8301 converter mounted on the TI LaunchXL F28069M Piccolo™ and connected to a PMDC motor, as shown in Figure 15.9 Regarding the Full-Bridge configuration, legs A and B are used for both unipolar and bipolar voltage switching. The cable connections from the BOOSTXL-DRV8301 to the motor terminals are shown by the connection scheme reported in Figure 15.23. In the following, the step-by-step procedures for programming the microcontroller for both modulation approaches are reported.

Firmware Environment (Unipolar Approach)

c28069_openDC_fbF_uni (solver: fixed step—ODE4, step size: $T_s = 50\,\mu s$)

Open a new blank Simulink® project and configure the environment as shown in Chapter 9. Then:

- Open the **Model Configuration Parameters** window and use the same settings adopted for the **Example 1** in Section 10.5.
- Insert an **ADC** block to read the potentiometer and set it as shown in the previous Section 15.4.1. Indeed, the potentiometer reading is the same. Hence, add **Data Type Conversion** blocks to manipulate the signal (i.e., `int16` and `single`). Place a **Gain** block set to $1/2^{12}$ to scale the digitalized signal within $[0, 1]$.
- Insert two **ePWM** blocks; use the settings reported in Section 13.5:

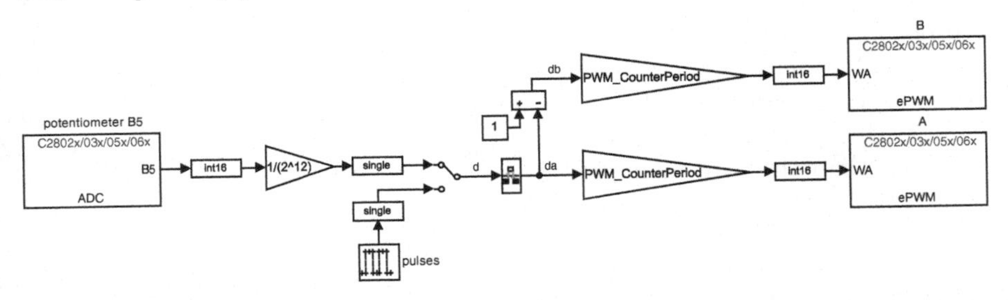

Figure 15.24 Simulink® scheme included in c28069_openDC_fbF_uni.

- **ePWM1** module is used with TBPRD corresponding to $f_c = 20\,\text{kHz}$, i.e., $\text{PWM}_{\text{counter-period}} = 2250$;
- Remember to unflag Enable ePWM1B and tick Inverted version of ePWMxA. This option allows to drive 1_{B} as the complement of 1_{A};
- Keep all the other settings as reported in Section 13.5;
- Set **ePWM2** analogously.

Since the same settings are used for both ePWM blocks (i.e., same triangular carrier and comparison instances), the modulating signals $d_{\text{A}}(k)$ and $d_{\text{B}}(k)$ are built as follow:

- Add a **constant** and an **add** block to bias the input signal. Set the value of the constant to 1 and the list of signs of the add block to $+-$. Thus, the modulating signal which is fed into ePWM1 is $d_{\text{A}}(k)$ while $d_{\text{B}}(k)$ is obtained as $d_{\text{B}}(k) = 1 - d_{\text{A}}(k)$ and fed into ePWM2.
- Add **gain** blocks with value $\text{PWM}_{\text{counter-period}} = 2250$ to both path for the ePWM modules to normalize the modulation signals.
  * $\widehat{d}_{\text{A}}(k) = d_{\text{A}}(k) \cdot \text{PWM}_{\text{counter\_period}}$
  * $\widehat{d}_{\text{B}}(k) = (1 - d_{\text{A}}(k)) \cdot \text{PWM}_{\text{counter\_period}}$
- Add other two **data type conversion** blocks (one for each ePWM block) set to `int16` to represent the duty cycle within a 16-bit base.

The aforementioned blocks should be connected as shown in Figure 15.24. Even in this case, it is recommended to set the power supply at $V_{\text{DC}} = 10\,\text{V}$ with a current limitation of 2 A. After the firmware deploy on the MCU, it is possible to manually vary the motor speed through the potentiometer as done previously. Since the system is running in open loop, the user is recommended to verify the maximum voltage and current which is fed into the system to avoid any damage or premature aging of the components.

Generally, it is preferable not to fully apply $d(k) = 0$ (0%) or $d(k) = 1$ (100%) because they could be potential causes of leg short-circuits.

Remark: the BOOSTXL-DRV8301 converter operates with a *bootstrap circuit*, which requires $d(k) < 1$ to keep the converter switching safely. Considering leg 1, the drain pin of 1_A is connected to the positive terminal of the of dc link while switching. It may happen that the voltage applied on the switch is higher than the threshold needed to close the switch, i.e., the gate-source voltage V_{gs} (e.g., $V_{gs} \approx 15\,\mathrm{V}$ with $V_\mathrm{dc} \approx 48\,\mathrm{V}$). Therefore, the gate circuit is using a diode to charge a small capacitor C_{gd} placed between gate and drain pins, designed to operate 1_A for a given value of V_{gs}. In practice, when $d(k) = 1$, only the switch 1_A is closed. However, keeping $d(k) = 1$, C_{gd} is slowly discharging in time until the energy is not enough to keep the switch closed (i.e., voltage becomes lower than V_{gs}). Hence, it is necessary to open it and close 1_B, which means $d(k) < 1$. Thus, some small interval is added to recharge C_{gd}. It is important to note that for low voltage applications (as in this case, being $V_\mathrm{SC} = 10\,\mathrm{V}$) it is enough to add a narrow margin on the duty cycle, e.g., $d(k) \in [0.01, 0.99]$.

Firmware Environment (Bipolar Approach)

c28069_openDC_fbF_bip (solver: fixed step - ODE4, step size: $T_s = 50\,\mu\mathrm{s}$)

Take the previous Simulink® project about the unipolar voltage switching There, the PWM actuation must be modified. In particular, the ePWM peripherals of the TI LaunchXL F28069M Piccolo™ allow *swap* features between ePWMxA and ePWMxB. By selecting *swap the ePWMA and ePWMB outputs*, the computed ePWMxA output signal is redirect/provided at the ePWMxB output terminal, and vice-versa. For the bipolar approach, this feature allows to simplify the PWM actuation since the duty cycles are set $d_\mathrm{A}(k) = d_\mathrm{B}(k)$. Two modulation stages are considered, ePWM1 (which provides $S_1(k)$, $S_2(k)$) and ePWM2 (which provides $S_3(k)$, $S_4(k)$), both fed with the same modulating signal. If ePWM2 is swap-enabled, then ePWM2A=$S_3(k)$ and ePWM2B=$S_3(k)$. Thus, if ePWM1 has standard settings, it follows that $S_1(k) = S_4(k)$ and $S_2(k) = S_3(k)$. This latter correspond to the bipolar voltage switching principle and it is exactly equivalent to make ePWM2 operating with a triangular carrier shifted by 180° with respect to ePWM1 and having $d_\mathrm{B}(k) = (1 - d_\mathrm{A}(k))$.

Therefore, the previous scheme has to be updated as follows:

- Remove the $d_\mathrm{B}(k)$ derivation, keeping one duty cycle only, i.e., $d_\mathrm{A}(k) = d_\mathrm{B}(k)$, by deleting the constant and the add block.
- Use the same settings adopted in the unipolar case for module ePWM1: TBPRD corresponds to $f_c = 20\,\mathrm{kHz}$, i.e., $\mathrm{PWM}_\mathrm{counterperiod} = 2250$.
- Keep the same settings for ePWM2 apart from the following:
  - Thick enable swap module A and B at the bottom of the General tab.
- Keep the rest of the scheme set as before.

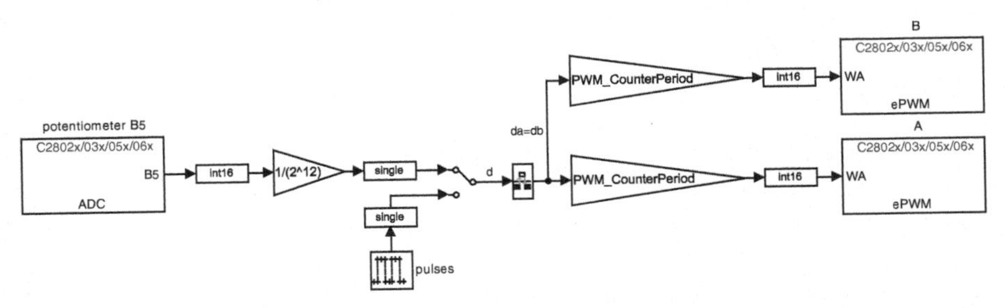

Figure 15.25 Simulink® scheme included in c28069_openDC_fbF_bip.

Finally, connect all these blocks as reported in Figure 15.25. It is recommended to set the power supply at $V_{\mathrm{DC}} = 10\,\mathrm{V}$ with a current limitation of $2\,\mathrm{A}$. After the deploy the firmware on the MCU board, it is possible to manually vary the motor speed in absolute value and sign according to the position of the potentiometer. Since the system is running in open loop, the user is recommended to verify the maximum voltage and current which is fed into the system to avoid any damage or premature aging of the components.

16

Low-Side Shunt Current Sensing

In order to close a control loop with the LaunchPad™ F28069M board, different measurements have to be managed and processed. Current, voltage and speed sensors (e.g., encoders or hall effect sensors) are widely adopted in electrical drive applications. These devices are usually placed at the power converter terminals or mechanically coupled with the load. The resulting measurements are fed into the MCU through ADC or GPIO channels. This section focuses on how to manage current measurements.

For instance, the TI™ BOOSTXL-DRV8301 converter has both voltage and current sensors on-board. In particular, this board includes a low-side shunt current sensor, which is a resistor placed below the low side MOSFET (e.g., 1_B) of each leg. A differential operational amplifier (op-amp) takes the differential voltage across the resistor as input, which varies depending to the current flow, to compute a filtered voltage value scaled for the peripheral inputs (i.e., ranging in $[0, 3.3]$ V). The corresponding equivalent circuit is shown in Figure 16.1 (a). It is worth noting that shunt resistors provide an indirect current measurements since, actually, a voltage drop across them is measured.

The main advantage of low-side sensing is the simple implementation. Moreover, being on the low side of the load, the common-mode voltage at the shunt is approximately zero; therefore, the robustness of the amplifiers is not an important factor. However, this solution comes at the cost of ground variations and it foresees a resistance between in the path to ground. Indeed, this resistance removes the ability to detect faults in the load if it is shorted to ground. Moreover, the op-amp circuit present several gains, which can be activated by acting on dedicated registers to adjust its output according to the magnitude of the measurement, as shown in Figure 16.1 (b). Hereafter, to simplify the notation, the terms op-amp or amplification stage directly refer to the differential structure shown in Figure 16.1 (b).

In order to design a firmware able to manage a current feedbacks, an accurate ADC conversion chain must be built taking into account the tolerances of the components mounted on sensoring circuits mounted on the BOOSTXL-DRV8301 converter as well as the ADC peripheral resolution of the LaunchPad™ F28069M board. In practice, this means to set some gains that allow a proper calibration of the conversion chain. To this aim, the following sections propose and compare two approaches: a theoretical and an experimental calibration targeted on the available hardware.

DOI: 10.1201/9781003196938-16

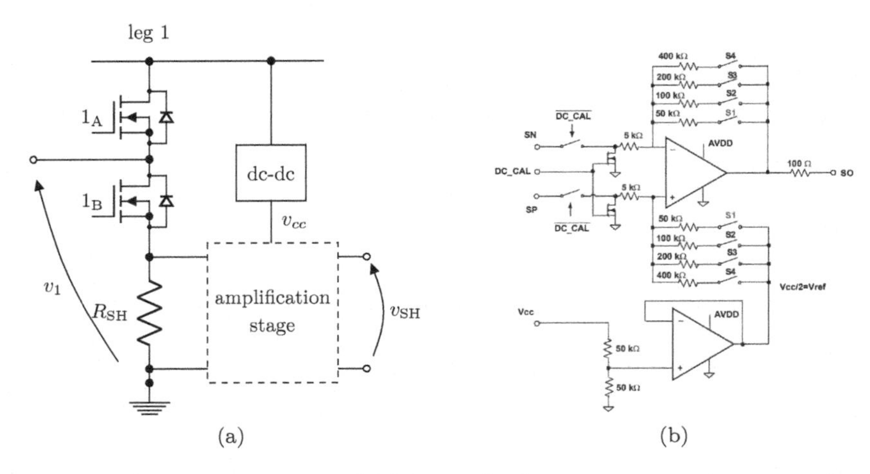

Figure 16.1 Equivalent circuit of (a) low-side shunt resistor and op-amp circuit for leg 1 and a (b) detailed view of the op-amp differential structure and selectable gains through switches $S1 - S4$. In this last schematic $v_{\mathrm{SO}} = v_{\mathrm{sh}}$ [14].

16.1 Sensor Characterization: Theoretical Approach

Referring to 16.1 (b), the op-amp circuit requires a reference voltage $v_{\mathrm{ref}} = 1.65\,\mathrm{V}$ to process bidirectional (positive/negative) current measurements while keeping its output bounded between $[0, 3.3]\,\mathrm{V}$. The theoretical characteristic of the current sensor is reported in Figure 16.2. In general, a $v_{\mathrm{ref}} = 1.65\,\mathrm{V}$ is not directly available in 3.3 V systems, but it can be derived through an internal voltage divider. This is depicted at the bottom of Figure 16.1 (b), where a voltage follower, i.e., voltage divider and amplification stage, is used to improve the signal stability and noise immunity. Provided that the op-amp circuit has several selectable gains, the default configuration refer to switch S1 closed (see Figure 16.1 (b)). Therefore, it follows:

$$v_{\mathrm{SO}}(k) = \frac{R_1}{R_{12}}\left(1 + \frac{R_1}{R_2}\right) v_{\mathrm{SP}}(k) - \frac{R_1}{R_2} v_{\mathrm{SN}}(k) + \frac{R_2}{R_{12}}\left(1 + \frac{R_1}{R_2}\right)\frac{v_{\mathrm{cc}}}{2} \quad (16.1)$$

where $R_1 = 50\,\mathrm{k\Omega}$, $R_2 = 5\,\mathrm{k\Omega}$ and $R_{12} = R_1 + R_2$, while $v_{\mathrm{cc}}/2 = v_{\mathrm{ref}}$. This equation can be rewritten as:

$$v_{\mathrm{SO}}(k) = v_{\mathrm{ref}} - 10\underbrace{(v_{\mathrm{SN}}(k) - v_{\mathrm{SP}}(k))}_{R_{\mathrm{SH}}i(t)} = v_{\mathrm{ref}} - 10 R_{\mathrm{SH}} i\,(k) \quad (16.2)$$

The BOOSTXL-DRV8301 board mounts a $R_{\mathrm{SH}} = 0.01\,\Omega$ shunt and it is designed for a maximum current peak of $\pm 14\,\mathrm{A}$. Therefore, the voltage values

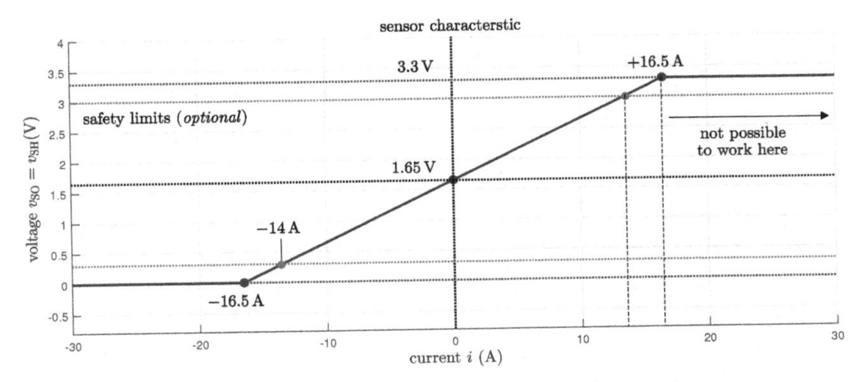

Figure 16.2 Theoretical characteristic of the current measurement through low-side shunt resistor.

provided at the ADC peripheral inputs are inside a $[0.25, 3.05]$ V range. This means that the amplification circuit has a safety margin for current management, since the magnitude of the current can be theoretically pushed up to ± 16.5 A. This last range allows to exploit the all the possible voltage values, that is, $[0, 3.3]$ V. Namely:

$$i_{\text{max}} : i(k) = +16.5\,\text{A} \quad \rightarrow \quad v_{\text{max}} : v_{\text{SH}}(k) = 3.3\,\text{V} \tag{16.3}$$

$$i_{\text{min}} : i(k) = -16.5\,\text{A} \quad \rightarrow \quad v_{\text{min}} : v_{\text{SH}}(k) = 0\,\text{V} \tag{16.4}$$

It is important to note that zero current corresponds to an output voltage of $v_{\text{SO}} = v_{\text{ref}} = 1.65\,\Omega$, as a result of the sensor characteristic. The first conversion gain is defined to translate the measured current $i\,(k)$ into voltage v_{SO}, including the offset v_{ref}. Considering v_{SO} and $i(k)$ at their maximum values, see equation (16.2), it follows that:

$$g_i = 10 R_{\text{SH}} = \frac{v_{\text{max}} - v_{\text{ref}}}{i_{\text{max}}} = 0.1\,\Omega = 100\,\frac{\text{mV}}{\text{A}} \tag{16.5}$$

Then, v_{SO} is processed and scaled over the 12 bit ADC peripheral through a quantization process, which can by accounted for by the gain:

$$g_{\text{v}} = \frac{v_{\text{max}}}{2^{12} - 1} = \frac{3.3}{4095} \approx 0.81\,\frac{\text{mV}}{\text{bit}} \tag{16.6}$$

The output of ADC channels is a voltage value in bit, $v_{\text{SO}}^{\text{bit}}$, for which the interval $[0, 2047]$ refers to negative currents, whereas $[2049, 4095]$ to positive ones. The value $v_{\text{SO}}^{\text{bit}} = 2048$ means to zero current. Therefore, g_i and g_v can be interpreted as the current and voltage conversion steps of the quantizied measurement chain, respectively.

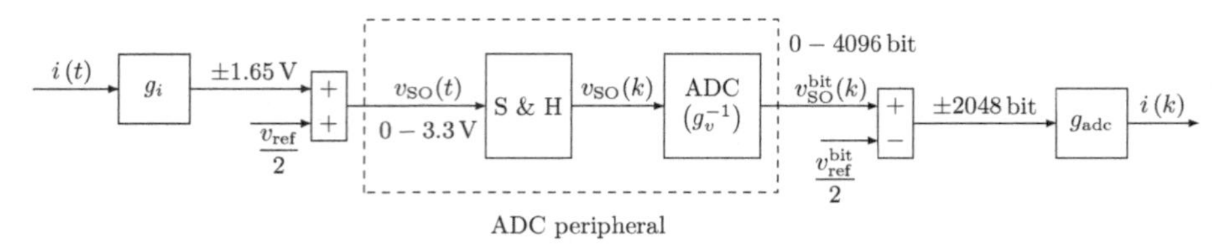

Figure 16.3 ADC conversion chain for current measurement through the BOOSTXL-DRV8301 converter board.

If the programming environment is required to work in SI units, the bit representation has to be re-scaled to provide values in Ampere. Therefore, combining g_i and g_v, a peripheral gain g_{adc} can be defined as follows:

$$g_{\text{adc}} = \frac{g_v}{g_i} = \frac{v_{\max} - v_{\text{ref}}}{i_{\max}} \frac{v_{\max}}{2^{12} - 1} \approx 8.1 \, \frac{\text{mA}}{\text{bit}} \tag{16.7}$$

The conversion chain of a current measurement from a shunt resistor to the ADC peripheral is reported in Figure 16.3, where $i(k)$ represent a variable ready to be processed to close a loop. This result is achieved through the following steps:

$$i(t) \underset{g_i, +v_{\text{ref}}/2}{\longrightarrow} v_{\text{SH}}(t) \underset{g_v}{\longrightarrow} v_{\text{SO}}^{\text{bit}}(t) \underset{S\&H}{\rightarrow} v_{\text{SO}}^{\text{bit}}(k) \underset{-v_{\text{ref}}^{\text{bit}}, g_{\text{adc}}}{\longrightarrow} i(k) \tag{16.8}$$

In particular, focusing on $i(t) = 16.5\,\text{A}$:

$$i(t) = 16.5\,\text{A} \rightarrow v_{\text{SH}}(t) = 3.3\,\text{V} \rightarrow v_{\text{SO}}^{\text{bit}}(k) = 4095 \rightarrow i(k) = 16.5\,\text{A} \tag{16.9}$$

The same process can be used to compute all the other current values.
Remark: it is convenient to synchronize the conversion procedure of the ADC peripheral with the ePWM module as reported in Section 13.9.

16.2 Locked Rotor Test

One of the easiest way for verifying the accuracy of the theoretical characterization of the current measurement is to perform a locked rotor test (also know as short circuit test) with a PMDC motor. This test is carried out by mechanically locking the rotor, thus, forcing the back-emf to zero. Considering the mathematical model presented in Chapter 15, the steady state equivalent circuit of the PMDC motor during a locked rotor test simplifies to that one shown in Figure 16.4. At setady state, the current I_{A} flows in the armature resistance only since d/dt = 0, i.e., the inductance behaves like a short circuit. The rated value of the DC motor resistance R_a can be computed as $R_{\text{a}} = V_{\text{a}}/I_{\text{a}}$, or it can be measured through an ohmmeter. It follows that $R_{\text{a}} \approx 0.6\,\Omega$.

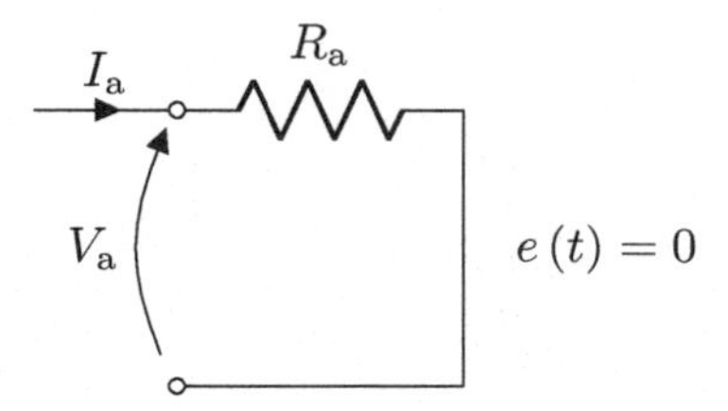

Figure 16.4 Equivalent circuit of a DC motor during a locked-rotor test at steady state.

This test can be carried out driving the PMDC motor in open-loop with a converter in half-bridge configuration (see the equivalent circuit in Figure 15.5). The power supply is set to $V_{DC} \approx 10\,\text{V}$ with current limitation at 2 A. Differently from the exercises reported in Chapter 15, the duty cycle is now constant and internally settled into the MCU instead of using a potentimeter. This is aimed to have a constant and stable voltage supplied at the motor terminal, which implies a constant current absorption as well. Hence, the duty cycle is set to $d = 0.2$ so that $V_a \approx 2\,\text{V}$ at armature windings. It follows that:

$$I_a = \frac{V_a}{R_a} = \frac{2\,\text{V}}{0.6\,\Omega} \approx 3.33\,\text{A} \tag{16.10}$$

Since the motor size is small and the applied voltage low, the rotor can be hand locked with the help of a clamp, as reported in Figure 16.5. A simple test is to measure the motor current at steady-state with an ADC peripheral and visualize it through the Serial Port (e.g., SCIA). If the conversion chain is correctly set, using g_v and g_{adc}, the result should be close to the value reported in equation (16.10). Regarding the MCU programming, it is suggested to start from the firmware developed in Section 15.4.1.

Firmware Environment

c28069_lockedDC_F.slx (solver: fixed step - ODE4, step size $T_s = 50\,\mu\text{s}$)

Open the example of the open-loop control of a PMDC motor (half-bridge converter configuration) presented in Section 15.4.1. Modify the scheme as follows:

- Edit the **ADC** block parameters:
  - Select SOCx **trigger source** ePWM1_ADCSOCA;
  - Set **sample time**: $T_{adc} = 0.001\,\text{s}$;
  - Select **Conversion channel** ADCINA0 in the **Input channel** tab.
- Insert a **data type conversion** block set to `int16`.

Figure 16.5 Example of hand locked rotor by using a clamp

- Insert a **Constant** equal to 2048 (as `int16`) and an **Add** block with list of signs +− to remove the offset at the ADC output.
- Add a **data type conversion** block set to `double` (or `single`) for safer mathematical computations. See see Chapter 7 for reference.
- Insert a **Gain** block set on $g_{\mathrm{adc}} = 8\,\mathrm{mA/bit}$ to map the ADC reading over a range of values in Ampere.
- Add a **data type conversion** block set to `single` combined with a **Rate Transition** block with sample time $T_{\mathrm{SCI}} = 0.01\,\mathrm{s}$ to prepare the data exchange.
- Insert a **Mux** block to send two inputs via the SCI Transmit block.
- Insert a **SCI Transmit** block and set it as shown in Section 10.5 (i.e., SCIA).
- Place a **pulse generator** block and double click on it:
  - Set **pulse type** on Sample based;
  - Insert an **Amplitude** equal to 0.2;
  - Choose a **period** of 250 samples;
  - Type a **pulse width** of 125 samples;
  - Insert a **sample time** of 0.01 s.

  The duty cycle is now a train of pulses with amplitude 0.2. This block is included to limit the steady-state condition. Indeed, even if the current is low, being related to the voltage input (which is 2 V while the rated value for this motor is 24 V), the temperatire of the armature windings may rise too much. The latter may lead to significant variations in the windings resistance, thus, affecting the measurement. Hence, the pulse generator block prevents overheating.

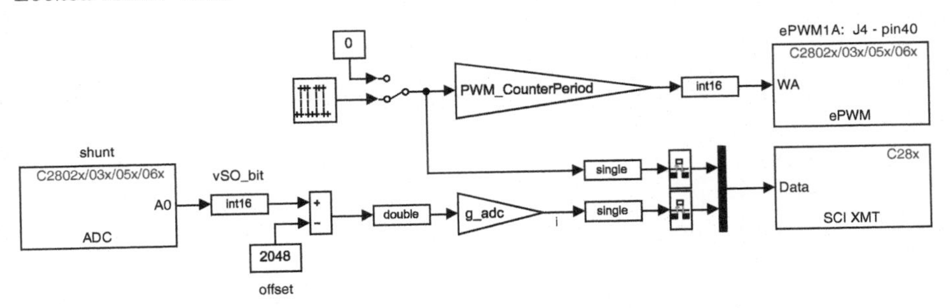

Figure 16.6 Simulink® scheme included in c28069_lockedDC_F.slx.

- Add a **data type conversion** block set to `single` combined with a **Rate Transition** block ($T_{\mathrm{SCI}} = 0.01\,\mathrm{s}$), to prepare the output data of the pulse generator. Connect it to the **Mux** block
- For the ePWM settings the reader is referred to Section 15.4.1 ($f_c = 20\,\mathrm{kHz}$)

The arrangement of these blocks is shown in Figure 16.6. Since $f_c = f_{\mathrm{sw}}$ holds, the sampling time for the current is $T_s = T_{\mathrm{sw}} = 50\,\mu\mathrm{s}$. A up-down couting mode is used with a $\mathrm{TBPRD} = \mathrm{PWM}_{\mathrm{counter-period}} = \mathrm{CPU}_{\mathrm{frequency}}/2 \cdot \mathrm{PWM}_{\mathrm{frequency}} = 2250$. An initialization script either in *Model Properites/Callbacks/InitFcn* or as a separate `m`-file should be edited to set the system parameters. The latter approach is used in this exercise (`lockDC_init.m`) and the code is reported here in the following:

```
%% F28069M clock
CPU_frequency        = 90e6;
%% carrier frequency definition
PWM_frequency        = 20e3;
Tsw                  = 1/PWM_frequency
%% TBPRD (counting mode) definition
PWM_counterperiod    = CPU_frequency/(2*PWM_frequency)
%% sampling time definition
Ts                   = Tsw;
%% ADC theoretical gain
g_adc                = 33/4095;
```

Testing Environment

c28069_lockedDC_T (solver: fixed step - ODE4, step size $T_{\mathrm{sim}} = 100\,\mu\mathrm{s}$)

Open a blank Simulink® project and set the **Model Configuration Parameters** as reported in Section 10.5. Then:

- insert a **Serial Configuration** block and set it as reported in Section 10.5.

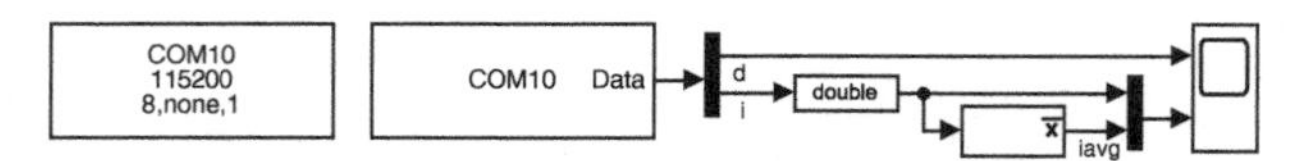

Figure 16.7 Simulink® scheme included in c28069_lockedDC_T.

- Insert a **Serial Receive** block and double-click on it:
  - Select the right COMx as **serial port**;
  - No **header** or **terminator** should be inserted;
  - Set **data size**: [1 2] since two signals have to be visualized;
  - Set **data type**: `single`;
  - Set **sample time**: 0.01 s.
- Insert a **Demux** block to split the two transmitted signals: $d(k)$ and $i(k)$.
- Insert a **Scope** block and set the number of input ports equal to two.
- *Optional:* a **Mean** block could be added to compute the mean value of $i(k)$. This can be used if $i(k)$ presents high noise or the average value is not clearly readable from the received data. Set:
  - Fundamental frequency: 10 Hz or 50 Hz;
  - Initial input: 0;
  - Sample time: 0.

  This block receives $i_{\text{SCI}}(k)$. A **data type conversion** block is added and set on `double`. The resulting variable is $i_{\text{avg}}(k)$.

Finally, connect these blocks as reported in Figure 16.7. Since the testing environment is used for a debug purposes only, the simulation step-size is set equal to $T_{\text{sim}} = T_s/2 = 100\,\text{ms}$.

The serial communication (here SCIA) is able to manage multiple data at the same time. They are ordered in columns, e.g., as $[1\ n]$ vectors for any commmunication involving n signals. In this case, the parameters of interest are the armature voltage $v_{\text{a}}(k)$ (which is equivalent to $d(k)$) and the armature current read by the ADC (channel A0) $i_{\text{SCI}}(k)$. It is up to the reader decide to further include the computation of the average current $i_{\text{avg}}(k)$ with a Mean block.

In a locked rotor test, a $d(k) \neq 0$ implies $v_{\text{a}}(k) \neq 0$, which causes a positive current $i_{\text{a}}(k)$ flowing in the armature windings and a positive motor torque as well. Note that no back-emf is acting on the machine. Thus, V_{DC} should be much lower than the nominal voltage rating of the electrical machine not to exceed the rated current of the motor. Given a repetitive step-wise duty cycle at $d(k) = 0.2$, it follows $V_{\text{a}} = v_{\text{a}}(k) \approx 2\,\text{V}$, as shown in Figure 16.8.

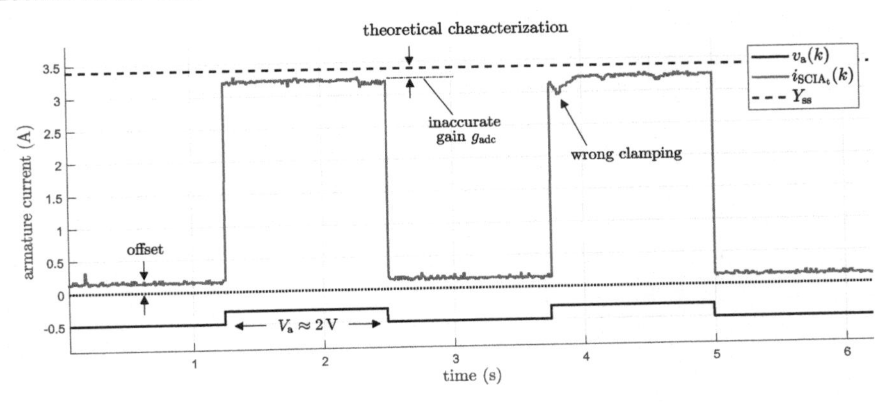

Figure 16.8 Comparison between the current measured through the low-side shunt resistor and the expected steady state values. Those current levels were evaluated thorough a current probe.

To specify that a theoretical characterization is used in this exercise, a subscript t is added. Figure 16.8 shows that $i_{SCI_t}(k)$ is affected by an offset in the zero value. Moreover, the gain g_{adc} is slightly inaccurate, since the steady-state value Y_{ss} is not reached (dash line) as well. These inconsistencies are due to the physical component tolerances of the measurement circuit, which influence the gain accuracy. Figure 16.8 also shows the effect of a wrong clamping, which can occur since the locking is done by hand.

To include a more accurate ADC conversion chain in the firmware environments, the gain g_{adc} and the offset must be corrected by performing an experimental characterization.

Remark1: by blocking the rotor, the mechanical power is entirely dissipated as Joule losses, increasing the temperature of the whole dc motor. If the temperature is too high, the armature resistance increases, which means $i_{SCI}(k)$ decreasing. Considering the previous example, this effect could be visible if the duty cycle is increased, e.g., $d(k) = 0.6$, thus, increasing the power to be dissipated.

Remark2: an higher Rate Transition sampling improves the visualization of the signals. However, if the number of samples is too large (i.e., a low sample time is set) the SCIA buffer might be saturated and the communication blocked. It can be noted that this saturation depends on how demanding is the executed control algorithm as well. If the firmware requires a high computational effort in terms of, e.g., processing time and memory usage, the sampling time must be lowered to keep the data exchange running.

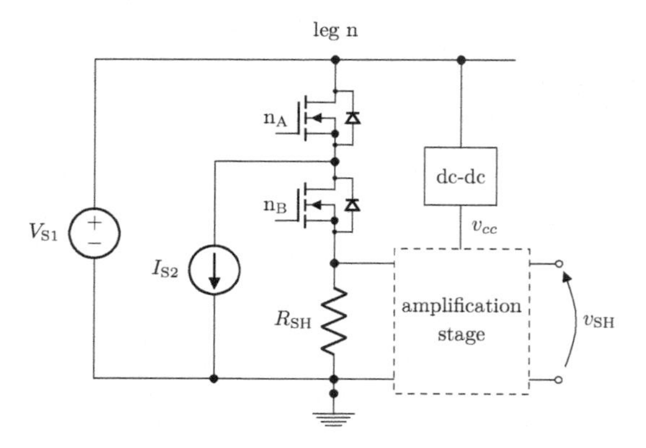

Figure 16.9 Equivalent circuit of the test setup for the characterization of the current sensor for the generic leg n, where n = 1, 2 or 3.

16.3 Sensor Characterization: Experimental Approach

The theoretical characterization presented in the previous section is based on datasheet parameters only. These latter are subjected to intrinsic tolerances, which may impact on the amplification circuit gain. In order to achieve accurate ADC conversions, an experimental characterization of the measurement channel is necessary. Such procedure is also helpful in many other cases, e.g., when the datasheet does not contain all the required parameters (e.g., R_{in}, R_{SH}) or when custom board are used. In addition, a proper offset at zero current should be identified as well. Focusing on the half-bridge configuration, a widely used procedure foresees a dc current injection into the low-side of a converter leg. This injection is done by using an external power supply which forces the current to flow from ground up to the output terminal of each leg through the free-wheeling diode of the low-side switches. In this case, the switches must be not operated, i.e., no gating signals must be provided. However, the dc-link has to be supplied with another power supply because the amplification stage circuit requires a voltage of 3.3 V, which is derived from the dc-link voltage. The amplification stage can not work without it. Figure 16.9 reports an equivalent circuit of the test setup.

Once the setup is correctly supplied, the ADC channel returns a value dependent to the current injected. Different current levels may be used to map the op-amp behavior. Since this is a forced test, the current flowing in the free-wheeling diodes should not be too high (they are not designed for this operation). For the sake of simplicity, only positive currents are tested, and a symmetrical behavior of the sensors is assumed.[1]

[1]This is not always the case and a characterization even for negative currents is recommended, especially for sensors operated with ac currents.

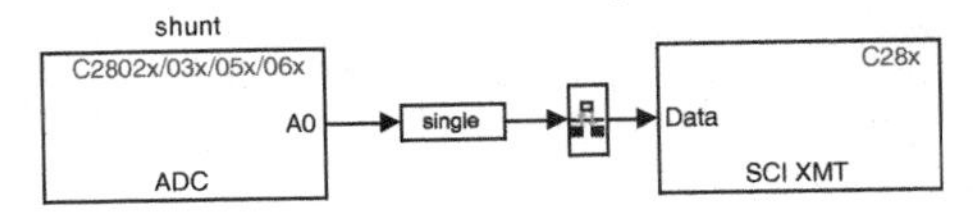

Figure 16.10 Simulink® scheme included in c28069_offset_F.slx.

Offset calibration: the computation of the offset value is relatively easy to perform. The idea is to send through the serial port the bit value corresponding to zero current. The settings for the firmware file are reported in the following.

Firmware Environment

c28069_offset_F.slx (solver: fixed step - ODE4, step size: $T_s = 50\,\mu s$)

Build a new firmware environment starting from the previous one:

- One **ADC** block is needed only. This time, no synchronzation with any ePWM module is required. Thus, set SOCx trigger source on CPU0_TINT0n. The input channels to be used are ADCINA0 for leg 1, ADCINB0 for leg 2 and ADCINA1 for leg 3. Here in the following, leg 1 only is characterized only.
- Keep a **data type conversion** block set to `single` (for data exchange).
- Add a **Rate Transition** block with sample time 0.01 s.
- Send the output of the ADCA0 (converted in `single`) directly to the **SCI Transmit** block.
- Delete or comment out the remaining blocks.

The connection scheme is reported in Figure 16.10.

Testing Environment

c28069_offset_T (solver: fixed step - ODE4, step size: $T_{\text{sim}} = 100\,\mu s$)

Build a new testing environment simplifying the previous one. Indeed, only one signal is sent trough SCI:

- Keep the **Serial Configuration** and the **Serial Receive** blocks. Change the data size back to [1 1].
- Use a **Scope** or a **Display** block to visualize the current reading translated in a bits.
- Delete or comment out the remaining blocks.

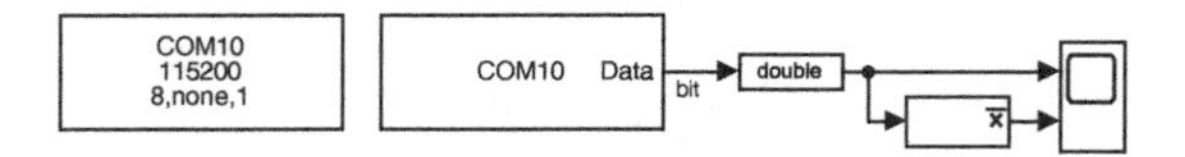

Figure 16.11 Simulink® scheme included in c28069_offset_T.slx.

- *Optional*: a **Mean** block could be added to filter the noise, which may be present in the bit reading. Use the settings presented in the previous exercise for this block.

Arrange these blocks as reported in Figure 16.11.

By feeding only the dc-link with a given voltage (e.g., $V_{DC} = 10\,V$) without connecting the second power supply (i.e., no forced current flow), the SCIA returns the bit value of v_{SH}^{bit} corresponding to zero current. Theoretically, this value should be 2048 for 12 bit ADC. However, due to the components tolerance, the resulting value is $v_{SH}^{bit} = 2067$ in this case. The latter value is strictly related to the hardware. Therefore, every reader performing such test will find (slightly) different results. The new bit value is used to compensate the bias highlighted in Figure 16.8.

Experimental Characterization of g_{adc}: this second experimental characterization relates the different recirculating current values (forced by the second power supply) to the bit values read through vCOM port. The settings for the **Firmware Environment** and the **Testing Environment** are the same as for the offset characterization.

Considering again leg 1 (phase A) only, the characteristic curve of the measurement channel is derived by injecting different currents and visualizing the corresponding bit values. The procedure is:

1. Connect the positive terminal of the power supply to the ground terminal of the DRV8301. Use another power supply to correctly supply the boosterpack dc-link, thus, the op-amp circuit. If the latter is not supplied, it does not return any bit value even in presence of a current.
2. Inject different current values $i_f(k)$ to understand which are the corresponding readings for $i(t) \longrightarrow v_{SH}(t) \longrightarrow v_{SO}^{bit}(k)$, i.e., in bits. The Table reported in Figure 16.12 shows all these readings.
3. The values shown in Figure 16.12 are assumed to be linearly distributed. Therefore, they are linearly interpolated in MATLAB® (e.g., through function polyval,) as shown in the same Figure. Keeping the same characterization even for negative current values, the resulting gradient of the characteristic is the experimental gain g_{adc}:

$$g_{adc} = 8.8\,\frac{\text{mA}}{\text{bit}} \neq 8.1\,\frac{\text{mA}}{\text{bit}} \tag{16.11}$$

which is slightly different with respect to that one resulting from the

current $i_f(k)$	ADC value $v_{SO}^{bit}(k)$
0 A	2067
240 mA	2097
495 mA	2122
800 mA	2161
1 A	2182

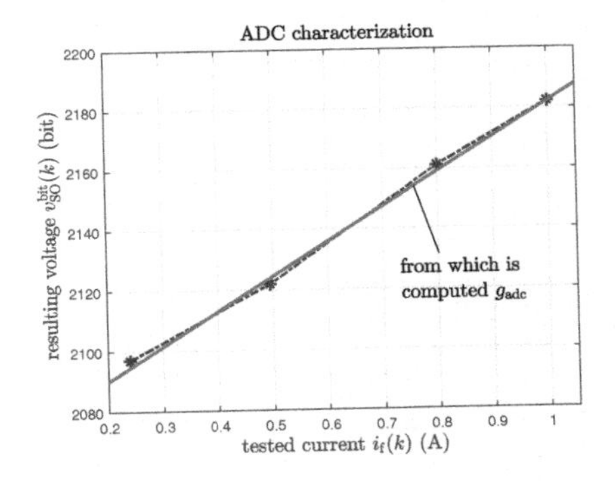

Figure 16.12 Characterization of the current measurement channel for leg 1.

theoretical characterization. It is worth noting that if more points can be investigate, the characterization procedure results to be more accurate.

In addition, it must be underlined that the same results can be achieved by using a firmware which already compensates the previously computed offset, as shown in Figure 16.13.

Repeating the locked-rotor test (i.e., coming back to the full firmware settings), the experimental characterization can be compared with the theoretical one, i.e., by applying $d(k) = 0.2$, thus, $V_a = v_a(k) \approx 2\,\text{V}$. The result is shown in Figure 16.14. The mismatches observed in Figure 16.8 are almost fully compensated. However, it must be noted that the proposed data interpolation is not accounting for possible non-linear behaviour of the amplification stages. Thus, the current measured through the ADC channel may overestimate the actual value. Since an experimental approach is considered, the current is now labeled as $i_{SCI_e}(k)$. This latter still has some thermal influence due to current flowing and it is sensitive to the rotor clamping, see Figure 16.14.

Validation: the experimental characterization seems to nicely approximate the steady-state value Y_{ss}. However, the whole measurement can be validate by directly comparing it with an exteranl measurement (a current probe

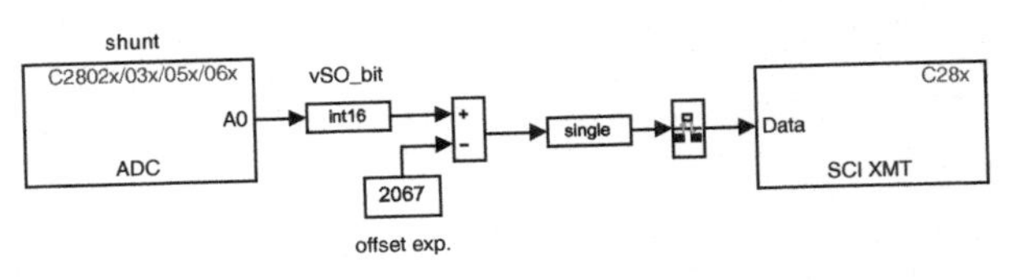

Figure 16.13 Firmware for ADC channel characterization as an alternative to c28069_offset_F.slx.

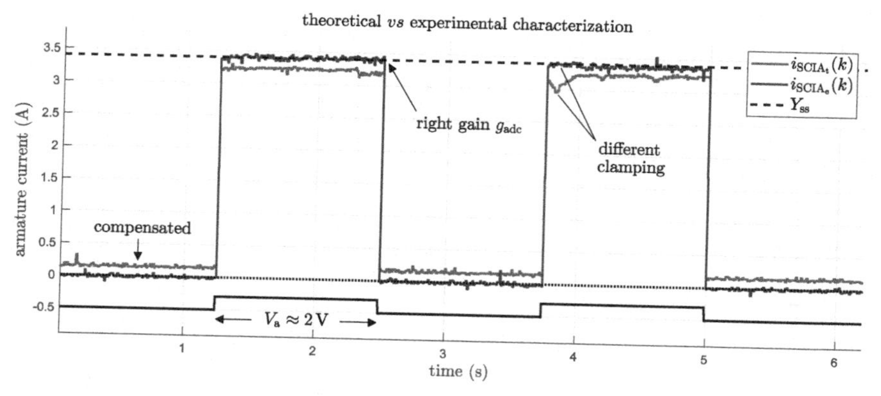

Figure 16.14 Locked-rotor test: comparison between theoretical and experimental characterizations. The same current is evaluated by the two approaches.

connected to an oscilloscope, for example), which has better accuracy. The results are reported in Figure 16.15. The $i_{SCI_e}(k)$ values are now superimposed with $i_{osc}(k)$ given by the oscilloscope, validating the use of the refined g_{adc}.

It is important to note that the serial communication as a debugging tool is satisfactory at steady-state only. However, due to the limited hardware resources for communication, it is not possible to visualize fast transients, which are typical of voltages and currents. In particular, serial communication is affected by computational/communication delay which translate in a discrepancy between the solid blue and red lines in Figure 16.15. Since $i(k)$ and

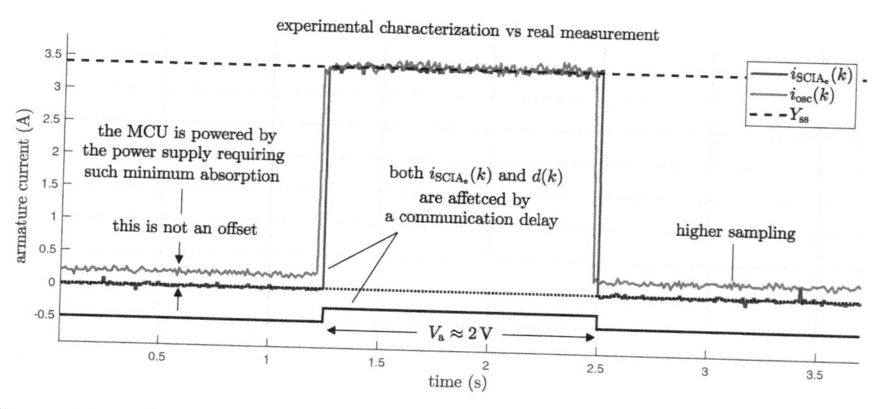

Figure 16.15 Comparison between the current waveforms measured through ADC channel (and displayed through SCI) and a current probe connected to an oscilloscope.

$d(k)$ are read via the SCIA, they are both affected by the same delay. This latter can be reduced by using a faster sampling rate (in this case 0.01 s should be reduced) until the MCU capabilities are reached. For instance, if a current probe is able to measure a transient with a classical exponential waveform, the serial communication tend to approximate it to a step-wise signal as a consequence of the lack of samples. Figure 16.15 also shows an offset in $i_{osc}(k)$. This is a consequence of the connection between the LaunchXL F28069M board and the BOOSXL - DRV8301 converter. Indeed, the MCU is powered by the power supply connected to the dc-link, requiring a minimum current absorption. It can be noted that the same locked-rotor test can be performed considering a H-bridge configuration.

Be careful: integer numbers are used to perform the sum aimed to remove the offset. Therefore, `int16` is chosen as ADC output data type to allow computations involving negative values. As seen previously, it is always possible to use (directly) `single` or `double` numbers to increase accuracy.

17

Current Control of an RL Load

After the explanation on how to manipulate measurements with the LaunchPad™ F28069M board, the following Sections show how to realize a closed-loop current control of a passive RL load. The on-board shunt resistors of the Boosterpack TI BOOSTXL-DRV8301 converter is used to measure the load current (read via ADC peripheral), and one or two converter legs can be involved to realize a Half-Bridge or a Full-Bridge configuration. In particular, this Section focuses on the controller implementation (PI-based) to meet the application requirements using a Half-Bridge configuration. Aiming to follow a given current reference $i^*(k)$, the resulting PI controller regulates the dc output voltage of the converter by manipulating the command signal(s) of the ePWM peripheral(s). This is a first and simple example of a closed-loop system implementation, which is equivalent (in practice) to the current control of a DC motor in a locked rotor condition (i.e. back-emf $e = 0$). The reference value of the output current can be provided by an external command, which may come from the SCIA/B (i.e., serial communication), or by the potentiometers mounted on the extPot3 or internally generated (e.g., through a pulse generator/constant block).

The whole system is simulated and analyzed first through Simulink® simulations. Then, a firmware is designed and implemented based on these results. The controller design for this application follows the methodology presented in Chapter 6. Nevertheless, this Section focuses more on the converter stage and how it can operate. In particular, the extRL(C) expansion board adopted for the implementation, it is possible to vary the inductance value L. Thus, some experimental tests are proposed to evaluate the effectiveness/robustness of the designed controller.

17.1 Required Hardware

- TI LaunchPad™ Piccolo™ F28069M and BOOSTXL-DRV8301 boards;
- extPot3 custom board (see Appendix B for technical details);
- extRL(C) board in RL configuration (see Appendix B for technical details);
- Power supply set on 15 V, 2 A.

DOI: 10.1201/9781003196938-17

Figure 17.1 Test-bench including a TI BOOSTXL-DRV8301 converter mounted on the TI LaunchXL F28069M Piccolo™ board and an extRL(C) bard configured in *RL* mode (see connection for Half-Bridge configuration in Figure 17.10). This setup includes the extPot3 board connected to the LaunchXL F28069M Piccolo™. The reader is referenced to Appendix B for further details.

This hardware should be arranged as shown in Figure 17.1. Both Half-Bridge and Full-Bridge configurations can be used as converter stage topologies. The corresponding equivalent circuits are shown in Figure 17.2 (a) and (b), respectively. The DC link voltage is provided by a power supply set to $V_{\mathrm{DC}} = 15\,\mathrm{V}$, with the output current limited to 2 A for safety reason. It can be noted that the extRL(C) board is designed to sustain up to 4 A currents. Different configurations can be set on this board by simply moving the on-board switches, as shown in Appendix B. This operation physically connects or disconnects different components on the printed circuit board. The RL configuration defines a circuit with resistor $R = 6.8\,\Omega$ and inductor $L = 860\,\mu\mathrm{H}$. A further resistance might be included to model the series-resistance of the inductor, which is $R_\ell = 150\,\mathrm{m}\Omega$ based on the manufacturer data-sheet, resulting in an equivalent resistive parameter equal to $R_{\mathrm{eq}} = R + R_\ell = 6.95\,\Omega$. The two circuits shown in Figure 17.2 differ from those reported in Chapter 15 for the absence of the back emf. Moreover, the relationship among duty cycle $d(k)$, output voltage of the converter $v_{\mathrm{A}}(k)$ (and $v_{\mathrm{B}}(k)$) and its averaged value are discussed in details in this Chapter, since it is a point of paramount importance in the design of the MCU firmware. In the following, the analytical model of the *RL* system is analyzed and a PI controller is designed based on it. Then, some simulations of the controlled system are reported for the two arrangements of the power electronics. Finally, experimental tests are carried out once the simulations returns satisfactory results.

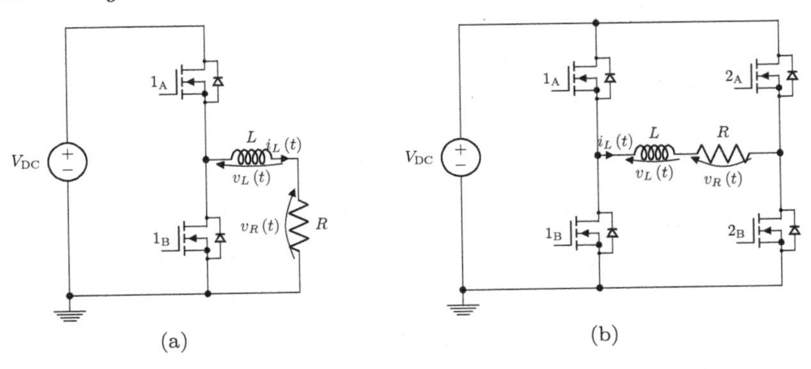

Figure 17.2 Equivalent circuits of the test setups involving the RL load driven by A DC/DC converter in Half-Bridge (a) and Full-Bridge (b) configurations.

17.2 Linear Average Model and Controller Design

Chapter 6 already presented the model derivation of a simple RL load circuit fed by a controllable voltage source (i.e., 1st order linear system) and a controller design approach. In particular, a closed-loop control regulates the current flowing into the winding $i(t) = i_L(t)$ by manipulating the voltage $v(t)$ applied on the RL load, i.e., $v(t) = v_L(t)+v_R(t)$. This latter is the controllable variable $u(t) = v(t)$. It is important to note that for a Half-Bridge configuration $v(t) = v_A(t)$, while for a Full-Bridge one $v(t) = v_A(t) - v_B(t)$. Besides the description provided in Chapter 6, the circuit physics can be analyzed more in detail during transients. In case a step-wise voltage is applied on the circuit, the current $i(t)$ cannot vary instantaneously, but it changes based on an exponential law. Applying the Kirchhoff's voltage law, it follows that:

$$v(t) - Ri(t) - L\frac{di_L(t)}{dt} = 0 \tag{17.1}$$

where $v(t)$ is equal to $v(t) = V_{DC}$ or $v(t) = 0$ depending on the current switching status of the converter. This non-linear behavior can be graphically represented by the two simplified, equivalent circuits reported in Figure 17.3 (a) and (b), where the switch is an equivalent representation of the whole converter. Solving the first order differential equation reported in (17.1) for the two scenarios, the obtained solution which represents the current transient going on in the circuit is:

$$i_L(t) = (i_{L0} - i_{L\infty})\, e^{-\frac{t}{\tau}} + i_{L\infty} \tag{17.2}$$

where the constant value i_{L0} refers to initial condition of the current for the considered equivalent circuit (i.e., the steady state value before the operation of the switch $u(t-1)$), while $i_{L\infty}$ is the steady state value resulting from the

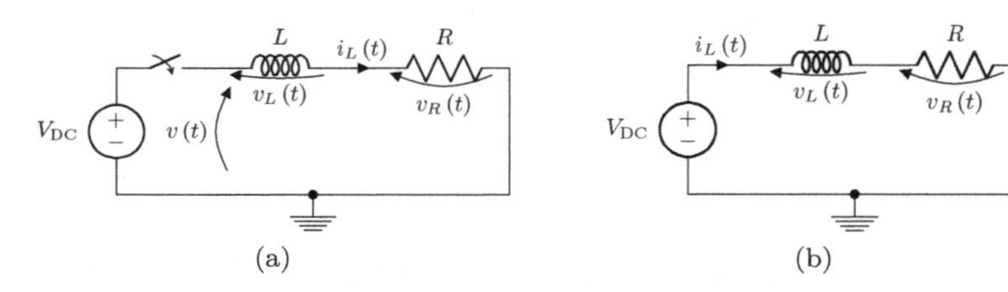

Figure 17.3 Equivalent circuits used for studying the transients caused by a switching operation in an *RL* circuit. The switches can cause (a) zero or (b) full voltage on the load.

application of $u(t)$ and ideally kept over an infinitely long time interval. This latter equals Y_{ss} defined in Chapter 6. Note that, in open-loop situation, $i_{L\infty}$ (or Y_{ss}) is reached after a time interval which is dependent on the natural dynamics of the circuit, that is the time constant $\tau = L/R$. For instance, for a step-wise supply voltage, e.g., varying from 0 up to V_{DC}, the evolution of the current in time domain can be described by settling a zero initial condition $i_{L0} = 0$ and $i_{L\infty} = V_{DC}/R$, which are the solutions of the circuits reported in Figure 17.3. These results lead to:

$$i_L(t) = -\frac{V_{DC}}{R}e^{-t\frac{R}{L}} + \frac{V_{DC}}{R} = \frac{V_{DC}}{R}\left(1 - e^{-\frac{t}{\tau}}\right) \tag{17.3}$$

Considering $x(t) = i(t)$ and $y(t) = x(t)$, both a state-space representation and system transfer functions can be derived based on the solution proposed in Chapter 6. Then, a PI controller $R(s) = k_p + k_i/s$ is used to close the loop and designed with pole/zero cancellation method. Here in the following the system $G(s)$ and open loop $L(s)$ transfer functions are simply recalled:

$$G(s) = \frac{i(t)}{v(t)} = \frac{y(s)}{u(s)} = \frac{1/R}{1 + s\tau_G} \qquad L(s) = \frac{i(t)}{e(t)} = R(s)G(s) = \frac{k_p/R}{sT_i} \tag{17.4}$$

where $L(s)$ is computed by setting $T_i = k_i/k_p = L/R$. Alternatively, $G(s)$ can be computed considering the equivalent impedance $Z(s) = R + sL$.

As previously mentioned, the modeling can consider the parasitic resistance of the inductor, which is summed up to the load resistor. An equivalent resistance $R_{eq} = 6.95\,\Omega$ is obtained. The uncontrolled system is characterized by a small time constant $\tau_G = L/R = 124\,\mu s$, which leads to a settling time of $T_{a,G} = 619\,\mu s$. For a better visualization of the result, the designed controller is slowing down the natural response by choosing $T_{a,F} = 2.67 \times T_{a,G} = 1.7\,ms$. This choice implies $\tau_F = T_{a,F}/5 = 333\,\mu s$ and a controller bandwidth equal to $\omega_c = 1/\tau_F = 3000\,rad/s$. The values of k_p and k_i are explicitly determined as:

$$k_p = \omega_c L = 2.58\,\Omega \tag{17.5}$$
$$k_i = \omega_c R = 20\,850\,\Omega/s \tag{17.6}$$

The dc voltage $V_{DC} = 15\,V$ provided by the power supply is considered an upper bound for the regulator, thus, the control variable $u(t)$ is saturated to this value. Therefore, an anti-windup strategy must be included in the PID controller block. Based on Chapter 6, a back-calculation approach is used with $k_{aw} = k_i/k_p = 8081\,s^{-1}$.

The discrete-time PI controller $R(z)$ is derived based on the continuous-time one and it is integrated in the closed-loop scheme together with a Rate Transition block in simulation to mimic the discretization process. The sampling time of the controller is $T_s = T_{sw} = 1/f_{sw}$, where the switching frequency is $f_{sw} = 10\,kHz$. Backward Euler is chosen as integration method.

17.3 System Simulations

The closed-loop control of the RL load is simulated in Simulink® to analyze its design and check the resulting performance of the regulator. This step is carried out for both converter configurations. To this purpose, the power electronics and the *RL* load are modeled through Simscape™ Electrical elements, referring to Specialized Power Systems ™ library (*Simscape/Electrical/Specialized Power Systems/*) as shown in Chapter 6.

In particular, the **Mosfet** blocks available in the subset */Fundamental Blocks/Power Electronics* are used to represent the switch mounted on the BOOSTXL-DRV8301 converter board. The semiconductor characteristic taken from the converter datasheet, e.g., R_{on} and R_d, are set into the Simulink® blocks as shown in Chapter 15. The two-level Half-Bridge configuration is obtained by rearranging these block based on the equivalent circuits reported in Figure 17.2 (a). The switches are driven by a PWM modulator having a carrier frequency of 10 kHz, i.e., $f_{sw} = 10\,kHz$. The load is modeled through a series **RLC branch** block set as *RL branch type* with $L = 860\,\mu H$ and $R_{eq} = R = 6.95\,\Omega$.

17.3.1 Detailed modeling of the actuation variables

Chapter 6 details how to design a closed-loop current control for a generic *RL* load, where the control variable $u(t) = v(t)$, that is, a dc voltage, is applied to a Transfer Fcn block which emulates the load. This operation is equivalent to a direct application of $u(t)$ (or, equivalently $u(k)$) on the *RL* terminals. This approach is a high-level analysis which considers an *averaged model* of the system, i.e., without including the converter stage or modeling it in a third nested loop operating at high bandwidth. As already said, the converter represents a non linear device, since it operates in discontinuous (switching) mode. Since linear controllers are used, the system model has to be linear too, thus, not including non-linear elements. Therefore, by using average models, the controller design can be based on standard approaches from linear control

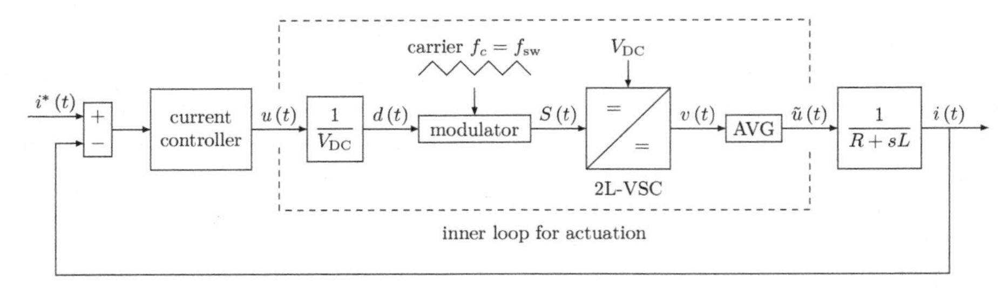

Figure 17.4 Closed-loop current control scheme including the inner loop for the converter actuation and its relationship with the average modeling.

theory,[1] leading to compute key parameters and figures of merit, e.g., k_p and k_i. Even if the controller parameters are necessary, the firmware architecture aimed to close a loop has to cope with the driving elements of the converter stage (modulator, scaling and protections) and the sensing stage as well (including measurements and sampling). Getting closer to the system physical behavior, Chapter 13 and 15 show how to include the converter topology and the ePWM peripheral working principle (which generates the switch positions) into the simulation environment. In particular, the actuation variable is the modulating signal $d(t)$, which is directly a duty cycle when referring to dc quantities.[2] Even though the controller design should be based on a linear average model, the controller output $u(t)$ must be related to $d(t)$ and the switch positions $S(t)$. This is detailed in the scheme reported in Figure 17.4. Based on the controller design and voltage limits (i.e., $u_{\max} = V_{\mathrm{DC}}$), the manipulated variable for a 2L-VSC is defined within $u(t) \in [0, V_{\mathrm{DC}}]$. This last interval divided by V_{DC} becomes the duty cycle range of variation $d(t) \in [0, 1]$. This is the input of the modulator operating at carrier frequency $f_c = f_{\mathrm{sw}}$ for each converter leg, which operates according to the switching pattern $S(t) \in [0, 1]$ (see Section 13.1). Therefore, $S(t)$ defines how long the dc-link voltage V_{DC} is applied to the load inside the time interval T_{sw}. The voltage $v(t)$ is then a squared waveform approximated by $v(t) = S(t)V_{\mathrm{DC}}$. The average value of $v(t)$ over T_{sw} is $\widehat{u}(t)$, which is $u(t) \approx \widehat{u}(t)$, see Figure 17.4. Such approximation becomes more and more accurate as f_{sw} increases. In practice, the current $i(t)$ is affected by a ripple which is influenced by f_{sw} and the resulting closed-loop dynamic. However, the ripple should not be included (or at least limited) in the feedback path used to compute the current error, otherwise the controller is not able to reach a zero error at steady-state, i.e. a proper command-tracking

[1] *Linear control theory* deals with systems governed by linear differential equations. These systems can be analyzed by powerful frequency-domain mathematical techniques of great generality, such as Laplace transform, Fourier transform, Z-transform, Bode plot, root locus, and Nyquist stability criterion. These tools lead to a system description using terms like bandwidth, frequency response, eigenvalues, gains, resonant frequencies, zeros and poles, and to provide simple/explicit design solutions to control the system response. An important subset of systems like these is the linear time invariant (LTI) ones.

[2] This is not the case for ac quantities.

must be ensured.[3] Indeed, the measurements sensing should be synchronized with the ePWM peripheral as discussed in Chapter 13.

The whole approach can be easily extended to discrete-time domain and represented by a Simulink® scheme. It can be noted that the models based on the Simscape™ Electrical elements were already including these features in Chapter 6. In particular, the use of the Controlled Voltage Source block was mimicking the inner actuation loop.

17.4 Half-Bridge Configuration

As shown in Chapter 15, a Half-Bridge configuration refers to a converter which uses a single-leg (or single-phase) of the boosterpack, requiring one ePWM module only, e.g., ePWM1. Hence, the switch 1_A is driven by the signal $S_1(k)$ and 1_B by $S_2(k)$, operating in a complementary way. The RL load is connected between the central point of leg A and ground. The *average* output voltage of the converter leg is function of the duty cycle $d(k)$ (or $d_\mathrm{A}(k)$):

$$V(k) = v_\mathrm{avg}(k) = d(k)V_\mathrm{DC} \tag{17.7}$$

where the variables are already considered in the discrete-time domain k, i.e., moving the simulation toward the implementation. The voltage V_DC is set by the power supply (i.e., it is the voltage on the DC bus) while $d(k) \in [0, 1]$.

As already mentioned, the system is characterized by $R = 6.95\,\Omega$, $L = 860\,\mu\mathrm{H}$. The supply voltage is $V_\mathrm{DC} = 15\,\mathrm{V}$ and the switching frequency is $f_\mathrm{sw} = 10\,\mathrm{kHz}$. These parameters can be initialized in *Model Properites/Callbacks/InitFcn* or in a separate `m`-file as shown here in the following:

```
%% power supply
Vdc     = 15;
%% carrier frequency definition
fsw     = 10e3;
Tsw     = 1/fsw;
%% sampling time definition
Ts      = Tsw;
%% step-size definition
Tsim    = Tsw/400;
%% load parameters
R       = 6.95;
L       = 860e-6;
```

[3]This is another fundamental principle in linear control theory.

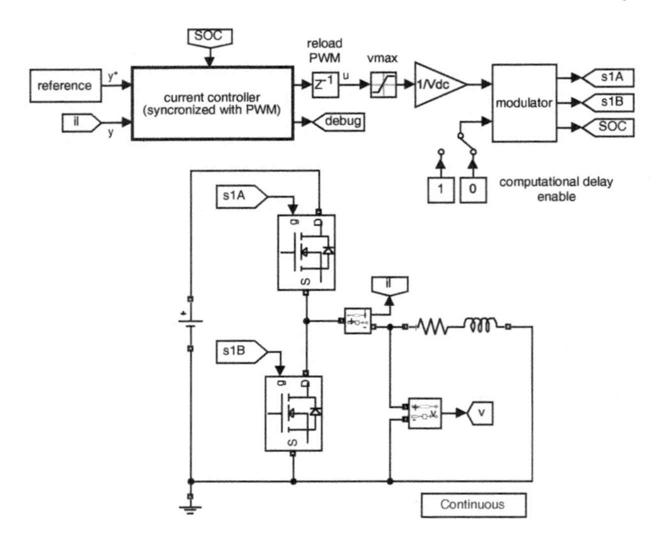

Figure 17.5 Simulink® scheme to simulate the closed-loop current control of an RL load driven in half-bridge configuration.

```
%% controller design
tauG   = L/R;
TaG    = 5*tauG;
wc     = 3000;
TaF    = 5/wc;
kp     = wc*L;
ki     = wc*R;
```

The equivalent circuit shown in Figure 17.2 (a) is then created in Simulink® shown in Figure 17.5. This scheme is included in the file:

HBsimRL_closed.slx (solver: fixed step - ODE4, step size: T_{sim})
(powergui: simulation type *Continuous*)

The simulation time step T_{sim} is particularly small to well approximate the switching behavior of current $i(k)$ and voltage $v(k)$. The reference subsystem includes different kinds of input signals to operate the system in several working conditions.

Considering the PWM stage, the duty cycle $d(k)$ is compared to a triangular carrier at $f_c = 10\,\text{kHz}$, generating a switching pattern $S(k)$. This is realized on the MCU by using the ePWM modules. Since the gate signals of the 2L-VSC topology are $S(k) = S_1(k)$ for 1_A and $S_2(k) = \text{not}(S_1(k))$ for 1_B, the instantaneous output voltage $v(k)$ is a pulsed waveform that can be computes as:

$$v_\text{a}(k) = S(k)V_\text{DC} \tag{17.8}$$

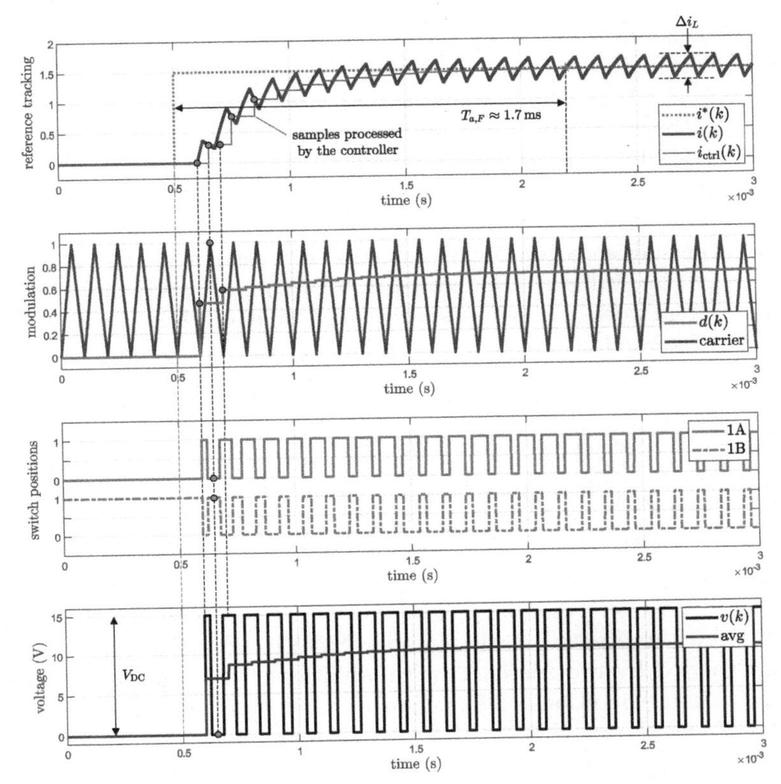

Figure 17.6 Closed-loop dynamics considering a step in the current reference (Half-Bridge configuration).

which is discontinuous and bounded within two voltage values, i.e., 0 and V_{DC}. The aim of this exercise is to design a *current control loop*, i.e., to control the current $i(k)$ flowing into the load based on a step-wise reference $i^*(k)$ by using a PI controller while achieving a settling time of $T_{a,F} = 1.7$ ms (see Section 17.2). To check these performances, a step-wise current reference $i^*(k)$ ranging from 0 A to 1.5 A (i.e., $Y_{\text{ss}} = 1.5$) at 0.5 ms is applied.

Figure 17.6 shows the current flowing into the load $y(k) = i(k)$ (blue line), which is correctly following the step-wise reference $y^*(k) = i^*(k)$ (red dotted line). The current dynamic follows an exponential behavior as described by Equation (17.3). Based on the design guidelines reported in Chapter 5, a $\epsilon \approx 1\%$ is considered as reference accuracy. Thus, the transient is completed once the current enters in the region $[(1 - 0.01\epsilon)Y_{\text{ss}}\ (1 + 0.01\epsilon)Y_{\text{ss}}]$, where $T_a \approx 5\tau$ holds. Indeed, it results that $T_{a,F} = 1.7$ ms, so the closed-loop control system is behaving as expected. The steady-state value Y_{ss} can be also be predicted thanks to the final value theorem (see Section 6.1.3).

The effects of the step variation in the reference are reflected into the switching patterns of 1_A and 1_B, which are aimed to make the average voltage $v_{avg}(k)$ increase by enlarging the pulse width of $v(k)$. Moreover, the switching behavior implies a current ripple equal to $\Delta i_L \approx 359\,\text{mA}$ at steady-state.

Based on Figure 17.6, the command-tracking performances are achieved thanks to the sudden regulation of the converter output voltage $u(k) = v(k)$, where its average value is approximated by $d(k)$. At steady-state, $u(k) \neq 0$ since the system works in a new operating point and a certain voltage is needed to keep the required current flow. This value can be computed thanks to the Kirchhoff's voltage law, where at steady-state the current derivative can be neglected. Namely, assuming $d(k) = D_{ss}$ at steady state, it follows:

$$v(t) - Ri(t) - L\cancel{\frac{\mathrm{d}i(t)}{\mathrm{d}t}} = 0 \quad \rightarrow \quad V_{ss} = D_{ss}V_{DC} = RI_{ss} \quad \rightarrow \quad D_{ss} = \frac{RI_{ss}}{V_{DC}} \tag{17.9}$$

which results in $D_{ss} = 0.7$. This value is reached approximately within the settling time $T_{a,F}$.

Focusing now on power supplies, they are limited by upper and lower bounds, which represent static non-linearity. Hence, the control variable $u(k)$ is bounded by the energy flow capabilities of a power supply. For this case study, a unidirectional DC source is considered, which works between $u_{min} = 0\,\text{V}$ and $u_{max} = V_{DC} = 15\,\text{V}$. An anti-windup PI controller based on the back-calculation method is considered. The coefficient k_{aw} is computed as $k_{aw} = 1/T_i = k_i/k_p$ to define how quickly the integrator of the PI controller is reset by the anti-windup loop. As reported in Section 6.2, the anti-windup parameters should be defined inside the PID block. In particular, Limit output must be selected and the saturation limits u_{min} and u_{max} have to be specified. Then, select back-calculation in the Anti-windup method menu and specify the Back-calculation coefficient Kb equal to k_{aw}. The verification of the settling time is just the first step of a closed-loop system evaluation. Simulations should provide an useful insight into the system operations, including as many implementation aspects as possible. The synchronization between ADC and ePWM modules to obtain an *average* current measurement is an example of how important it is to choose the specific sampling time to avoid issues related to peripheral and task scheduling early during the design stage. This directly reflects on the calculation and actuation updates that the controller should provide, i.e., *integrate MCU scheduling and peripherals in power electronics-based applications*, as discussed in Section 13.10. To this purpose, an embedded closed-loop control scheme (with MCU target) involving current reading (ADC module), 2L-PWM acutation (ePWM module) and a PI-based controller which use the ADC reading to compute the duty cycle for the ePWM is subjected to a *sequential* events execution.

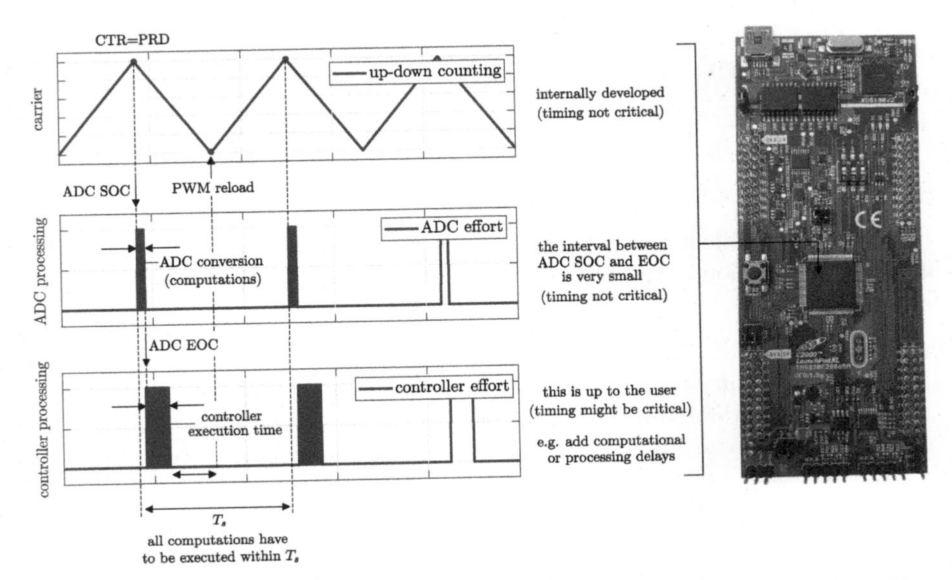

Figure 17.7 Example of computations to be performed within the sampling period T_s.

The current controller is scheduled to be executed within the sampling period $T_s = T_{sw} = 1/f_{sw}$. Therefore, in the target hardware, the controller routines are synchronized with the ADC interrupt every T_s. Figure 17.7 shows the event sequence, interrupt trigger, and software execution time for the control algorithm running in the target hardware. The scheduled event sequence is:

- At the beginning of T_s, when the value of the PWM counter equals the PWM counter period (TBCR=TBPRD, that simply equals 1 in Simulink® simulations), the ePWM peripheral, which is center-aligned in this case (up-down counting mode), triggers the start-of-conversion (SOC) event for the ADC module;
- The ADC module converts the sampled analog signal into digital counts and triggers the end-of-conversion (EOC) event;
- The EOC triggers the ADC interrupt for the controller, which reads the phase current value;
- Given the PI structure, the controller processes the current error to compute the modulating signal for the ePWM module. The number as well as the complexity of these mathematical operations define the execution window length require by the controller to have a data ready. This time interval is expected to be sufficiently lower than T_s;

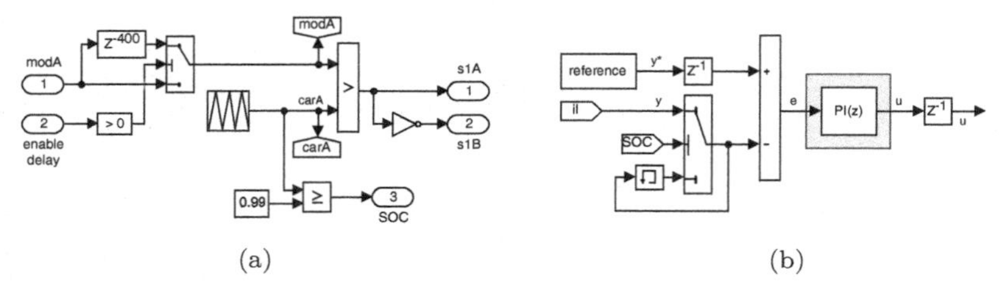

Figure 17.8 Detailed view of (a) the modulator subsystem and (b) the equivalent representation of the triggered subsystem for the current controller (see Figure 17.5).

- The resulting duty cycle is held until the end of the sampling interval and it is sent to the adopted ePWM module. Therefore, even if the updated duty cycle value is ready before the end of T_s, the actuation happens only at the reload of the ePWM peripheral. During the remaining time in-between, the MCU is assigned to other tasks (e.g., communication) or enters in sleep mode.

Therefore, the controller adopted in the Simulink® simulation is expected to be consistent with such event sequence. To this end, the scheme reported in Figure 17.5 includes a modulator subsystem, in which a trigger is generated every time the carrier equals 1. Namely, a simulated `SOC` signal is created. The implementation is achieved by using a Relational Operator block which returns one when the carrier is greater than 0.99 or zero otherwise, i.e., a boolean squarewave is obtained. A detailed scheme of this subsystem is shown in Figure 17.8 (a). The `SOC` signal is used as trigger source for a triggered subsystem[4] which realizes a current controller synchronized with the carrier peaks.

Assuming to adopt the previously mentioned step-wise current reference $i^*(k)$, both a Pulse Generator block (sample based, with gain $A = 1.5$, number of samples $n_s = 120$, pulse width 50 %, phase delay = 5, and sample time T_s) and a square-wave of period $120T_s = 12\,\text{ms}$ can be used. Focusing on the first rising edge after the step shown in the top plot of Figure 17.9, the current reference varies at the zero value of the triangular carrier. Since the ADC SOC trigger is generated at the carrier peak, $i^*(k)$ is delayed by $T_s/2$, thus, generating a current error at the controller input. This delay can be included through a Delay block, where the sample time is set equal to $T_s/2$ and the delay length is 1. The resulting reference current is called $i^*_{\text{ctrl}}(k)$. This is detailed in the second plot of Figure 17.9 . Even if $i(k)$ shows a superimposed ripple,

[4]A triggered subsystem is a conditionally executed atomic subsystem that runs each time the trigger signal assume one of the selected triggering condition, i.e., rising, falling, rising-falling trigger type. For more information see `https://www.mathworks.com/help/simulink/ug/triggered-subsystems.html`.

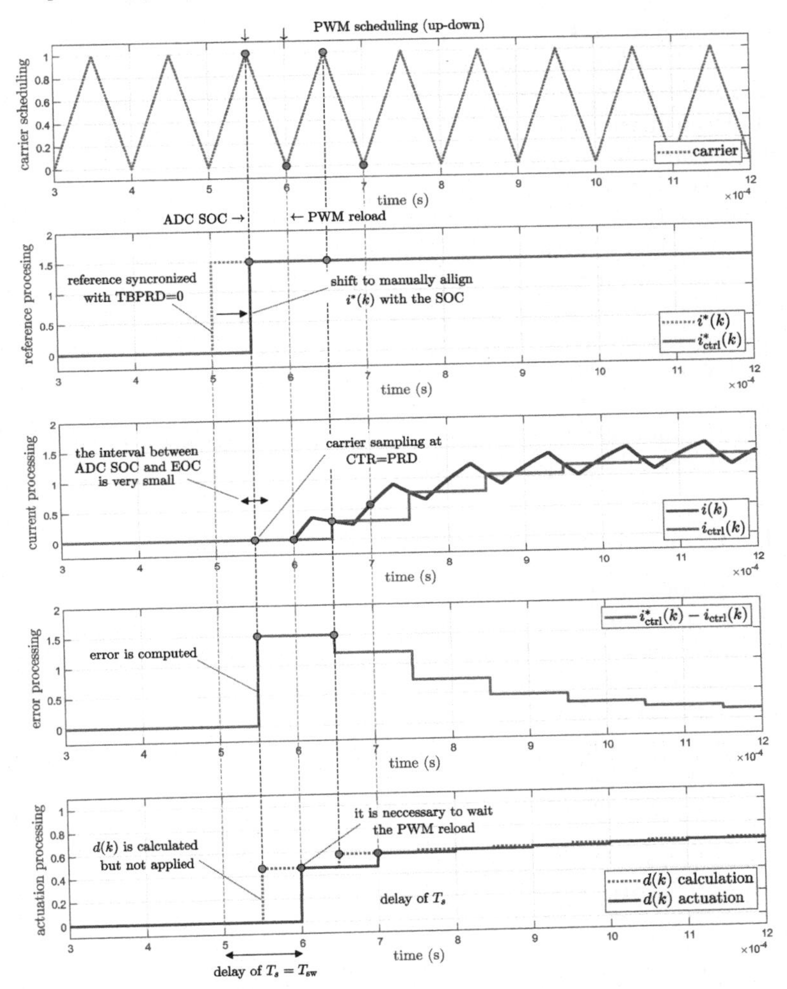

Figure 17.9 Example of MCU scheduling and latency between sensing and actuation.

the center-aligned carrier ensures an ADC SOC which allows to measure the average value of the current, which is hold until the next SOC, i.e., for the hwole sampling period T_s. The resulting measured current is called $i_{\text{ctrl}}(k)$. The subscript ctrl refers to the quantity that are processed by the controller, which is expected to read the current error $i_{\text{ctrl,err}}(k) = i^*_{\text{ctrl}}(k) - i_{\text{ctrl}}(k)$ and compute the resulting duty cycle $d(k)$. It can be noted that, the controller starts processing $i_{\text{ctrl,err}}(k)$ and the related computations at the ADC EOC. The time interval between ADC SOC and EOC is small but not zero, i.e., both $i_{\mathrm{ctrl,err}}(k)$ and $d(k)$ data values are not evaluated strictly simultaneously with the carrier peak (at TBPRD). In Figure 17.9, the time interval between SOC and EOC is not visible due to the limited figure resolution, but it is still present. Even if the controller computations are relatively easy to perform, providing a value for $d(k)$ before the end of the switching period $T_s = T_{\mathrm{sw}}$, the controller has to wait the next reload PWM action to feed the resulting duty cycle value at the ePWM input. This happens when CTR=zero. Therefore, another Delay block, with sample time $T_s/2$ and data length 1, is included at the controller output to synchronize the actuation with the zero value of the carrier. This event sequence is fulfilled by the implementation of the current controller as a triggered subsystem. To understand better its behavior, the involved blocks are showed in Figure 17.8 (b), which is used to analyze the controller scheduling only at simulation level.

Given the aforementioned parameters and performances, the designed PI controller meets all the specifications.

17.4.1 Control implementation

The closed-loop dynamics are tested on a test-bench which includes a TI BOOSTXL-DRV8301 converter mounted on the TI LaunchXL F28069M Piccolo™ board and connected to the extRL(C) board in *RL* mode (see Figure 17.1). The extPot3 board is also used to manipulate analog inputs. In Half-Bridge configuration, only leg (phase) A is used, as shown in the connection scheme reported in Figure 17.10. Therefore, one ePMW module is used only and it is operated in up-down counting mode at $f_{\mathrm{sw}} = \mathrm{PWM}_{\mathrm{frequency}} = 10\,\mathrm{kHz}$. Reminding that for the F28069M board $f_{\mathrm{ck}} = \mathrm{CPU}_{\mathrm{frequency}} = 90\,\mathrm{MHz}$, it follows:

$$\mathrm{TBPRD} = \mathrm{PWM}_{\mathrm{counter-period}} = \frac{\mathrm{CPU}_{\mathrm{frequency}}}{2 \cdot \mathrm{PWM}_{\mathrm{frequency}}} = 4500 \qquad (17.10)$$

where the sampling time is $T_s = T_{\mathrm{sw}} = 1/\mathrm{PWM}_{\mathrm{frequency}}$.

The reference current can be internally generated in the firmware or externally provided either using the available potentiometers from extPot3 board or through COM port (e.g., SCIA). The analog solution is adopted to add versatility to the control scheme, avoiding the slow communication issue through serial communication. The needed parameters can be initialized in *Model Properites/Callbacks/InitFcn* or in a separate `m`-file as done for simulation.

A step-by-step procedure for programming the microcontroller for this task is reported here in the following.

Firmware Environment

c28069_RLcontrol_hbF.slx (solver: fixed step - ODE4, step size: $T_s = T_{\mathrm{sw}}/2 = 50\,\mu\mathrm{s}$)

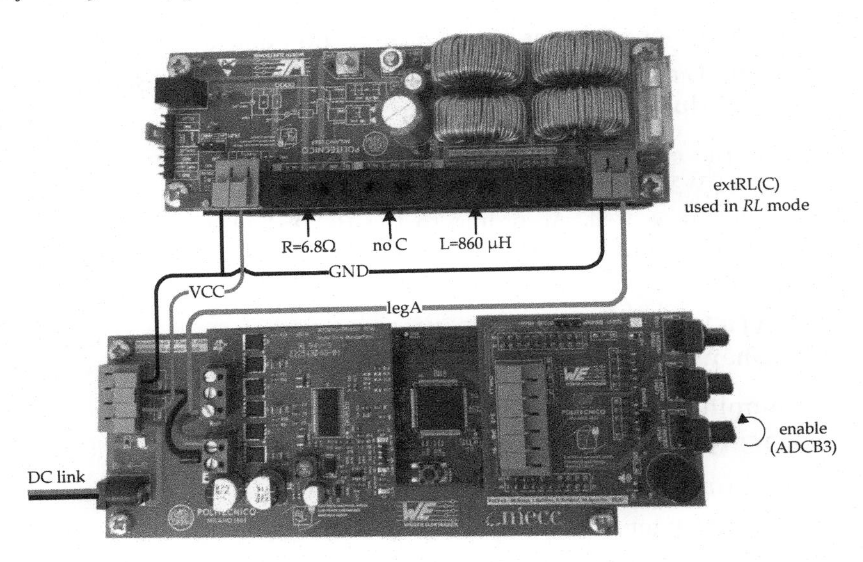

Figure 17.10 Connections of the extRL(C) board with the BOOSTXL-DRV8301 converter realizing a Half-Bridge configuration.

Open a new blank Simulink® project and configure the environment as shown in Section 9. Remember that the BOOSTXL-DRV8301 is enabled by settling **GPIO50** high (see Section 3.2).

- Insert an **ePWM** block and use the same settings reported in Section 13.5:
  - Select **ePWM1** module with TBPRD related to $f_c = 10\,\text{kHz}$, i.e., $\text{PWM}_{\text{counter-period}} = 4500$;
  - Remember to unflag Enable ePWM1B and tick Inverted version of ePWMxA. This option allows to drive 1_B as the complement of 1_A;
  - Keep all the other settings as reported in Section 13.5.
- Add a **gain** block to normalize the modulation signal over the peak value of the carrier signal $\text{PWM}_{\text{counter-period}}$. Include a **data type conversion** set on `int16` to represent the duty cycle within a 16-bit base.

For up-down counting mode, the measurement of an average current $i(k)$ is feasible thanks to the ePWM module which sends a synchronization signal that triggers the SOC of an ADC module when the carrier equals TBCTR=TBPRD, as shown previously. To this aim, the following settings are summarized here below:

- In the **Event Trigger** (Tab) of ePWM1
  - Select **Enable ADC start of conversion for module A**;

  - Set **Number of event for SOCA to be generated** to First event;
  - Select Counter equals to period (CTR=PRD) as **Start of conversion for module A event selection.**

Every time the carrier signal reaches the maximum value TBPRD, a SOC input (i.e., ePWM1_ADCSOCA) is sent to the ADC. This last block has to be configured as well. According to the BOOSTXL-DRV8301 datasheet (see Section 3.2) and Chapter 16, the shunt-resistor of leg A is connected to ADCA0 channel:

- Add a **ADC** block and select ADCINA0 as **Conversion channel** in the **Input Channels** tab. Then, go back in the **SOC Trigger** tab;
  - **Sampling mode**: Single sampling mode;
  - **SOC trigger number**: SOC0;
  - **SOCx acquisition window**: 7;
  - **SOCx trigger source**: ePWM1_ADCSOCA;
  - **ADC will trigger SOCx**: No ADCINT;
  - **Sample time**: T_s;
  - **Data type**: uint16;
  - Flag **Post interrupt at EOC trigger** (optional);
  - **Interrupt selection**: ADCINT1 (optional);
  - Flag **ADCINT1 continous mode** (optional).
- Add a **Data Type Conversion** block set to int16 (since positive signal only are used) or `single` to exploit 32 bit.
- Place a **Gain** block for scaling the digitalized signal within $[0, 1]$ since a 12-bit reading returns values ranging from 0 (0 V) up to $2^{12} - 1 = 4095$ (3.3 V).
- Build a proper conversion chain for the current sensing by removing the estimated offset (e.g., 2064) as well as using the scaling factor g_{adc} to obtain Ampere values after moving to `single` data type. See the characterization procedure in Chapter 16.
- Use the same discrete **PID controller** block adopted in simulation; set the controller type as PI (Parallel) with sample time T_s; set P=k_p and I=k_i. Select an external source for the **reset** with **rising** condition. Saturate the controller output to keep the voltage in the range $[0 - V_{\text{DC}}]$. A more conservative choice might be to consider $0.98 \cdot V_{\text{DC}}$ as upper bound.[5] Set a back-calculation method as anti-windup strategy with Kb=k_{aw}.

[5]This choice implies to keep a sufficiently good current sensing, since the shunt-resistor measures the current only when $S_2(k) = 1$ only, that occurs at every switching period when $d(k) < 1$.

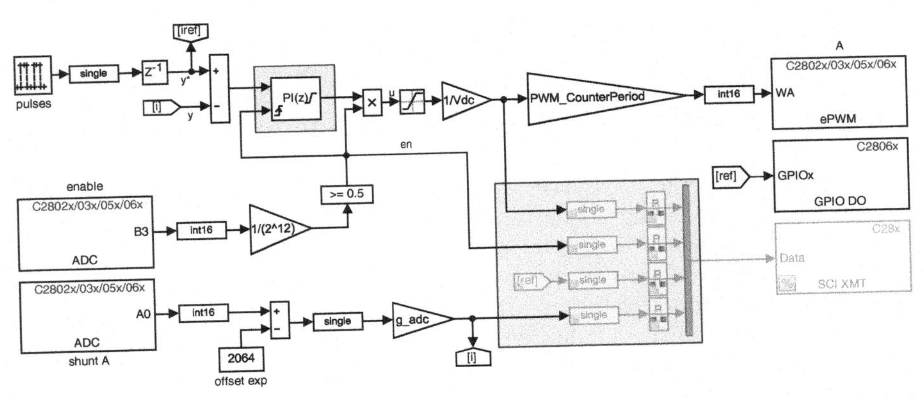

Figure 17.11 Simulink® scheme included in c28069_RLcontrol_hbF.slx.

- A redundant saturation block can be used to protect the system by limiting the required voltage.
- Add a **Gain** block to normalize the controller output (which varies in the $[0, V_{\mathrm{DC}}]$ range) with a scaling factor $1/V_{\mathrm{DC}}$. Then, the signal spans a $[0, 1]$ range, being consistent with the definition of $d(k)$.
- Add a **Pulse Generator** (sample based) block to create the reference $i^*(k)$. Set the amplitude equal to $A = 1.5$, number of samples $n_s = 120$, pulse width 50 %, phase delay 5, and sample time T_s. It is equivalent to have $i^*(k)$ generated by a square-wave of period $120T_s = 12\,\text{ms}$. It can be noted that $i^*(k)$ is synchronized with the zero value of the triangular carrier
- Add a **Delay** block to be consistent with the ADC SOC trigger generated at the carrier peak. Set sample time $T_s/2$ and delay length 1.

Arrange these aforementioned blocks as shown in Figure 17.11.

Resetting the Integrator

Every time the firmware is downloaded into the microcotnroller, the PI controller (as well as the the other functions/routines) is constantly executed. Since a closed-loop structure involves feedbacks which are processed by the peripherals and the conversion chains, the resulting measurement values are affected by noise, being nonzero. Therefore, even without a reference or with its value equal to zero, during the first seconds of firmware executions the PI controller tries to react providing a control action to change the system state. This latter is particularly important. Indeed, solvers in Simulink® sequentially computes/solves states and outputs from block-to-block. When a reference is applied but it is not synchronized with the controller action, the closed-loop system may see a state discontinuity. This lead to a break down of the computation process due to numerical conditions, which in this case means to feed the integral part (integrator) of the controller with an infinite value. So, even if the closed-loop system is well designed, this issue could make the control not operating, e.g., I/O freezing or even entering in fault mode (`nFault`) condition, causing the controller action to be not suitable for the load.

To avoid this issue, the PID controller block has an external reset that can be selected/activated in the Initialization tab, as described above. By enabling this feature, an additional input port is created which allows to specify the trigger condition that causes the block to reset the integrator to initial conditions, e.g., Source: internal - Integrator: 0. By default, the output of the integrator is set to initial conditions when the scheme/code is initialized, while it is not during the simulation/firmware execution. When the trigger occurs, i.e., sending a signal to the External reset port, the integral part of the PI controller is reset on a rising-, falling-edge, or level of such signal. The port icon indicates the selected trigger type. For this exercise, an external reset on the rising-edge is chosen.

In order to manually manage the activation of the PI controller, the external reset (rising) is connected to a potentiometer (ADCINB3/pin68). Without moving the latter, the control algorithm is forced not to provide any actuation signal. To do that:

- Add an **ADC** block and select ADCINB3 as **Conversion channel** in the **Input Channels** tab. Then, go back in the **SOC Trigger** tab:
  - **Sampling mode**: Single sampling mode;
  - **SOC trigger number**: SOC1;
  - **SOCx acqusition window**: 7;
  - **SOCx trigger source**: Software;
  - **ADC will trigger SOCx**: No ADCINT;
  - **Sample time**: $T_{sig} = 0.0011$,s;
  - **Data type**: uint16.
- Add a **Data Type Conversion** block set to int16 (since positive signal only are used) and the 12-bit scaling to move from a range (0 V) up to $2^{12} - 1 = 4095$ (3.3 V) to $[0, 1]$.
- Add a **Compare to Constant** block which is used to evaluate a greater or equal condition referred to Constant value 0.5. If this condition is true, the output switches from zero to one resetting the integrator of the PI controller. The enable signal is defined as $\text{en}(k)$.
- Additionally, $\text{en}(k)$ can be used to add a further protection redundancy. Indeed, by using a Product block, the enable signal is multiplied by the actuation variable. As long as $\text{en}(k) = 0$ the duty cycle is forced to zero $d(k) = 0$, i.e., without switching.

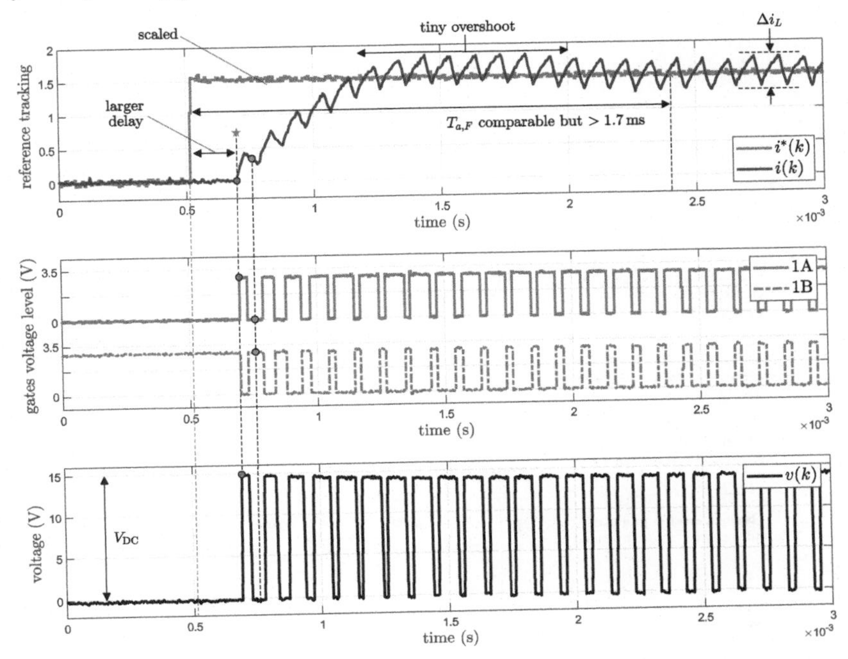

Figure 17.12 Closed-loop dynamics measured from the F28069M board pins through an oscilloscope (Half-Bridge configuration).

- To visually evaluate the pulsed reference $i^*(k)$, the Pulse Generator can be connected to a **GPIO** block (e.g., GPIO7) with a proper scaling of 1/1.5 to normalize the signal in a $[0, 1]$ range. This allow to visualize through an oscilloscope the reference signal ranging in $[0, 3.3]$ V at pin 79.

Optional: the firmware could be also prepared to a debug through the serial COM. In this case:

- Add **Data Type Conversion** and **Rate Transition** (e.g., $T_{TX} = 0.01$ s) blocks to prepare the data exchange via **SCI Transmit** block. Figure 17.11 shows how those blocks should be connected. See Section 10.5 or Chapter 16 for further details.
- In this case, even a Testing Environment must be arranged.

Figure 17.12 reports the closed-loop behavior of the system obtained by executing the given firmware. The system dynamics are validated through an oscilloscope. In particular, the current $i(k)$ is measured by using a current probe, while the converter output voltage $v(k)$ and the gating signals $S_1(k)$, $S_2(k)$ (considered as voltage levels at the related pins) are measured by voltage probes. These latter are also used to check the current reference available at pin 79.

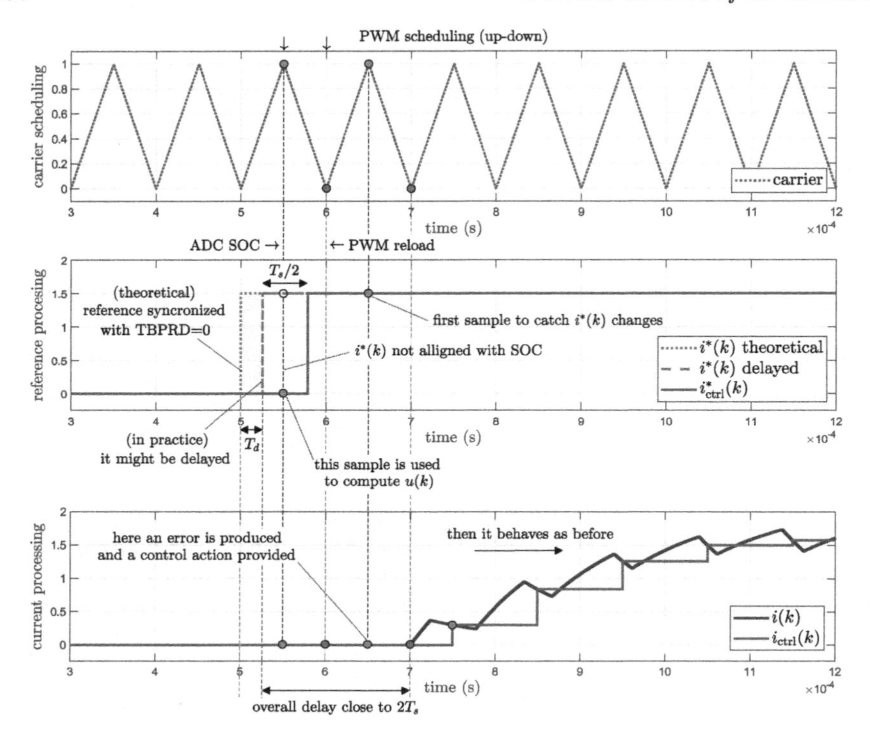

Figure 17.13 Effects of a processing delay on the current.

As done in simulation, the analysis of Figure 17.12 starts from the evaluation of the command tracking performances. The resulting $i(k)$ (blue line) is following the internally generated $i^*(k)$ (red line), which is re-scaled from 3.3 V on the actual reference value in Ampere. The settling time is slightly larger than 1.7 ms. This is due to the presence of a delay in the converter actuation. Indeed, both the gating signals of 1A, 1B switches and the voltage $v(k)$ are operating late compared to the dynamics shown in Figure 17.6. Moreover, this delay also implies a (small) overshoot in $i(k)$ that modifies the shape of the current response. The current ripple is $\Delta i_L \approx 540$ mA. Thus, it is slightly larger than that one computed in simulation, but it is still comparable. This value is influenced by the inductor parameters as well as the probe accuracy. For instance, the series-resistance R_ℓ might be lower than the value reported in the datasheet. The Pulse generator which provides the internal reference $i^*(k)$ is synchronized with the zero value of the triangular carrier. Nevertheless, such synchronization is not as strict as other tasks, e.g., counter comparison in ePWM, which is based on hard synchronization (hard-sync) event through interrupts. Therefore, it might happen that the rising edge of the step-wise $i^*(k)$ is not perfectly synchronized with the ADC SOC (i.e., carrier peak). This effect can be simulated by using the Simulink® scheme in

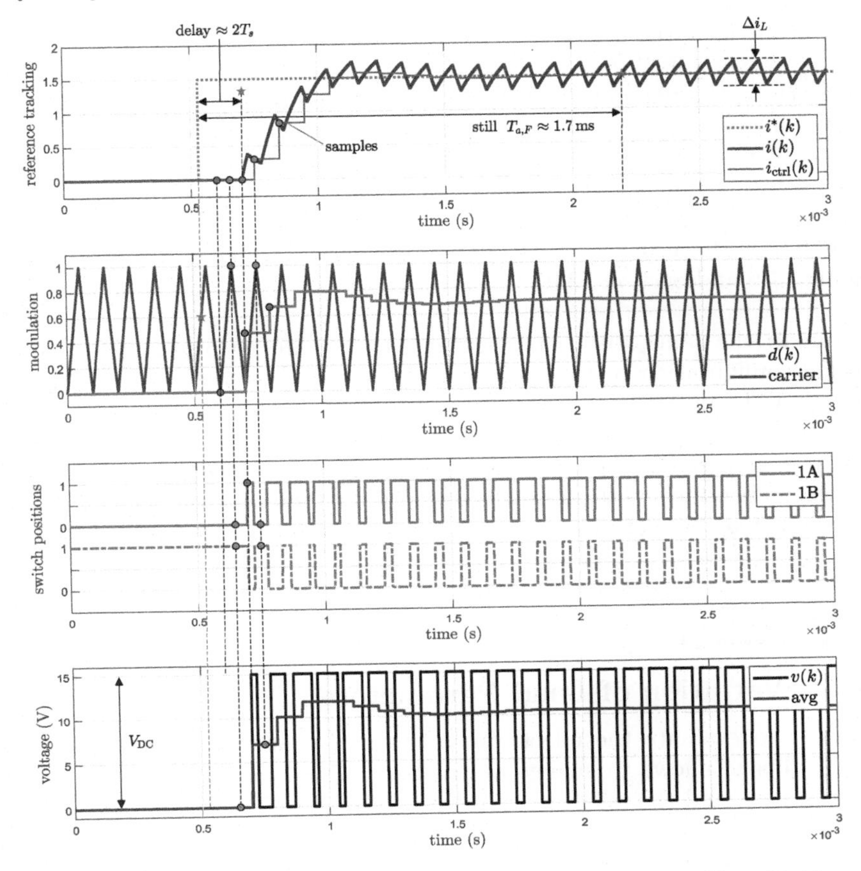

Figure 17.14 Closed-loop dynamics considering a delay $T_d = T_s/4$.

Figure 17.5 which already includes the possibility to add a delay T_d up to $T_d = 2T_s$. The results are analyzed through Figure 17.13. Theoretically, $i^*(k)$ is synchronized with TBPRD=0. By adding a fixed shift of $T_s/2$, $i^*_{\text{ctrl}}(k)$ is coincident with ADC SOC, thus, sampling a value of $i^*_{\text{ctrl}}(k) = 1.5\,\text{A}$. If a delay is present (here quantified by T_d), the current sample processed at the SOC is $i^*_{\text{ctrl}}(k) = 0\,\text{A}$, which implies a zero error $i_{\text{ctrl,err}}(k)$ and no need for any control action. It is necessary to wait until the next carrier peak to detect the value $i^*_{\text{ctrl}}(k) = 1.5\,\text{A}$, after which an error is computed and $d(k) \neq 0$ is applied. Therefore, in a worst case scenario, the overall interval between the rising edge of $i^*(k)$ and the starting of the system dynamics (i.e., when $i(k)$ starts to increase due to $v(k) \neq 0$) reaches $2T_s$. Figure 17.14 shows the simulation results by considering a delay of $T_d = T_s/4$, which implies an actuation delay close to $2T_s$. The system dynamics and the width of the ripple are compatible with those obtained by the experimental results shown in Figure

17.12, validating the firmware.

Remark: if the reference $i^*(k)$ is particularly close to the maximum current limit, which is related to the maximum voltage provided by the power supply, i.e., V_{DC}, the current sensing may be quite inaccurate because $d(k)$ is approximately 1.[6] Indeed, the low side shunt resistor is reading when $S_1(k) = 0$, thus, $d(k)$ should be sufficiently lower than 1 to allow a off-time window[7] large enough to perform the current sampling, which is centered on the off-time (see Figure 17.6 and Section 13.9).

Remark: the same study and implementation can be repeated for a Full-Bridge configuration of the converter, which requires two ePWM modules, e.g., ePWM1 and ePWM2. The same procedure is suggested, that is to start from simulation first and, then move to implementation. Moreover, the considerations on the processing delay are still valid. The ripple on the current changes depending on the adopted strategy for the converter operation, that is, unipolar or bipolar voltage switching. For more information see Chapter 15.

17.5 Variation of Load Parameters

In order to evaluate the effectiveness/robustness of the controller, the ex-tRL(C) board allows to change on-the-fly the load parameters, such as the inductance L and the resistance R values. The main purpose of the designed controller is to guarantee a stable current regulation despite load changes. Nevertheless, even if the tracking of $i^*(k)$ is guaranteed, load changes imply different settling time and current waveforms compared to the previous standard case. Such analysis is useful to understand which parameter variations may lead to potential damage or instability in the system.

The control architecture is left equal to the previous ones as for the step-wise current reference. The inductance can be varied from $L = 860\,\mu\text{H}$ up to $L = 4 \cdot 860\,\mu\text{H}$, while the resistance from $R = 6.8\,\Omega$ down to $R = 6.8/2\,\Omega$ (without considering R_ℓ). The implementation still refers to a Half-Bridge configuration.

17.5.1 Effects on the transient response

The same parameters adopted for the previous exercise are kept as reference, i.e., $L = 860\,\mu\text{H}$ and $R = 6.8\,\Omega + R_\ell$, as well as the resulting dynamical responses, e.g., $\Delta I_L = 540\,\text{mA}$.

An increase in L by a factor of 4 is performed while keeping the resistance value constant. As theory predicts, a bigger inductance L implies a lower

[6]Dead-time protection and signal saturation are always present.

[7]If the duty cycle $d(k)$ is defined as the on-time over the switching period, the off-time represents the remaining time interval.

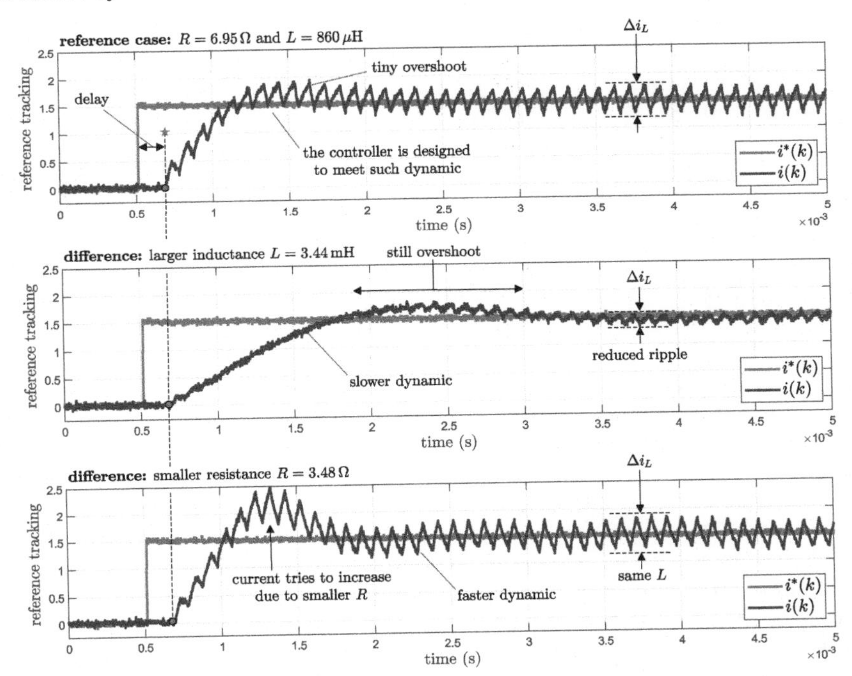

Figure 17.15 Closed-loop dynamics measured from the LaunchXL F28069M board pins through an oscilloscope in correspondence of a variation of the load parameters (Half-Bridge configuration).

current ripple ΔI_L. At steady state, an increase of four times in the inductance, $L = 3.44\,\text{mH}$ results in a quite relevant current ripple reduction, i.e., $\Delta I_L = 259\,\text{mA}$. As a matter of fact, given the same steady-state value for the current (i.e., average value), the continuous-time domain current waveform and, thus, the ripple can be computed considering the integral:

$$i(t) = i_L(t) = \frac{1}{L}\int_{t_1}^{t_2} v_L(t)\,\mathrm{d}t \tag{17.11}$$

which is defined within a generic switching period. For instance, $t_1 = 0$ and $t_2 = T_{\text{sw}} = T_s$. Keeping constant f_{sw}, the voltage applied on the inductor $v_L(t)$ changes based on the variations of L, allowing to achieve the same average value of the load current, but with a lower fluctuation than the reference case. Such effects are visible in Figure 17.15, top and central plots. It can be noted that the integral is not considering thermal variations or magnetizing effects on the inductor, which may further influence the current ripple. Nevertheless, since the closed-loop current controller is designed on the dynamic subject to $L = 860\,\mu\text{H}$, the settling time results to be larger than before. This is also motivated by the fact that the controller is expecting a quicker current

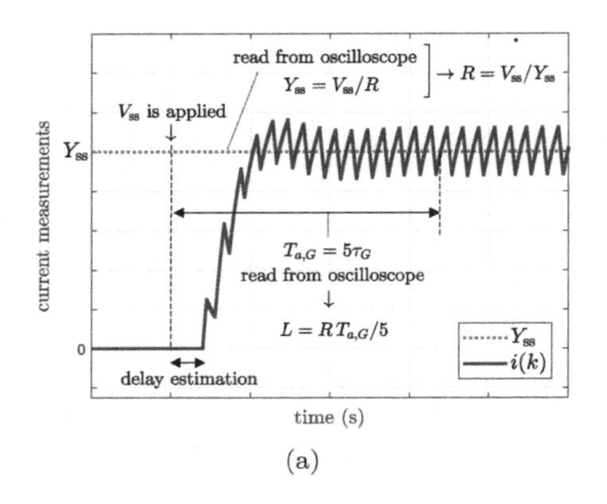

(a)

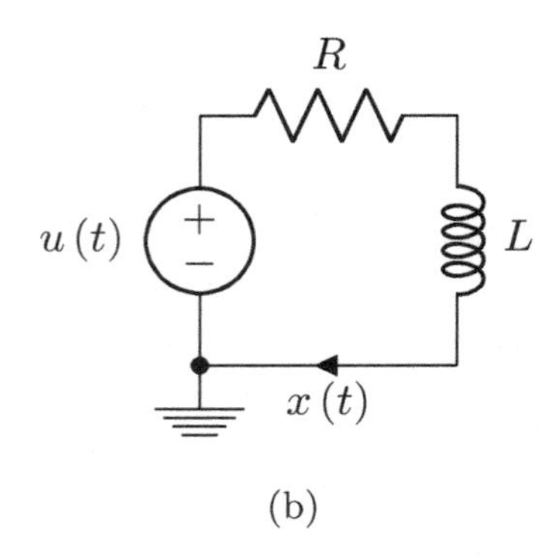

(b)

Figure 17.16 Open-loop identification of an RL load: measured current (a) and equivalent circuit (b).

reaction, which is now not possible due to a greater energy stored in four inductors (insted of one), that implies an higher electrical inertia. Despite the different settling time, the command tracking is still guaranteed.

On the other hand, decreasing R, e.g., moving to $R = 3.48\,\Omega$, while keeping the reference inductance value does not have any impact on ΔI_l. Due to this low resistance, the current rapidly increase enlarging the overshoot if the same control action applied in the reference case (i.e., designed considering a larger R) is provided. This imply a faster dynamic, thus, a smaller settling time, while the command tracking is still guaranteed. However, the current overshoot might be dangerous for the system, which now have to sustain an higher current and possible overheating.

Both cases are simple but effective test which could represent practical scenarios that may lead to components degradation or load revamp.

Remark: the same effects caused by variations of load parameters are valid for the Full-Bridge configuration, both for unipolar and bipolar voltage switching.

17.5.2 Parameters estimation

In case of large parameter uncertainty, different identification techniques may be used to compute the load values. As an example, the simplest technique consist in a open-loop identification of the system to be controlled, i.e., the RL load (see Figure 17.16).

Based on the definition of the transfer function $G(s)$ and ϵ from Chapter 5 and 6, the settling time is $T_a, G = 5\tau_G$. Then, the identification procedure follows three steps:

- Drive the system in open-loop (connection scheme as shown in Figure 17.10) evaluating the system response with an oscilloscope.

- By applying a constant duty cycle, the load is subject to a constant (average) voltage V_{ss} at steady state. It follows that $i(k) = Y_{ss} = V_{ss}/R$. Since the voltage is know, the resistance can be simply estimated as $R = V_{ss}/Y_{ss}$ (see Figure 17.16 (a)).

- The constant duty cycle forces the system to react with its natural dynamic. The open-loop settling time can be directly read from the oscilloscope. Therefore, provided the resistance estimation, the inductance can be computed as $L = R\tau_G = RT_{a,G}/5$ (see Figure 17.16 (a)).

Different operating points (i.e., different values of $d(k)$, thus, V_{ss}) can be tested in order to refine the parameters computation. This operation might be useful when the system is sensitive to physical or environment variations. As an example, in case the TI LaunchPad™ Piccolo™ F28069M board, the BOOSTXL-DRV8301 converter together with the extRL(C) load are operating in a closed box where the ambient temperature is different from 25 °C, the data-sheet values could be corrected for the considered working condition with an identification procedure. It can be noted that, even in a standard environment, a high current flowing in the RL load affects the component temperature, impacting on the values of R and R_ℓ.

The same identification procedure can be also used to estimate the processing delay introduced by the control algorithm.

Remark: in order to limit the components stress, it is recommend not to apply the full dc-link voltage during the open-loop test. Set V_{DC} as the upper bound, suitable voltages for this test may be $V_{DC}/2$ or $V_{DC}/3$.

18

Voltage Control of an RLC load

This chapter shows how to realize a closed-loop voltage control of a passive *RLC* load. An extRL(C) board is adopted. The isolated output voltage sensing stage included in the latter (i.e., it is external to the MCU board) is used to measure the voltage across the resistor through the ADC peripheral. A Half-Bridge converter configuration is considered, thus, using one leg of the TI BOOSTXL-DRV8301 converter only. This aims, in practice, to realize a (non-isolated) bidirectional step-down DC-DC circuit, i.e., two switches connected to a *LC* filter. The latter is aimed to decrease the input voltage to a lower and stable output voltage with a low harmonic content.

Even if a simple resistive load is considered, comments on the design of the closed loop scheme proposed here in the following and on its closed-loop dynamics are quite general. Aiming to follow a given voltage reference $v_o^*(k)$, the resulting PI controller regulates the dc output voltage of the converter by manipulating the command signal of the ePWM peripheral, that is, $d(k)$. The reference value of the output voltage can be provided by an external command, which may come from serial communication, or by the extPot3 potentiometers, or internally (e.g., through a pulse generator/constant block).

The whole system, which modeling is based on the general state-space averaging method, is simulated and analyzed first through Simulink® simulations. Then, a firmware is designed and implemented on the LaunchPad™ F28069M board based on these results. Moreover, guidelines for the hardware design of an *RLC* load are also provided. This Section focuses on the components selection of the conversion stage based on given requirements, such as power rating, current/voltage ripple and stability. Since the extRL(C) board is a configurable load, the effectiveness/robustness of the closed-loop control scheme is experimentally tested through online variations of the inductance L, capacitance C and resistance R values.

18.1 Required Hardware

- TI LaunchPad™ Piccolo™ F28069M and BOOSTXL-DRV8301 boards;
- extPot3 custom board (see Appendix B for technical details);

DOI: 10.1201/9781003196938-18

Figure 18.1 Test-bench including a TI BOOSTXL-DRV8301 converter mounted on the TI LaunchXL F28069M Piccolo™ board and an extRL(C) bard configured in *RLC* mode. The reader is referenced to Appendix B for further details.

- extRL(C) board in *RLC* configuration (see Appendix B);
- Power supply set on 15 V, 2 A.

This Chapter requires the same hardware adopted in Chapter 17, as shown in Figure 18.1 and 17.1. Half-Bridge configuration is used as converter stage topology. The corresponding equivalent circuit is shown in Figure 18.2. The latter shows that the system to be controlled is a resistive load R connected to a LC filter which smooth the discontinuous voltage supplied by the two-level converter leg. The DC link voltage is again provided by a power supply set to $V_{\mathrm{DC}} = 15\,\mathrm{V}$, with the output current limited to 2 A. It can be noted that the extRL(C) board is designed to sustain up to 4 A.

There is *no unique control paradigm* that solves this power electronic-based control problem [40]. Only one among the classical DC-DC converter control approaches is presented in this book. Namely, the so called *voltage-mode control* is outlined together with its related embedded implementation. The extRL(C) board has a voltage sensing stage connected in parallel to the resistive load, allowing to measure $v_o(k)$. This voltage is scaled by a voltage divider in a $[0, 3.3]\,\mathrm{V}$ range, so that it can be directly read by an ADC peripheral.

Differently from Chapter 17, the relationship among duty cycle $d(k)$, output voltage of the converter $v_{\mathrm{A}}(k)$ and its averaged value are described by using the general state-space averaging method. In particular, the control proposed in Chapter 17 is a *current-mode control*, which considers the output voltage $v(k)$ as the controlled variable. Thus, there is no need to specify the duty cycle in the derivation of the system transfer functions both in continuous- or discrete-time domain. For a *voltage-mode control*, the duty cycle is required to

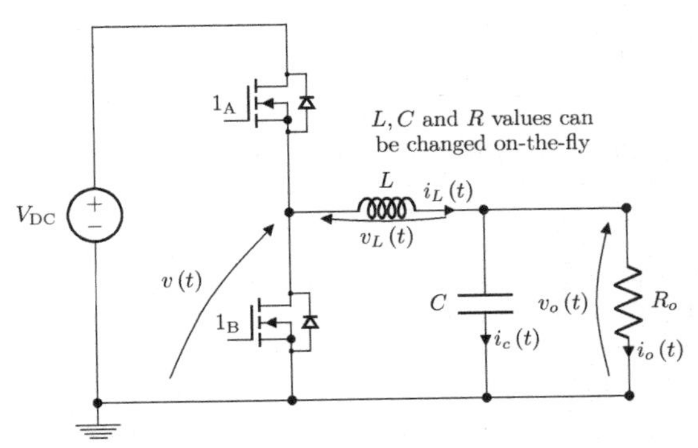

Figure 18.2 Equivalent circuit of the test setup involving the *RLC* load, which is connected to a bidirectional step-down DC-DC converter.

be directly considered in the transfer function. This is achieved by applying the general state-space averaging method, which is a procedure to compute a low frequency model of the system to be controlled.

Different configurations can be set on this extRL(C) board by simply moving the on-board switches, as shown in Appendix B. This operation physically connects or disconnects different components on the printed circuit board. The *RLC* configuration has a $R = 6.8\,\Omega$ resistor, a $L = 860\,\mu\text{H}$ inductor and a $C = 100\,\mu\text{F}$ capacitor. Given the manufacturer data-sheet, the system modeling can be refined by including further (parasitic) resistances, such as the series-resistance of the inductor, which is $R_\ell = 150\,\text{m}\Omega$ at $20\,°\text{C}$ and the equivalent series-resistance of the capacitor (ESR), e.g. $\text{ESR} = R_c = 66\,\text{m}\Omega$ at $f_{\text{sw}} = 10\,\text{kHz}$ (according to the ESR-f characteristic).

As detailed in Appendix B, the extRL(C) is a configurable load that allows to modify the L, C and R parameters through on-board manual switches. For instances, the switch S1 can be used to put in parallel two equal $6.8\,\Omega$ resisitors, thus, resulting in a $R_{\text{eq}} = 3.4\,\Omega$ value seen at the output terminals (i.e. accross the capacitor). Such manual resistance variation may be useful to test the robustness of the control scheme. Indeed, the control algorithm should stabilize the dc output voltage ensuring either good command-tracking perfomance and sufficient disturbance rejection against load variations (i.e. output voltage variations). In the following, the analytical model of the *RLC* system is analyzed and a PI controller is designed based on it. Then, some simulations of the controlled system are reported and considered as reference case for the implementation on hardware. Finally, experimental tests are carried out once the simulations returns satisfactory results.

Table 18.1 Starting data for the extRL(C) load design.

V_{DC}	V_o	R_o	f_{sw}	i_{max}	i_{pk}	P_{max}
15 V	6 V	6.8 Ω	10 kHz	2 A	4 A	7 W

Remark: a bidirectional step-down DC-DC converter can operate in continuous conduction mode (CCM) or discontinuous conduction mode (DCM), depending on the waveform of the inductor current. In CCM the inductor current never reaches zero values in a switching period, whereas in DCM it may even go towards value lower than zero for some time interval. Only CCM operations are considered in this Chapter.

18.2 Guidelines for the Hardware Design of a *RLC* Load

As detailed in Chapter 5, the control design depends on a good knowledge of the load system capabilities. In particular, this means to deal with the power electronics design in terms of passive components sizing, magnetic and saturation effects, evaluation of power losses and heat exchange as well as many more aspects.[1] Therefore, before entering in the details of the control, this Section provides some guidelines to approach the *LC* output filter design in a step-down DC-DC converter configuration, that is, the procedure leading to the design of the extRL(C) board.

The starting parameters are reported in Table 18.1. Uppercase letters are referring to steady-state quantities. The passive components have to be selected aiming to make the system fulfilling a set of constraints. For example:

- The ripple of the current flowing into the inductor must be $\Delta i_L \leq 500\,\text{mA}$
- The ripple of the voltage measured across the load resistor (or equivalently across the capacitor branch) must be $\Delta v_o \leq 100\,\text{mV}$

These two values are related to target power losses and energy quality. The hardware design refers to the worst-case scenario, i.e., passive components are chosen based on the most stressful condition to which the system is subjected.

Steady-State Analysis

Assuming that the capacitor C is sufficiently large to keep $v_o(t) = V_o$, which implies $i_c(t) = \mathrm{d}v_c(t)/\mathrm{dt} = 0$, the average value of the output quantities at steady-state[2] are:

[1] *"An engineer who deals with power electronics applications has to know the basics of power electronics design in order to figure out system physical limits, later used as information for the control structure"* Prof. Ralph Kennel and Prof. Jose Rodriguez.

[2] In order to keep a simple notation, the subscript $_{ss}$ is omitted from variable definitions.

$$I_o = I_L = \frac{V_o}{R_o} = 882.4\,\text{mA} \tag{18.1}$$

$$P_o = R_o I_o^2 = 5.3\,\text{W} \tag{18.2}$$

where the power dissipated on the load is lower than $P_{\text{max}} = 7\,\text{W}$.

A Half-Bridge configuration foresees the use of one leg of the TI BOOSTXL-DRV8301 converter (e.g., leg A), for which switches 1_A, 1_B are considered. The switch positions are defined as $S_1(k)$ and $S_2(k)$, respectively. During commutations, the overall switching state is defined by the vector $\boldsymbol{S}(k) = [S_1(k)\ S_2(k)]^T$. Due to the two-level converter topology, the inductor is subjected to the following average voltages:

ON state $\rightarrow$ $\boldsymbol{S}(k) = [1\ 0]^T$	OFF state $\rightarrow$ $\boldsymbol{S}(k) = [0\ 1]^T$
$V_L = V_{\text{DC}} - V_o - \cancel{R_\ell I_L} \approx 9\,\text{V}$	$V_L = V_o - \cancel{R_\ell I_L} \approx 6\,\text{V}$

where the dc-link voltage V_{DC} is assumed to be constant and the inductor voltage drop negligible, i.e., $R_\ell I_L \approx 0$. The MOSFETs included in the TI Boosterpack[3] have a low gate-source voltage threshold and low drain-source on resistance. Hence, even the MOSFET voltage drop is negligible.

From an energetic point of view, the energy stored in the inductor is $E_L = (1/2)L\,I_L^2$, which increases during ON states and decreases during OFF states. The inductor voltage can be related to the current variation within the switching period $T_{\text{sw}} = 1/f_{\text{sw}}$ and it can be computed as:

$$v_L(t) = L\frac{di_L(t)}{\text{d}t} \quad \rightarrow \quad V_L = L\frac{\Delta i_L}{T_{\text{sw}}} \tag{18.3}$$

where the inductor current variation Δi_L can be computed for the ON and OFF states analytically as:

ON state $\rightarrow$ $\boldsymbol{S}(k) = [1\ 0]^T$	OFF state $\rightarrow$ $\boldsymbol{S}(k) = [0\ 1]^T$
$\Delta i_{L(\text{on})} = \frac{V_{\text{DC}}-V_o}{L}\,DT_{\text{sw}}$	$\Delta I_{L(\text{off})} = \frac{V_o}{L}\,(1-D)\,T_{\text{sw}}$

At steady-state, the amount of energy stored inside the inductor at the beginning and at the end of the switching period is equal $E_{L(\text{on})} = E_{L(\text{off})}$. This equality is extended to the current ripple $\Delta i_{L(\text{on})} = \Delta i_{L(\text{off})}$. Therefore, the duty cycle at steady-state D can be computed considering the previous equations as:

$$\Delta i_{L(\text{on})} = \Delta i_{L(\text{off})} \quad \rightarrow \quad D = \frac{V_o}{V_{\text{DC}}} = 0.4 \rightarrow 40\% \tag{18.4}$$

As the name says, for a step-down DC-DC converter $V_o \leq V_{\text{DC}}$ since the output voltage has to be lower than the input.

---

[3]The TI BOOSTXL-DRV8301 use CSD18533Q5A N-Channel MOSFET [22].

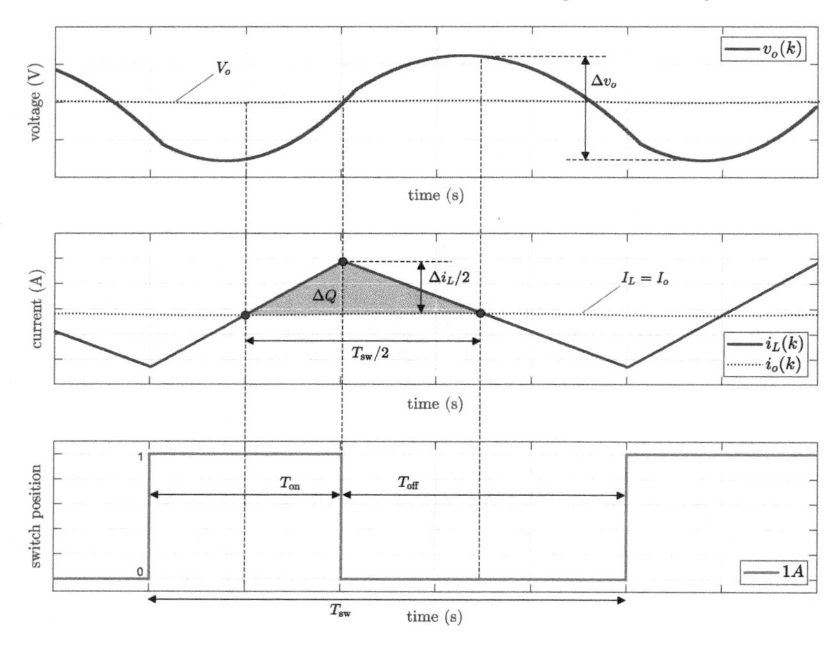

Figure 18.3 Waveforms of the load voltage, inductor current and position of switch 1A for a step-down DC/DC converter. The area highlighted in light blue is the charge ΔQ that is stored (and, then, released) on the capacitor in $T_{sw}/2$. The dotted lines refer to average values, e.g. $I_o = \Delta i_L/2 = DT_{\text{sw}}V_{\text{DC}}/(2L)$

Reminding the aforementioned requirement for the current ripple $\Delta i_L \leq 500\,\text{mA}$, $\Delta i_{L(\text{on})} = \Delta i_L$ can be assumed. Hence, the minimum inductance value L_{min} is:

$$L_{\text{min}} = \frac{V_{\text{DC}} - V_o}{\Delta i_{L(\text{on})}} D\,T_{\text{sw}} = 720\,\mu\text{H} \;\rightarrow\; L \geq L_{\text{min}} \tag{18.5}$$

Then, a first selection proposal is a commercial $L = 860\,\mu\text{H}$ leaded toroidal inductor from Würth Elektronik™ Group, with ±20% tolerance, rated current 3 A and series resistance $R_\ell = 150\,\text{m}\Omega$ at 20 °C. Considering R_ℓ, this component leads to the following current ripple:

$$\Delta i_{L(\text{on})} = \frac{V_{\text{DC}} - V_o - R_\ell I_L}{L} D\,T_{\text{sw}} = 412.4\,\text{mA} \tag{18.6}$$

$$\Delta i_{L(\text{off})} = \frac{V_o - R_\ell I_L}{L}\,(1 - D)\,T_{\text{sw}} = 409.4\,\text{mA} \tag{18.7}$$

which are both lower than 500 mA. Thus, this requirement is satisfied.

Nevertheless, the effects of the parasitic resistance R_ℓ on the load circuit are not easy to predict with simple equations. For instance, the larger R_ℓ, the lower the average value of the output voltage V_o. This result is due to a larger voltage drop on the inductance. However, a large R_ℓ is useful to reduce the current ripple Δi_L. Moreover, sufficiently large values of R_ℓ add passive damping in the load circuit to reduce the resonating effects created by the LC filter. This aspect is detailed in the next Section.

On the other hand, the ripple in the output voltage Δv_o strongly depends on the capacitance C. Assuming a negligible ESR, i.e., $R_c \approx 0$, then it can be considered that $v_c(t) = v_o(t)$. Similarly to the inductor voltage, the capacitor current $i_c(t)$ depends on the voltage variation inside the switching period T_{sw}:

$$i_c(t) = C\frac{\mathrm{d}v_o(t)}{\mathrm{d}t} \quad \rightarrow \quad I_c = C\frac{\Delta v_o}{T_{\text{sw}}} \tag{18.8}$$

During CCM operations, the ripple in the output voltage Δv_o can be computed considering the waveform relationships shown in Figure 18.3. Assuming that all the ripple component of the inductor current Δi_L flows through the capacitor only, while its average value I_L flows into the load resistor R_o, the filtering action of C provides an additional charge ΔQ that is stored and released on the capacitor layers (the light blue area reported in Figure 18.3). Based on Figure 18.3, the voltage ripple can be computed through a geometric approach:

$$\Delta v_o = \frac{\Delta Q}{C} = \frac{\left(\frac{\Delta i_L}{2}\right)\cdot\left(\frac{T_{\text{sw}}}{2}\right)}{2C} \tag{18.9}$$

Reminding the aforementioned voltage and current ripple requirements $\Delta v_o \leq 100\,\text{mV}$, $\Delta i_L \leq 500\,\text{mA}$ and considering equation (18.9), the minimum capacitance value C_{min} is computed as:

$$C_{\text{min}} = \frac{\Delta i_L T_{\text{sw}}}{8\Delta v_o} = 62.5\,\mu\text{F} \quad \rightarrow \quad C \geq C_{\text{min}} \tag{18.10}$$

A first selection proposal is a commercial electrolytic capacitor with $C = 100\,\mu\text{F}$. Considering the specific current ripples resulting from the implementation with the selected inductor, the resulting voltage ripples can be computed through equation (18.9), that is:

$$\Delta v_{o(\text{on})} = \frac{\Delta i_{L(\text{on})} T_{\text{sw}}}{8C} = 51.6\,\text{mV} \tag{18.11}$$

$$\Delta v_{o(\text{off})} = \frac{\Delta i_{L(\text{off})} T_{\text{sw}}}{8C} = 51.2\,\text{mV} \tag{18.12}$$

Nevertheless, the effects of the ESR on the voltage ripple should be verified. Indeed, the presence of a resistance R_c implies a further voltage drop in the capacitor branch, that leads to $v_o(t) = v_c(t) + R_c i_c(t)$. In terms of variations, it becomes:

$$\Delta v_o = \Delta v_c + R_c \Delta i_L \tag{18.13}$$

The ripple Δv_c can be computed similarly to Δv_o from equation (18.9) through a geometric approach. The Δv_c waveform equals Δv_o except for a different

Table 18.2 Selected components for the extRL(C) load design.

L	R_ℓ	Δi_L	C	R_c	Δv_o	D
860 µH	150 mΩ	≈ 412 mA	100 µF	66 mΩ	≈ 52 mV	0.4

scaling. In order to keep $\Delta V_o \approx \Delta V_c$, an easy and conservative approach[4] to estimate the maximum allowed value for the ESR $R_{c,\max}$ is to consider the upper bound $\Delta v_o/2$. If the voltage drop in the capacitor branch is lower than $\Delta v_o/2$, its contribution is negligible. Given $\Delta v_o \leq 100\,\text{mV}$ and $\Delta i_L \leq 500\,\text{mA}$, it follows that:

$$\frac{\Delta v_o}{2} = R_{c,\max}\Delta i_L \rightarrow R_{c,\max} = \frac{\Delta v_o}{2\Delta i_L} = 100\,\text{m}\Omega \rightarrow R_c \leq R_{c,\max} \quad (18.14)$$

By selecting a capacitor with R_c lower than $R_{c,\max}$, the parasitic resistance does not have a large impact on ΔV_o. Thus, a $C = 100\,\mu\text{F}$ aluminum electrolytic capacitor with ±20% tolerance, rated voltage 100 V, and ESR $= R_c = 66\,\text{m}\Omega$ at $f_{\text{sw}} = 10\,\text{kHz}$ from Würth Elektronik™ Group is selected.

Many manufacturers produce different versions of the same capacitor with equal capacitance, but different ESR values. Those with $R_c < 40\,\text{m}\Omega$ are called *low-ESR* type. A low R_c is useful to reduce the system power losses. However, R_c must be large enough to damp the resonating effects together with R_ℓ.

Once components are selected, the main system parameters are summarized in Table 18.2.

Remark: this chapter refers to a bidirectional step-down DC-DC converter topology (also called *synchronous buck converter*). This device uses two switches instead of a combination of one switch and one diode, which is the classical *buck converter* structure. Given the application targets (e.g., power, voltage, current ratings), a bidirectional topology may be used to reduce losses. For instance, if the diode in a buck converter is replaced by a MOSFET switch with low drain-source on resistance, the resulting power losses are reduced (especially for low switching frequencies). Moreover, a bidirectional topology allows a reverse current flow, which can be used in motor applications for regenerative braking and/or to avoid DCM operations.

18.3 General State-Space Average Modeling Method

Middlebrook and Ćuk introduced in [36] a state-space averaging technique aimed to derive linear models for power converters and modulation stages.

[4]The method proposed here is simple, but not the optimal one. Indeed, many nonlinear effects should be taken into account.

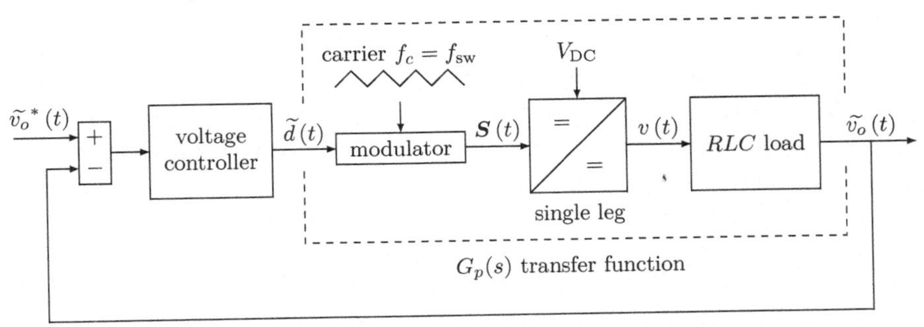

Figure 18.4 *Voltage-mode* control scheme based on the state-space average modeling method, where $G_p(s)$ describes the whole *power stage* dynamic, including the modulator, power converter and *RLC* load stage.

The basic idea is to focus only on small AC signals (*small signal control theory*), linearizing the system around a steady-state DC operating point.

The step-down DC-DC converter topology shown in Figure 18.2 is considered. A *voltage-mode* control scheme based on the state-space average modeling method can be represented as reported in Figure 18.4. All time domain variables can be represented by a steady-state DC value plus a small AC perturbation, e.g., $v_o(t) = V_o + \widetilde{v_o}(t)$. The goal of this approach is to compute the *small signal transfer function* $G_p(s) = \widetilde{v_o}(s)/\widetilde{d}(s)$ in the continuous-time domain by using the Laplace operator. In particular, $G_p(s)$ describes in a condensed manner the dynamics of the modulator, power converter and *RLC* load stage. The variables $\widetilde{v_o}(s)$ and $\widetilde{d}(s)$ are the small perturbations in the output voltage $v_o(t)$ and in the duty cycle $d(t)$ around their steady state DC operating values V_o and D, respectively.

Then, a state vector $\boldsymbol{x}(t) = [i_L(t)\ v_c(t)]^T$ is considered, which includes the current flowing in the inductor L and the voltage measured across the capacitor C, as well as the switch position vector $\boldsymbol{S}(t) = [S_1(t)\ S_2(t)]^T$. This latter defines the converter output voltage value $v(t)$. The general procedure can be outlined through the following steps:

1. **Compute dynamic models for each allowed circuit configurations**
   In a 2-level topology two cases are possible: ON state $\boldsymbol{S}(t) = [1,0]^T \rightarrow v(t) = V_{DC}$ and OFF state $\boldsymbol{S}(t) = [0,1]^T \rightarrow v(t) = 0$. For each scenario, the state equations can be determined as

ON state $\rightarrow d(t)T_s$

$$\frac{\mathrm{d}\boldsymbol{x}(t)}{\mathrm{dt}} = \boldsymbol{A_1}\boldsymbol{x}(t) + \boldsymbol{B_1}V_{DC}$$
$$v_o(t) = \boldsymbol{C_1}\boldsymbol{x}(t)$$

OFF state $\rightarrow (1 - d(t))\,T_s$

$$\frac{\mathrm{d}\boldsymbol{x}(t)}{\mathrm{dt}} = \boldsymbol{A_2}\boldsymbol{x}(t)$$
$$v_o(t) = \boldsymbol{C_2}\boldsymbol{x}(t)$$

where matrices A_1 and A_2 and vectors B_1, C_1 and C_2 are defined based on the system parameters.

2. **Average the state-variables using the duty cycle $d(t)$**
To achieve an average description of the dynamic models in T_{sw}, the equations corresponding to the two foregoing states are merged in a *time weighted* and *averaged* manner using $d(t)$:

$$\frac{\mathrm{d}\boldsymbol{x}(t)}{\mathrm{dt}} = \left[\boldsymbol{A_1}d(t) + \boldsymbol{A_2}\left(1 - d(t)\right)\right]\boldsymbol{x}(t) + \boldsymbol{B_1}d(t)V_{\mathrm{DC}} \tag{18.15}$$

$$v_o(t) = \left[\boldsymbol{C_1}d(t) + \boldsymbol{C_2}\left(1 - d(t)\right)\right]\boldsymbol{x}(t) \tag{18.16}$$

3. **Introduce small AC perturbations and split AC and DC dynamics descriptions**
Small AC perturbations, denoted by the hat $\tilde{}$, are superimposed to DC steady-state quantities, which are represented by capital letters:

$$\boldsymbol{x}(t) = \boldsymbol{X} + \widetilde{\boldsymbol{x}}(t) \qquad v_o(t) = V_o + \widetilde{v_o}(t) \qquad d(t) = D + \widetilde{d}(t) \tag{18.17}$$

where a constant dc-link voltage V_{DC} is assumed. After some mathematical commutations using the definitions reported in equation (18.17), (18.15) and (18.16) are rewritten as:

$$\begin{aligned}\frac{\mathrm{d}\widetilde{\boldsymbol{x}}(t)}{\mathrm{dt}} &= \boldsymbol{AX} + \boldsymbol{B}V_{\mathrm{DC}} + \boldsymbol{A}\widetilde{\boldsymbol{x}}(t) + \left[\boldsymbol{A_1}D + \boldsymbol{A_2}\left(1 - D\right)\right]\widetilde{\boldsymbol{x}}(t) \\ &\quad + \left[\left(\boldsymbol{A_1} - \boldsymbol{A_2}\right)\boldsymbol{X} + \boldsymbol{B_1}V_{\mathrm{DC}}\right]\widetilde{d}(t)\end{aligned} \tag{18.18}$$

$$\widetilde{v_o}(t) = \boldsymbol{CX} + \boldsymbol{C}\widetilde{\boldsymbol{x}}(t) + \left(\boldsymbol{C_1} - \boldsymbol{C_2}\right)\boldsymbol{X}\widetilde{d}(t) - V_o \tag{18.19}$$

where terms containing products of $\widetilde{v_o}(t)$ and $\widetilde{d}(t)$ are neglected since they are second order terms. To simplify the notation, matrices $\boldsymbol{A} = \boldsymbol{A_1}D + \boldsymbol{A_2}\left(1 - D\right)$, $\boldsymbol{B} = \boldsymbol{B_1}D$ and $\boldsymbol{C} = \boldsymbol{C_1}D + \boldsymbol{C_2}\left(1 - D\right)$ are defined. The *steady-state contributions* shown in equation (18.18) and (18.19) can be computed by setting all the AC perturbation terms and their time derivatives equal to zero. It follows:

$$\boldsymbol{AX} + \boldsymbol{B}V_{\mathrm{DC}} = 0 \tag{18.20}$$

$$\boldsymbol{CX} - V_o = 0 \tag{18.21}$$

This operation leads to compute the steady-state voltage transfer function, which can be seen as a DC gain:

$$\frac{V_o}{V_{\mathrm{DC}}} = -\boldsymbol{CA}^{-1}\boldsymbol{B} \tag{18.22}$$

By removing the steady-state contributions from equations (18.18) and (18.19), a *small AC perturbation description* is finally achieved.

$$\frac{\mathrm{d}\widetilde{\boldsymbol{x}}(t)}{\mathrm{dt}} = \boldsymbol{A}\widetilde{\boldsymbol{x}}(t) + \left[\left(\boldsymbol{A_1} - \boldsymbol{A_2}\right)\boldsymbol{X} + \boldsymbol{B_1}V_{\mathrm{DC}}\right]\widetilde{d}(t) \tag{18.23}$$

$$\widetilde{v_o}(t) = \boldsymbol{C}\widetilde{\boldsymbol{x}}(t) + \left[\left(\boldsymbol{C_1} - \boldsymbol{C_2}\right)\boldsymbol{X}\right]\widetilde{d}(t) \tag{18.24}$$

4. **Rewrite AC equation in s-domain and compute the power stage transfer function**
   The power stage transfer function $G_p(s)$ can be computed using the Laplace operator, from the small signal AC model reported in equation (18.23) and (18.23):

$$G_p(s) = \widetilde{v}_o(s)/\widetilde{d}(s) = \\ = \boldsymbol{C}\left(s\boldsymbol{I} - \boldsymbol{A}\right)^{-1}\left(\left(\boldsymbol{A_1} - \boldsymbol{A_2}\right)\boldsymbol{X} + \boldsymbol{B_1}V_{\mathrm{DC}}\right) + \left(\boldsymbol{C_1} - \boldsymbol{C_2}\right)\boldsymbol{X} \tag{18.25}$$

18.3.1 Linear average model and controller design

The state-space average modeling presented in the previous section is specific for a 2-level converter topology and it can be extended to many different load stages. Given $\boldsymbol{x}(t) = [i_L(t)\ v_c(t)]^T$ and $\boldsymbol{S}(t) = [S_1(t)\ S_2(t)]^T$, the step-by-step modeling procedure starts considering:

ON state $\rightarrow$ $\boldsymbol{S}(t) = [1, 0]^T$

$$\frac{\mathrm{d}\boldsymbol{x}(t)}{\mathrm{dt}} = \boldsymbol{A_1}\boldsymbol{x}(t) + \boldsymbol{B_1}V_{\mathrm{DC}}$$
$$v_o(t) = \boldsymbol{C_1}\boldsymbol{x}(t)$$

OFF state $\rightarrow$ $\boldsymbol{S}(t) = [0, 1]^T$

$$\frac{\mathrm{d}\boldsymbol{x}(t)}{\mathrm{dt}} = \boldsymbol{A_2}\boldsymbol{x}(t)$$
$$v_o(t) = \boldsymbol{C_2}\boldsymbol{x}(t)$$

where the matrices $\boldsymbol{A} = \boldsymbol{A_1} = \boldsymbol{A_2}$, $\boldsymbol{B} = \boldsymbol{B_1}$ and $\boldsymbol{C} = \boldsymbol{C_1} = \boldsymbol{C_2}$ are defined as

$$\boldsymbol{A} = \begin{bmatrix} -\dfrac{R_oR_c + R_oR_\ell + R_cR_\ell}{L\left(R_o + R_c\right)} & -\dfrac{R_o}{L\left(R_o + R_c\right)} \\ \dfrac{R_o}{C\left(R_o + R_c\right)} & -\dfrac{1}{C\left(R_o + R_c\right)} \end{bmatrix}$$

$$\boldsymbol{B} = \begin{bmatrix} \dfrac{1}{L} \\ 0 \end{bmatrix}$$

$$\boldsymbol{C} = \begin{bmatrix} \dfrac{R_oR_c}{R_o + R_c} & \dfrac{R_o}{R_o + R_c} \end{bmatrix}$$

The DC gain of the steady-state voltage transfer function is:

$$\frac{V_o}{V_{\mathrm{DC}}} = D\frac{R_o + R_c}{R_o + R_c + R_\ell} \qquad \text{if } R_c, R_\ell \approx 0 \rightarrow \frac{V_o}{V_i} = D \tag{18.26}$$

Furthermore, the power stage transfer function $G_p(s)$ can be computed from equation (18.25), based on the small signal AC perturbation model:

$$G_p(s) = \frac{\widetilde{v}_o(s)}{\widetilde{d}(s)} = \boldsymbol{C}\left(s\boldsymbol{I} - \boldsymbol{A}\right)^{-1}\boldsymbol{B}V_{\mathrm{DC}} =$$

$$= V_{\mathrm{DC}}\frac{R_oR_c}{L\left(R_o+R_c\right)}\frac{s+\dfrac{1}{CR_c}}{\underbrace{s^2+s\dfrac{C\left(R_o\left(R_\ell+R_c\right)+R_cR_\ell\right)+L}{LC\left(R_o+R_c\right)}+\dfrac{R_o+R_\ell}{LC\left(R_o+R_c\right)}}_{\text{oscillating response}}} =$$

$$= V_{\mathrm{DC}}\frac{\omega_o^2}{\omega_z}\frac{s+\omega_z}{\underbrace{s^2+2\xi\omega_o s+\omega_o^2}} \tag{18.27}$$

The highlighted term of $G_p(s)$ includes a 2nd order denominator which models resonating effects. Therefore, the intrinsic behavior of a LC filter increases the model complexity. In particular, its oscillating response is identified by the resonance frequency ω_o (which is associated to the poles), the anti-resonance frequency ω_z (which is associated to the zeros), and the damping coefficient ξ. Based on equation (18.27), they are defined as

$$\omega_o = \sqrt{\frac{R_o+R_\ell}{LC\left(R_o+R_c\right)}} \qquad \xi = \frac{C\left(R_o\left(R_\ell+R_c\right)+R_cR_\ell\right)+L}{2\omega_o\left(LC\left(R_o+R_c\right)\right)} \qquad \omega_z = \frac{1}{CR_c} \tag{18.28}$$

Figure 18.5 (a) shows the Bode plot of $G_p(s)$ considering the numerical values reported in Table 18.2. The transfer function presents a fixed gain (about 23 dB) and a almost zero phase shift at frequency below $\omega_o \approx 3431$ rad/s. Beyond the resonance frequency ω_o, $|G_p(s)|$ is decreasing with a -40 dB/dec slope, whereas the phase approaches $-180°$. Beyond the anti-resonance frequency $\omega_z \approx 151.521 \times 10^3$ rad/s, the absolute value decreases with a -20 dB/dec slope and the phase starts rising toward $-90°$. The damping coefficient is $\xi \approx 0.249$, keeping the resonance peak limited below 30 dB. From equation (18.28) it is evident how important R_ℓ and R_c are to increase the damping factor.

As shown in Figure 18.4, a PI controller $R(s) = k_p + k_i/s$ is used to close the voltage loop. From now on the hat $\tilde{}$ is omitted. The open-loop transfer function is defined as:

$$L(s) = \frac{v_o(t)}{e(t)} = R(s)G_p(s) = k_p\frac{1+sT_i}{sT_i}V_{\mathrm{DC}}\frac{\omega_o^2}{\omega_z}\frac{s+\omega_z}{s^2+2\xi\omega_o s+\omega_o^2} \tag{18.29}$$

Since $G_p(s)$ is a second order transfer function, a simple pole-zero cancellation is not straightforward in this case. The explanation and analysis of the tuning techniques for this kind of systems are out of the scope of this book. Nevertheless, a first set of tuning parameters can be found using the MATLAB®

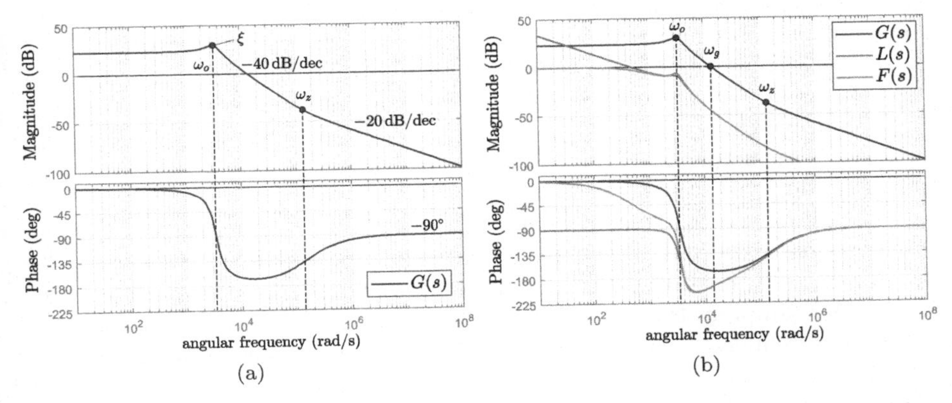

Figure 18.5 Bode diagrams of (a) $G_p(s)$ and (b) comparison between $G_p(s)$, $L(s)$ and $F(s)$ transfer functions for the designed controller. The amplitude of the overshoot strongly depends on the damping coefficient ξ.

interface **pidTuner()**, as shown in Chapter 5. In particular, **pidTuner()** allows to work with a user friendly interface in which several step responses generated with different values of k_p and k_i can be compared. Based on the Bode plot shown in Figure 18.5(a), the uncontrolled system is characterized by the angular frequency $\omega_g \approx 1.33 \times 10^4\,\text{rad/s}$. Due to the 2nd order denominator, the system produces an overshoot. Hence, the settling time is $T_{a,G} > 5/\omega_g = 376\,\mu\text{s}$. For a better visualization of the result, the designed controller is slowing down the natural response of the system by choosing $T_{a,F} = 23.9 \times T_{a,G} = 9\,\text{ms}$.[5] This value implies $\tau_F = T_{a,F}/5 = 1.8\,\text{ms}$ and a controller bandwidth equal to $\omega_c = 1/\tau_F = 555.56\,\text{rad/s}$. By setting this last value and a phase margin of $\phi_m = 90°$, **pidTuner()** returns:

$$k_p = \frac{\cos(\phi_m - \pi - \arg(G_p(j\omega_c)))}{|G_p(j\omega_c)|} = 0.0047\,\text{V}^{-1} \tag{18.30}$$

$$k_i = -\omega_c \frac{\sin(\phi_m - \pi - \arg(G_p(j\omega_c)))}{|G_p(j\omega_c)|} = 33.35\,\text{V}^{-1}\text{s}^{-1} \tag{18.31}$$

where the units of the gains are defined to be consistent with $L(s)$. The resulting bode diagrams are shown in Figure 18.5(b).

The control variable is $u(t) = d(t)$. However, the dc voltage $V_{\text{DC}} = 15\,\text{V}$ provided by the power supply is considered an upper bound. This is set as regulator saturation by limiting the control variable $u(t)$ in a range $[0\ 1]$.

[5]It is recommended to multiply the time constant T_a, G with a large coefficient. Indeed, the estimation of $T_{a,G}$ is performed from the ω_g computed based on the bode plot of $G_p(s)$. Hence, it does not consider the magnitude of resonance. Depending on the energy introduced in the system, the resonance may increase, implying a larger settling time than expected. This effect is investigated through open-loop tests.

Therefore, an anti-windup strategy must be included in the PID controller block. Based on Chapter 6, a back-calculation approach is used with $k_{aw} = k_i/k_p = 7096\,\mathrm{s}^{-1}$. It can be noted that the intrinsic nature of a step-down DC-DC converter is to achieve a steady-state output voltage much lower than the input, which implies $D \ll 1$. However, $d(t)$ could suddenly increase during transients to compensate for a voltage variation. Hence, this explains the need for bounding the control variable.

The discrete-time PI controller $R(z)$ is derived based on the continuous-time counterpart and it is integrated in the closed-loop scheme together with a Rate Transition block in simulation to mimic the discretization process. The sampling time of the controller is $T_s = T_{\mathrm{sw}} = 1/f_{\mathrm{sw}}$, where the switching frequency is $f_{\mathrm{sw}} = 10\,\mathrm{kHz}$. Backward Euler is chosen as integration method.

18.4 System Simulations

The closed-loop control of the *RLC* load is simulated in Simulink® to analyze its design and check the resulting performance of the controller. This step is carried out the Half-Bridge configuration of the converter. To this purpose, the power electronics and the *RLC* load are modeled through Simscape™ Electrical elements, thanks to the Specialized Power Systems ™ library (*Simscape/Electrical/Specialized Power Systems/*) as shown in Chapters 6 and 17.

In particular, the **Mosfet** blocks available in the subset */Fundamental Blocks/Power Electronics* are used to represent the switches mounted on the BOOSTXL-DRV8301 converter board. The semiconductor characteristics taken from the converter datasheet, e.g., R_{on} and R_{d}, are set into the Simulink® blocks as shown in Chapters 15 and 17. The two-level Half-Bridge configuration is obtained by rearranging these block based on the equivalent circuits reported in Figure 18.2. The switches are driven by a PWM modulator with a carrier frequency $f_{\mathrm{sw}} = 10\,\mathrm{kHz}$. The load is modeled through three series **RLC branch** block set as follows: *RL branch type* with $L = 860\,\mu\mathrm{H}$ and $R_\ell = 150\,\mathrm{m}\Omega$, *RC branch type* with $C = 100\,\mu\mathrm{F}$ and $R_c = 66\,\mathrm{m}\Omega$, and *R branch type* with $R_o = 6.8\,\Omega$. Moreover, a second *R branch type* with $R_o = 6.8\,\Omega$ is added in parallel to the latter component. A switch manages the parallel insertion of the two R_o to check the dynamical response of the system to load variations.

18.5 Half-Bridge Configuration

As shown in Chapter 17, a Half-Bridge configuration refers to a converter which uses a single-leg of the boosterpack, requiring one ePWM module only,

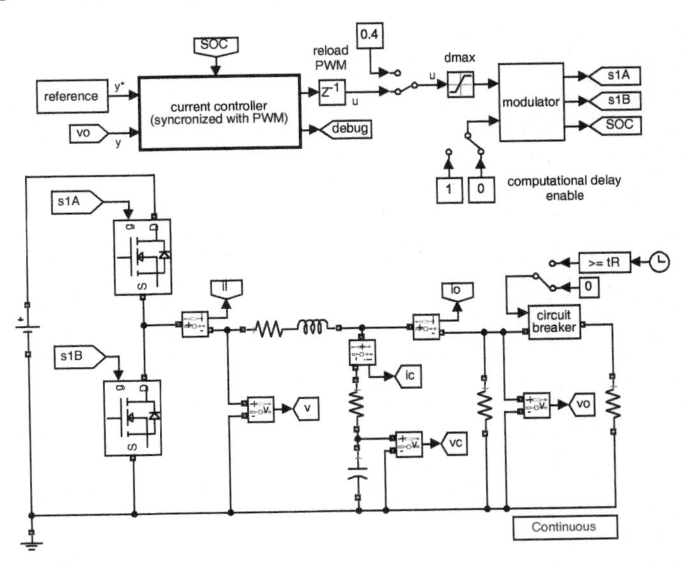

Figure 18.6 Simulink® scheme to simulate the closed-loop voltage control of an *RLC* load driven in Half-Bridge configuration.

e.g., ePWM1. Hence, the switch 1_A is driven by the signal $S_1(k)$ and 1_B by $S_2(k)$, operating in a complementary way. The *RLC* load is connected between the central point of leg A and ground. The *average* output voltage of the converter leg is function of the duty cycle $d(k)$ (or $d_A(k)$):

$$V(k) = v_{\text{avg}}(k) = d(k)V_{\text{DC}} \tag{18.32}$$

where the variables are already considered in the discrete-time domain k, i.e., moving the simulation toward the implementation. The voltage V_{DC} is set by the power supply, while $d(k) \in [0, 1]$. Considering the PWM stage, the duty cycle $d(k)$ is compared to a triangular carrier at $f_c = 10\,\text{kHz}$, which generates a switching pattern $S(k)$. This is realized on the MCU board by using ePWM modules. Since the gate signals of the 2-level topology are $S(k) = S_1(k)$ and $S_2(k) = \text{not}(S_1(k))$, the instantaneous output voltage $v(k)$ is a pulsed waveform that can be computes as:

$$v(k) = v_{\text{a}}(k) = S(k)V_{\text{DC}} \tag{18.33}$$

which is discontinuous and bounded within 0 and V_{DC}. This result allows to analyze the different states in the state-space average model.

As already mentioned, the system is characterized by the parameters reported in Tables 18.1 and 18.2. The control coefficients are defined in the previous Section. These parameters can be initialized in *Model Properites/-Callbacks/InitFcn* or in a separate `m`-file as shown here in the following:

```
%% power supply
Vdc    = 15;
%% carrier frequency definition
fsw    = 10e3;
Tsw    = 1/fsw;
%% sampling time definition
Ts     = Tsw;
%% step-size definition
Tsim   = Tsw/400;
%% load parameters
Ro     = 6.8;
L      = 860e-6;
Rl     = 150e-3;
C      = 100e-6;
Rc     = 66e-3;
imax   = 2;
%% controller (pidTuner)
%% wc = 555 - phiM=90
kp     = 0.0047;
ki     = 33.35;
```

The equivalent circuit shown in Figure 18.2 is then created in Simulink® as reported in Figure 18.6. This scheme is included in the file:

SDsimRLC_closed.slx (solver: fixed step - ODE4, step size: T_{sim})
(powergui: simulation type *Continuous*)

The simulation time step T_{sim} is particularly small to well approximate the switching behavior of currents $i_L(k)$, $i_c(k)$, $i_o(k)$ and voltages $v(k)$, $v_c(t)$, $v_o(k)$. The reference subsystem includes different kinds of input signals to operate the system in several working conditions.

The time constants τ_G and $T_{a,G}$ were estimated by analyzing the bode diagram of $G_p(s)$. Nevertheless, it is recommended to analyze the open-loop behavior of the system in simulation to verify both $T_{a,G}$ and the resonance effects on the variables. A constant duty cycle $d(k) = D = 0.4$ is directly applied avoiding any control, as shown in Figure 18.6. Based on the previous steady-state analysis, the output voltage is computed as $v_o(k) = V_o = DV_{\text{DC}} = 6\,\text{V}$. The resulting open-loop dynamics are reported in Figure 18.7. Both $v_o(k)$ and $i_L(k)$ shows oscillating responses influenced by the forcing action provided by $D \neq 0$ and the resonance excitation at ω_o. The peaks are attenuated thanks due to the damping ξ, as reported in equation (18.28). In particular, the steady state value V_o differs from the predicted one due to the voltage drop determined across the inductor, which is due to R_ℓ. Indeed, reminding

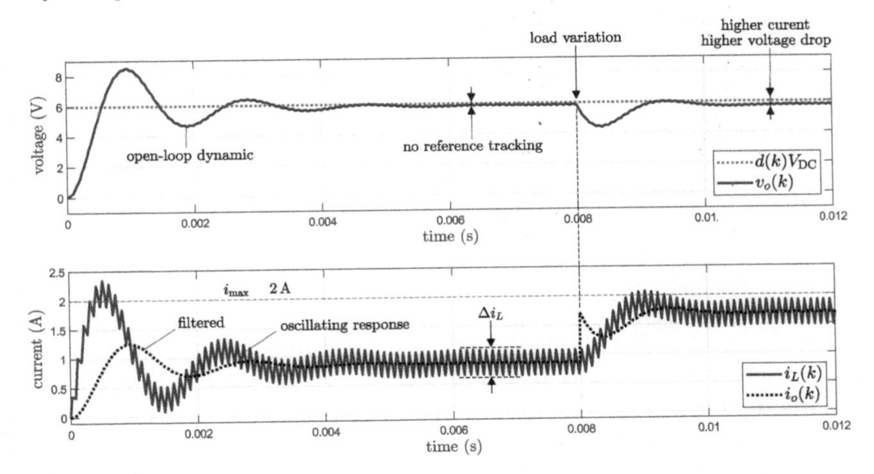

Figure 18.7 Open-loop dynamics considering a load variation at $t_R = 8\,\text{ms}$.

that this is a open loop test, the duty cycle $d(k)$ cannot be dynamically adjusted to compensate such voltage error, leading to $V_{o,\text{open}} \approx 5.718\,\text{V} < 6\,\text{V}$. As a consequence, at steady-state, $I_{o,\text{open}} \approx 865\,\text{mA} < I_o = 882.4\,\text{mA}$. Even the current ripple $\Delta i_{L,\text{open}}$ slightly differs from the predicted one too. The output current $i_o(k)$ presents a filtered dynamic compared to $i_L(k)$. However, the latter has a fast dynamic which exceed the limit $i_{\max} = 2\,\text{A}$. Similarly, the output voltage overshoot $S_{v\%}$ can be analytically identified through the coefficients reported in equation (18.28):

$$S_{v\%} = 100\mathrm{e}^{\left(-\frac{\pi\xi}{\sqrt{1-\xi^2}}\right)} \approx 44.63\% \rightarrow v_{o,\text{pk}} = V_o\left(1 + \frac{S_{v\%}}{100}\right) \approx 8.68\,\text{V} \tag{18.34}$$

The peak value $v_{o,\text{pk}}$ is in-line with the results provided in Figure 18.7. The settling time can be estimated as:

$$T_{a,G,\text{open}} = \frac{5}{\xi\omega_o} = 6\,\text{ms} \;>\; T_{a,G} \tag{18.35}$$

which is larger than the one computed through the bode plot. As already detailed, this behavior is due to the energy-related effects of the resonance, which influences the time decay of the voltage waveform.

Moreover, the scheme shown in Figure 18.6 allows to test load variations. At time instances t_R (which is a parameter to be initialized), the circuit breaker subsystem connects the second resistor R_o, which lead to an equivalent resistance $R_{\text{eq}} = R_o/2 = 3.4\,\Omega < R_o$. Setting $t_R = 8\,\text{ms}$, Figure 18.7 shows how the system reacts to such perturbation. Since $R_{\text{eq}} < R_o$, both $i_L(k)$ and $i_o(k)$ are higher than before. Based on Section 18.2, it follows:

$$I_o^{(R_{\text{eq}})} = \frac{V_o}{R_{\text{eq}}} = 1.76\,\text{A} \;>\; I_{o,\text{open}}^{(R_{\text{eq}})} \approx 1.69\,\text{A} > I_{o,\text{open}} \tag{18.36}$$

Table 18.4 Steady-state values of the currents and voltages in open loop simulations.

$V_{o,\text{open}}$	$I_{o,\text{open}}$	$\Delta\, i_{L,\text{open}}$	$\Delta\, v_{o,\text{open}}$	$V_{o,\text{open}}^{(R_{\text{eq}})}$	$I_{o,\text{open}}^{(R_{\text{eq}})}$
5.87 V	865 mA	≈ 420 mA	≈ 52 mV	5.75 V	1.69 A

Consequently, the new steady state value of $V_{o,\text{open}}^{(R_{\text{eq}})}$ is lower than the one before load variation due to the increase in the load current, which, in turns, causes a larger voltage drop in the inductor and in the load resistance. The steady-state values of the open-loop dynamics are summarized in Table 18.4.

The aim of this exercise is to design a *voltage-mode control loop*. Namely, this means to control and stabilize the output voltage $v_o(k)$ measured across the load R_o (which is effectively the output variable $y(k)$) based on a specific reference $v_o^*(k)$ (lower than V_{DC}) by using a PI controller while achieving a settling time $T_{a,F} = 9\,\text{ms}$ (see Section 18.3.1). To check these performances, a step-wise voltage reference $v_o^*(k)$ jumping from 0 up to 6 V (i.e., $Y_{\text{ss}} = 6$) at 0.5 ms is applied.

Figure 18.8 shows the output voltage $y(k) = v_o(k)$ (blue line), which is correctly following the step-wise reference $y^*(k) = v_o^*(k)$ (red dotted line). The current flowing in the inductor $i_L(k)$ is subjected to a more oscillating dynamic than $v_o(k)$, but always below the upper bound $i_{\text{max}} = 2\,\text{A}$. The resulting waveforms of $v_o(k)$ and $i_L(k)$ follow an exponential behavior with oscillations superimposed on them, which is a typical case in second order systems.

Based on the design guidelines reported in Chapter 5, $\epsilon \approx 1\%$ is considered as reference accuracy. Thus, the transient is completed once the voltage enters in the region $[(1 - 0.01\epsilon)Y_{\text{ss}}\ (1 + 0.01\epsilon)Y_{\text{ss}}]$, where $T_a \approx 5\tau$ holds. Indeed, it results that $T_{a,F} = 9\,\text{ms}$, so the closed-loop control system is behaving as expected. The steady-state value Y_{ss} can be verified by applying the concepts reported in Sections 18.2 and 18.3. Furthermore, the closed-loop dynamics lead to a voltage ripple $\Delta v_o \approx 56\,\text{mV}$ and current ripple $\Delta i_L \approx 423\,\text{mA}$, which are consistent with the predicted values (see Section 18.2 to check the requirements). The effects of the step variation in the reference are reflected into the switching patterns of 1_{A} and 1_{B}, which are aimed to make the average voltage $v_{\text{avg}}(k)$ increase by enlarging the pulse width of $v(k)$. This is evident from Figure 18.9, which shows a portion of the voltage transient to relate current and voltage dynamics to the switch positions and modulation behavior. The command-tracking performances are achieved thanks to the regulation of $u(k) = d(k)$ which effectively approximates $v_{\text{avg}}(k)$. In particular, the load current $i_o(k)$ is filtered thanks to the capacitor, which absorbs all the high-frequency components. Indeed, $i_c(k)$ shows the same ripple of $i_L(k)$.

Note that the control variable $u(k)$ (i.e., duty cycle dynamic) is indirectly bounded by the energy flow capabilities of a power supply, which is a static non-linearity. For this case study, a unidirectional DC source operating within

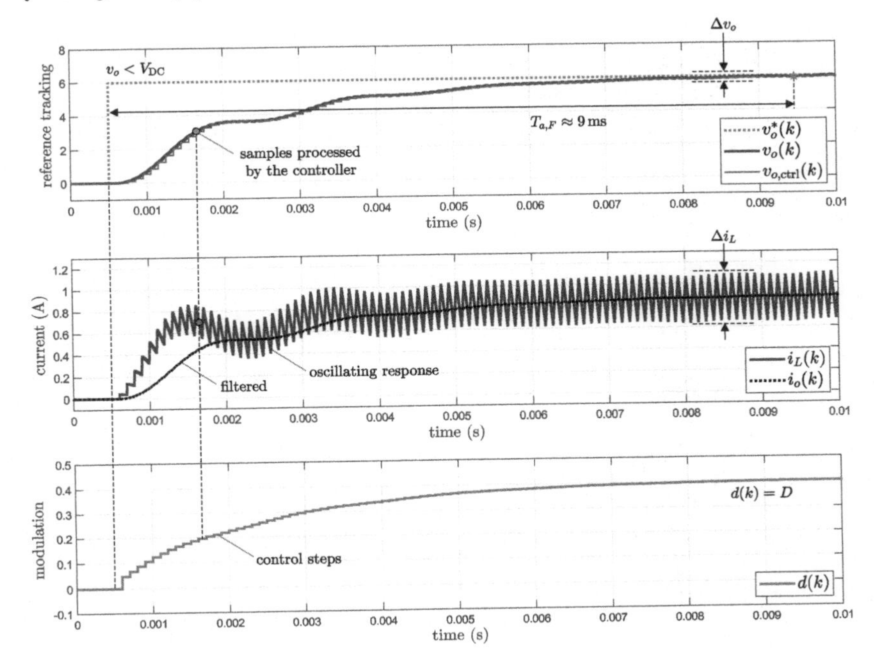

Figure 18.8 Closed-loop dynamics considering a step in the voltage reference.

[0 15 V] is considered. Thus, referring to $d(k)$, $u_{\text{min}} = 0$ and $u_{\text{max}} = 1$ holds. An anti-windup PI controller based on the back-calculation method is considered, with the back-calculation coefficient Kb equal to $k_{aw} = 1/T_i = k_i/k_p$. As reported in Section 6.2, the anti-windup parameters should be defined inside the PID block as done for the RL load in Chapter 16. It can be noted that, due to the *voltage-mode* control principle, there is no need to scale the controller output by $1/V_{\text{DC}}$, since it is a duty cycle already.

Besides the verification of the settling time of a closed-loop system, the choice of the sampling time to avoid issues related to peripheral and task scheduling, e.g., controller calculation and actuation updates, is investigated, as discussed in Section 13.10 and 17.4). To this purpose, the elements involved in the closed-loop control scheme loaded into the MCU, i.e., voltage reading (ADC module), 2L-PWM acutation (ePWM module) and a PI-based controller, are scheduled to be executed within the sampling period $T_s = T_{sw} = 1/f_{sw}$. Therefore, the target hardware is subjected to the following *sequential* events execution:

- At the beginning of T_s, when the value of the PWM counter equals the PWM counter period (TBCR=TBPRD, that simply equals 1 in Simulink® simulations), the ePWM peripheral is center-aligned (up-down counting mode) and it triggers the start-of-conversion (SOC) event for the ADC module;

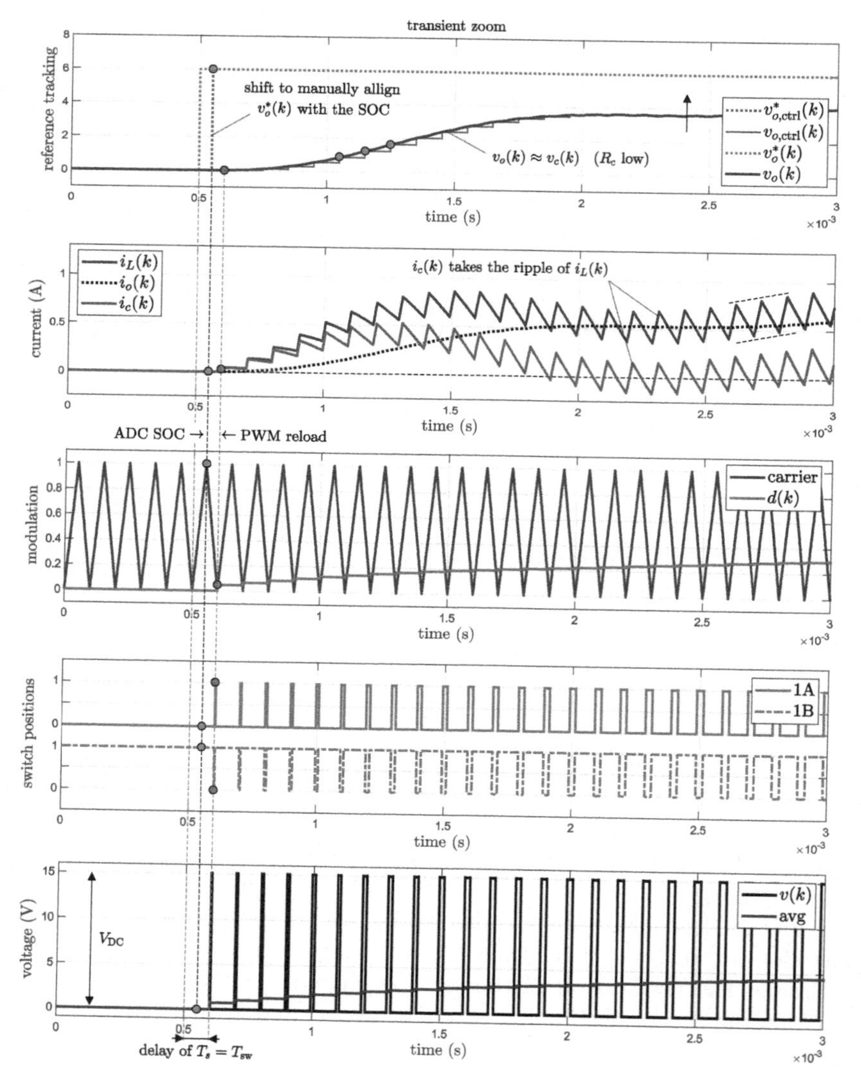

Figure 18.9 Detailed view of the closed-loop dynamics, which relates the current/voltage dynamics to the switch positions and modulation behavior.

- The ADC module converts the sampled analog signal into digital counts and triggers the end-of-conversion (EOC) event;
- The EOC triggers the ADC interrupt for the controller, which reads the output voltage value (sample);
- Given the PI structure, the controller processes the voltage error to compute

the modulating signal for the ePWM module. The computation interval required to obtain $d(k)$ is expected to be sufficiently lower than T_s;

- The resulting duty cycle is held until the end of the sampling interval and it is sent to the adopted ePWM module. As already explained, the actuation happens at the reload of the ePWM peripheral.

Therefore, the controller adopted in the Simulink® simulation is consistent with such event sequence. Indeed, the scheme reported in Figure 18.6 use the same approach adopted for the current control of an RL load (Section 17.4).

In particular, a trigger signal SOC (i.e., a boolean square-wave) is generated inside the modulator subsystem every time the carrier equals 1. The SOC is used as trigger source for a triggered subsystem which realizes a voltage controller synchronized with the carrier peaks. A voltage reference $v_o(k)$ internally generated by a Pulse Generator block is considered (sample based, with gain $A = 6$, number of samples $n_s = 500$, pulse width 50 %, phase delay = 5, and sample time T_s). This is a square-wave of period $500T_s = 25\,\text{ms}$. The Pulse Generator block is synchronous with the zero value of the triangular carrier. Since the ADC SOC trigger is generated at the carrier peak, $v_o^*(k)$ is delayed by $T_s/2$ to provide a correct voltage error at the controller input.[6] The resulting reference voltage is called $v_{o,\text{ctrl}}^*(k)$, where the subscript ctrl refers to the quantity that are processed by the controller. Hence, the processed voltage is called $v_{o,\text{ctrl}}(k)$. Even if the controller computations return a value for $d(k)$ before the end of $T_s = T_{\text{sw}}$, the controller has to wait the next reload PWM action to feed $d(k)$ to the ePWM input (this happens when CTR=zero). Therefore, another Delay block (sample time $T_s/2$ and data length 1) is included at the controller output to synchronize the actuation with the zero value of the carrier. The whole event sequence is denoted in Figure 18.9.

As did for the open-loop test, the closed-loop dynamics are also evaluated in presence of load variations. Figure 18.10 shows how the system reacts to such perturbation occurring at $t_R = 8\,\text{ms}$.

Immediately after the operation of the switch, the inductor current $i_L(k)$ cannot instantaneously increase since it is a state variable. Thus, the load current $i_o(k)$ increase (due to the lower resistance) is supplied by the capacitor, which starts to discharge. Then, $i_L(k)$ starts to increase too. The capacitor goes on discharging until the new steady-state value for $i_o(k)$ is reached. This current compensation is evident in Figure 18.10. In addition, it can be noted that the output voltage quickly stabilizes around $v_o(k) = v_o^*(k)$ after the transient.

Considering the achieved aforementioned performances, the designed PI controller meets all the specifications.

For the sake of completeness, the close-loop dynamics can be compared with those provided by a faster controller. Therefore, $T_{a,F} = 3.6\,\text{ms}$ which

[6]This delay can be included through a Delay block, where the sample time is set equal to $T_s/2$ and the delay length is 1.

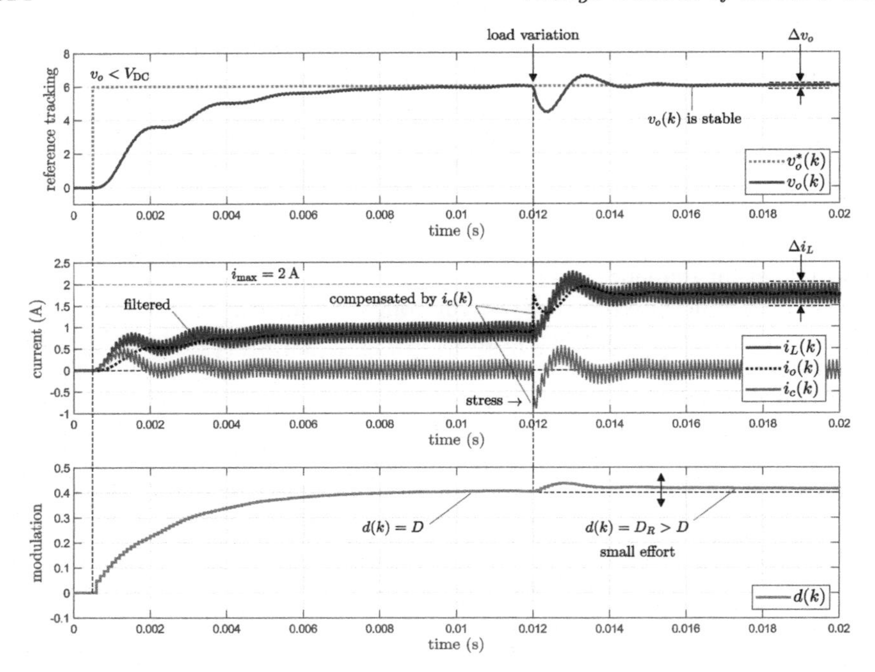

Figure 18.10 Closed-loop dynamics considering a load variation at $t_R = 8\,\mathrm{ms}$.

implies $\tau_F = T_{a,F}/5 = 714\,\mu s$ and a controller bandwidth equal to $\omega_c = 1/\tau_F = 1400\,\mathrm{rad/s}$ are considered. Moreover, by setting a phase margin $\phi_m = 90°$, `pidTuner()` returns $k_p = 0.0133\,\mathrm{V}^{-1}$ and $k_i = 79.68\,(\mathrm{Vs})^{-1}$. Keeping the other settings as before, the resulting plots are reported in Figure 18.11.

To achieve a faster voltage response $v_{o,f}(k)$, the controller feeds the system with higher energy (i.e., $d_f(k) > d(k)$) exciting the resonance at ω_o more than before and enlarging the oscillating effects. Likewise, the dynamics of $i_{L,f}(k)$ and $i_{o,f}(k)$ may reach more stressful values, even if the current ripple is comparable to the previous case.

18.5.1 Control implementation

The closed-loop dynamics are tested on a test-bench which includes a TI BOOSTXL-DRV8301 converter mounted on the TI LaunchXL F28069M Piccolo™ board and connected to the extRL(C) board in *RLC* mode (see Figure 18.1). The extPot3 board is also used to manipulate analog inputs.

Since only leg (phase) A is used for this implementation, the connection scheme is reported in Figure 18.12 accordingly. Therefore, one ePMW module is used only and it is operated in up-down counting mode at $f_{sw} = \mathrm{PWM_{frequency}} = 10\,\mathrm{kHz}$. Reminding that for the F28069M board

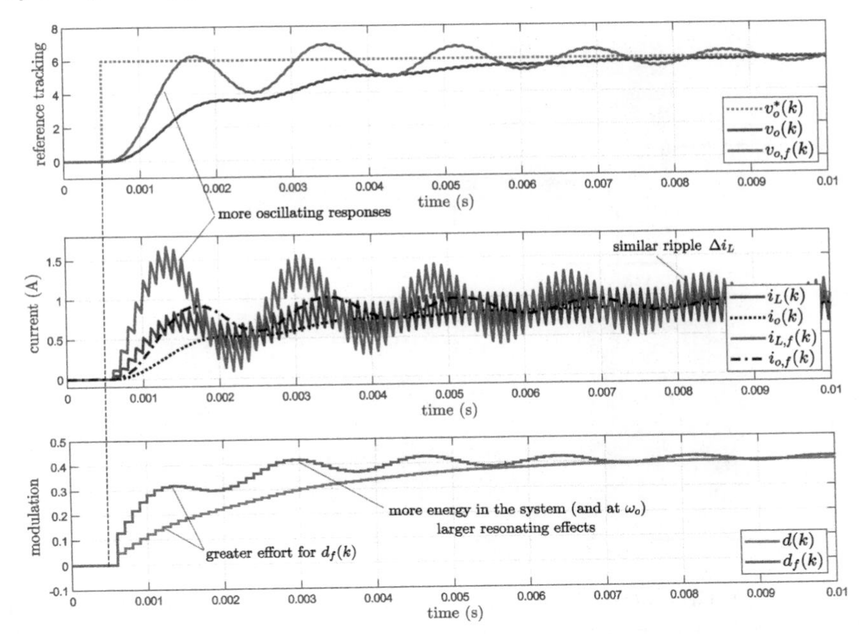

Figure 18.11 Comparison of the closed-loop dynamics for two different control design. The fastest one is denoted by subscript $_f$.

$f_{\text{ck}} = \text{CPU}_{\text{frequency}} = 90\,\text{MHz}$, it follows:

$$\text{TBPRD} = \text{PWM}_{\text{counter-period}} = \frac{\text{CPU}_{\text{frequency}}}{2 \cdot \text{PWM}_{\text{frequency}}} = 4500 \tag{18.37}$$

where the sampling time is $T_s = T_{\text{sw}} = 1/\text{PWM}_{\text{frequency}}$.

The reference voltage can be internally generated in the firmware or externally provided both using the available potentiometers on the extPot3 board or through COM port (e.g., SCIA). The analog solution is adopted to add versatility to the control scheme in the proposed solution. The needed parameters can be initialized in *Model Properites/Callbacks/InitFcn* or in a separate `m`-file as done for simulation.

A step-by-step procedure for programming the microcontroller for this task is reported here in the following.

Firmware Environment

c28069_RLCcontrolSD_F.slx (solver: fixed step - ODE4, step size: $T_s = T_{\text{sw}}/2 = 50\,\mu\text{s}$)

Open a new blank Simulink® project and configure the environment as shown

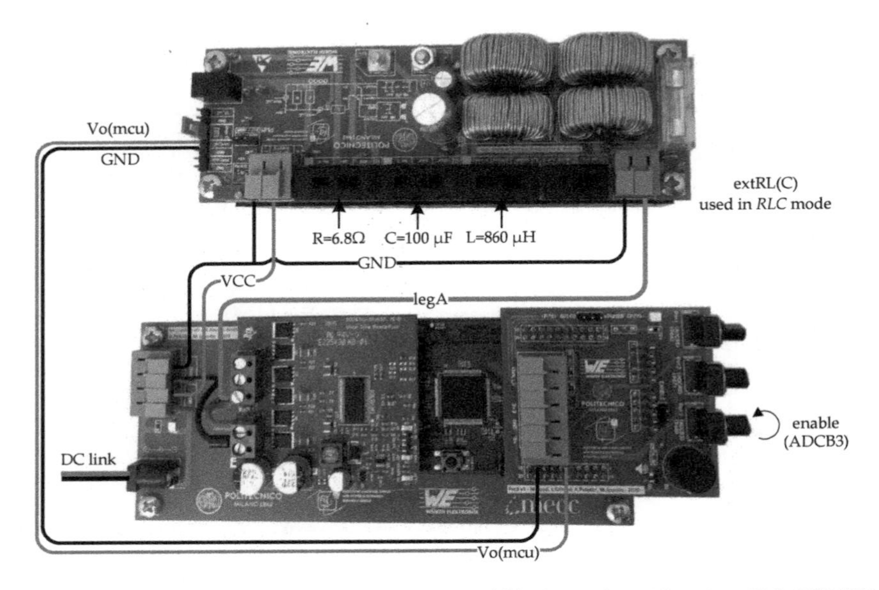

Figure 18.12 Connections of the extRL(C) board with the BOOSTXL-DRV8301 converter realizing a step-down DC-DC converter configuration.

in Section 9. Remember that the BOOSTXL-DRV8301 is enabled by settling **GPIO50** high (see Section 3.2).

- Insert an **ePWM** block and use the same settings shown in Section 13.5:
  - Select **ePWM1** module with TBPRD corresponding to $f_c = 10\,\text{kHz}$, i.e., $\text{PWM}_{\text{counter-period}} = 4500$;
  - Remember to unflag Enable ePWM1B and tick Inverted version of ePWMxA. This option allows to drive 1_B as the complement of 1_A;
  - Keep all the other settings as reported in Section 13.5.
- Add a **gain** block to normalize the modulation signal over the peak value of the carrier signal $\text{PWM}_{\text{counter-period}}$. Include a **data type conversion** set on `int16` to represent the duty cycle within a 16-bit base.

For up-down counting mode, the measurement of an average current $i_L(k)$ or voltage $v_o(k)$ is feasible thanks to the ePWM module which sends a synchronization signal that triggers the SOC of an ADC module when the carrier equals TBCTR=TBPRD, as shown previously. To this aim, the following settings are summarized here below:

- In the **Event Trigger** (Tab) of ePWM1
  - Select **Enable ADC start of conversion for module A**;

  - Set **Number of event for SOCA to be generated** to First event;
  - Select Counter equals to period (CTR=PRD) as **Start of conversion for module A event selection.**

Every time the carrier signal reaches the maximum value TBPRD, a SOC input (i.e., ePWM1_ADCSOCA) is sent to the ADC. This last block has to be configured as well. The extRL(C) board (see Appendix B) has an external voltage sensor, which has to be connected to one of the BOOSTXL-DRV8301 ADC channels.

- Add an **ADC** block and select **ADCINA5** as **Conversion channel** in the **Input Channels** tab. Then, go back in the **SOC Trigger** tab;
  - **Sampling mode**: Single sampling mode;
  - **SOC trigger number**: SOC2;
  - **SOCx acquisition window**: 7;
  - **SOCx trigger source**: ePWM1_ADCSOCA;
  - **ADC will trigger SOCx**: No ADCINT;
  - **Sample time**: T_s;
  - **Data type**: uint16.
- Add a **Data Type Conversion** block set to int16 (since positive signals only are used) or `single` to exploit 32 bit.
- Place a **Gain** block for scaling the digital signal over $[0, 1]$, since a 12-bit reading returns values ranging from 0 (0 V) up to $2^{12} - 1 = 4095$ (3.3 V).
- Build a proper conversion chain for the voltage sensing by removing the estimated offset as well as using a scaling factor $g_{v,\text{adc}}$ to obtain Volt values after moving to `single` data type.

  The external voltage sensor is essentially a voltage divider which scales $v_o(k)$ within a $[0, 3.3]$ V range (suitable for MCU inputs). Since the *RLC* board is designed to sustain higher voltage than the one used in this exercise, the voltage divider uses $R_{\text{high}} = 15\,\text{k}\Omega$ and $R_{\text{low}} = 1\,\text{k}\Omega$ to achieve:

$$v_{o,\texttt{mcu}}(k) = v_o(k)\frac{R_{\text{lo}}}{R_{\text{up}} + R_{\text{lo}}} \qquad \text{if } v_o(k) = 6\,\text{V} \rightarrow v_{o,\texttt{mcu}}(k) \approx 375\,\text{mV} \tag{18.38}$$

  Similarly, $15V \rightarrow v_{o,\texttt{mcu}}(k) \approx 937.5\,\text{mV}$, thus, a limited portion of the $[0, 3.3]$ V voltage range is used. This implies that the voltage conversion chain is characterized by the following gain:

$$g_{v,\text{adc}} = \frac{3.3}{4095}\frac{R_{\text{up}} + R_{\text{lo}}}{R_{\text{lo}}} \tag{18.39}$$

  As done for the low-side shunt (Chapter 16), the voltage sensor should be

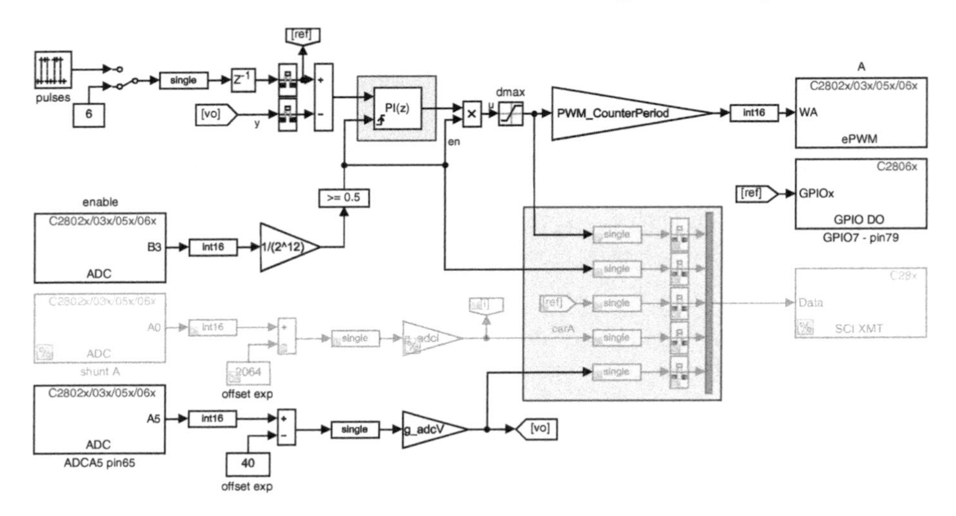

Figure 18.13 Simulink® scheme included in c28069_RLCcontrolSD_F.slx.

characterized from an experimental point of view to include the variations given by the resistors tolerance (e.g., the setup provides some extra wiring connections other than extra contact resistances) as well as the estimation of the experimental offset (e.g., 40), as shown in Figure 18.13.

- Use the same discrete **PID controller** block adopted in simulation. Set the controller type as PI (Parallel) with sample time T_s; set P=k_p and I=k_i. Select an external source for the **reset** with **rising** condition. Saturate the controller output to keep the duty cycle in the range $[0, 1]$. Set a back-calculation method as anti-windup strategy with Kb=k_{aw}.

- A redundant saturation block can be used to protect the system by limiting $d(k)$ besides the limits provided by the PID block.

- Add a **Pulse Generator** (sample based) block to create the reference $v_o^*(k)$. Set the amplitude equal to $A = 6$, number of samples $n_s = 500$, pulse width 50 %, phase delay 5, and sample time T_s. It is equivalent to have $v_o^*(k)$ generated by a square-wave of period $500T_s = 25\,\text{ms}$. It can be noted that $v_o^*(k)$ is synchronized with the zero value of the triangular carrier.

- Add a **Delay** block to be consistent with the ADC SOC trigger generated at the carrier peak. Set sample time $T_s/2$ and delay length 1.

Arrange these aforementioned blocks as shown in Figure 18.13.

In order to manually manage the activation of the PI controller, the external reset (rising) is connected to a potentiometer (ADCINB3/pin68). Without moving it, the control algorithm is forced not to provide any actuation signal. To do that:

- Add an **ADC** block and select ADCINB3 as **Conversion channel** in the **Input Channels** tab. Then, go back in the **SOC Trigger** tab:
  - **Sampling mode**: Single sampling mode;
  - **SOC trigger number**: SOC1;
  - **SOCx acqusition window**: 7;
  - **SOCx trigger source**: Software;
  - **ADC will trigger SOCx**: No ADCINT;
  - **Sample time**: $T_{sig} = 0.001\,\mathrm{s}$;
  - **Data type**: uint16.
- Add a **Data Type Conversion** block set to int16 (since positive signal only are used) and the 12-bit scaling to move from a range (0 V) up to $2^{12} - 1 = 4095$ (3.3 V) to $[0, 1]$.
- Add a **Compare to Constant** block which is used to evaluate if signals are greater or equal than a Constant value 0.5. If this condition is true, the output switches from zero to one resetting the integrator of the PI controller. The enable signal is defined as $\mathrm{en}(k)$.
- Additionally, $\mathrm{en}(k)$ can be used to add a further protection redundancy. Indeed, by using a Product block, the enable signal is multiplied by the actuation variable. As long as $\mathrm{en}(k) = 0$, the duty cycle is forced to stay equal to zero $d(k) = 0$, i.e., no switching is performed.
- To visually evaluate the pulsed reference $v_o^*(k)$, the Pulse Generator can be connected to a **GPIO** block (e.g., GPIO7) with a proper scaling of $1/6$ to normalize the signal in a $[0, 1]$ range. This processing allows to visualize the reference signal ranging in $[0, 3.3]$ V at pin 79 through an oscilloscope.

Optional: the firmware could be also prepared to visualize the duty cycle $d(k)$ through a DAC peripheral. In this case:

- Copy the **ePWM** block, select ePWM8 (which is connected to DAC3 in J8, pin 72), keep the other settings. Connect this block in parallel to the other ePWM (i.e., connect $d(k)$ just after the Data type conversion block). Hence, $d(k)$ can be observed by connecting a voltage probe to pin 72.

Figure 18.14 reports the closed-loop behavior of the system obtained by executing the firmware and observing the current and voltages of interest through an oscilloscope. In particular, voltages $v_o(k)$ and $v_o^*(k)$ are measured by using voltage probes (with $v_o^*(k)$ scaled on GPIO7), while the currents $i_L(k)$ and $i_o(k)$ are measured with current probes. In case DAC3 is used to check $d(k)$, another voltage probe connected to pin 72 is needed.

As done in simulation, the analysis of Figure 18.14 starts from the evaluation of the command tracking performances. The resulting $v_o(k)$ (blue line)

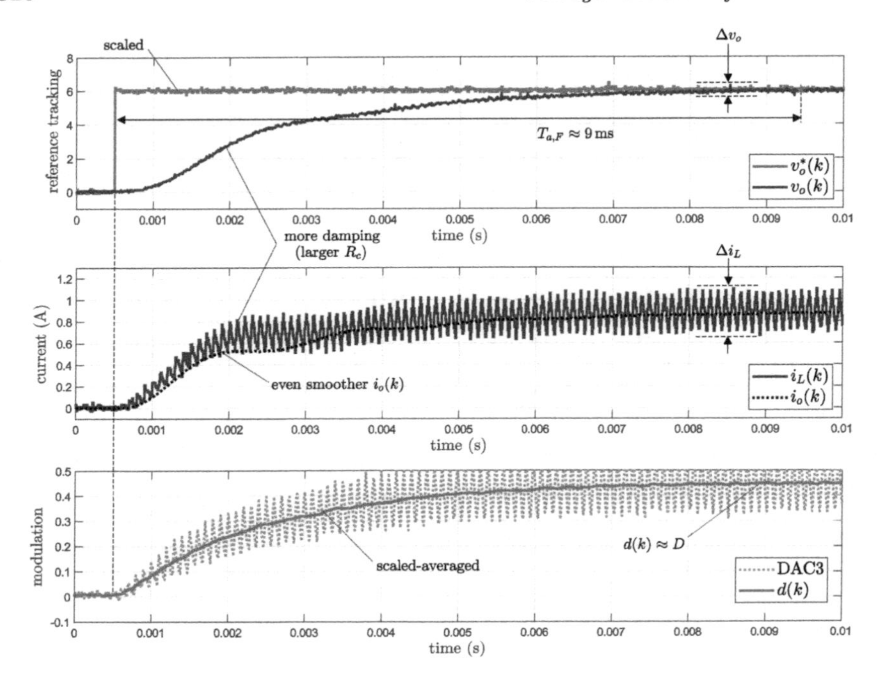

Figure 18.14 Closed-loop dynamics measured from the F28069M board pins through an oscilloscope.

is following the internally generated $v_o^*(k)$ (red line), which is re-scaled from 3.3 V on the actual reference value in Volt. The settling time is similar to the expected one $T_{a,F} = 9\,\mathrm{ms}$ even if both voltage and current dynamics looks smoother (i.e., reduced oscillating responses) during transient. Besides the probe accuracy, this effect is mainly due to the parasitic resistances R_L and R_c, which differ from the theoretical ones. In particular, R_c is a great matter of concern since even a slightly increase has a significant impact on the damping coefficient ξ, while the settling time $T_{a,F}$ is almost unaffected.

Consequently, the current ripple is $\Delta i_L \approx 382\,\mathrm{mA}$, which is lower than the value predicted in simulation. The voltage ripple is $\Delta v_o \approx 100\,\mathrm{mV}$, which is larger than the value predicted in simulation. However, both Δi_L and Δv_o are compliant with the requirements. This last discrepancy could be a consequence of a more noisy environment. The Pulse generator which provides the internal reference $v_o^*(k)$ is synchronized with the zero value of the triangular carrier (TBPRD=0). Nevertheless, such synchronization is not as strict as other tasks, e.g., counter comparison in ePWM, which is based on hard synchronization (hard-sync) event through interrupts. Therefore, it might happen that the rising edge of the step-wise $v_o^*(k)$ is not perfectly synchronized with the ADC SOC (i.e., carrier peak). By adding a fixed shift of $T_s/2$, $v_{o,\mathrm{ctrl}}^*(k)$ is coincident with ADC SOC, thus, sampling a value of $v_{o,\mathrm{ctrl}}^*(k) = 6\,\mathrm{V}$. This is the same

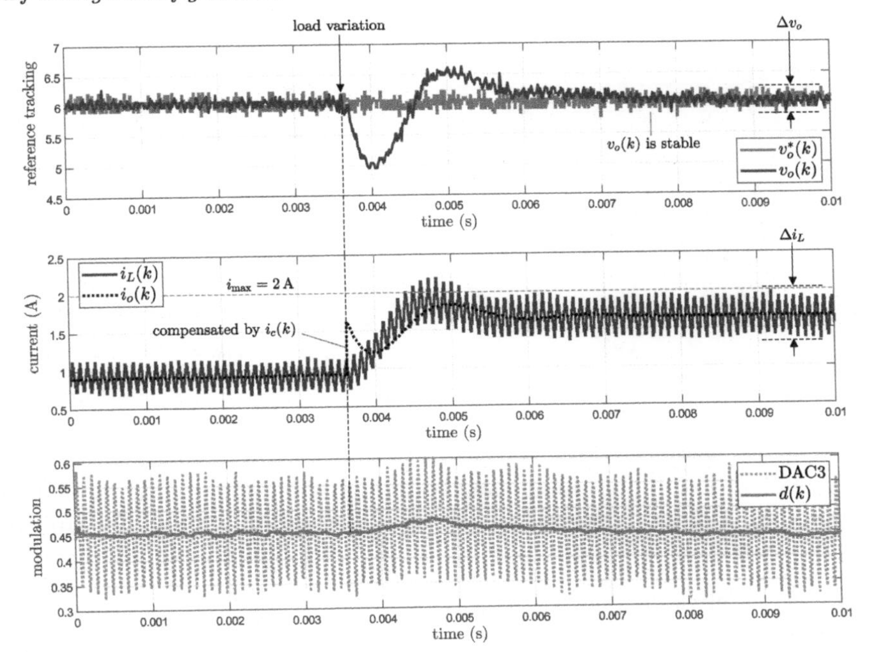

Figure 18.15 Closed-loop dynamics with a load variation at $t_R = 8$ ms observed through an oscilloscope.

processing obtained for the RL load, thus, the reader is referred to the detailed description reported in Section 17.4.1. Based on this part, the overall interval between the the rising edge of $v_o^*(k)$ and the start of the system dynamics[7] (i.e., when $v_o(k)$ and $i_L(k)$ start to increase due to $v(k) \neq 0$) reaches $2T_s$ in a worst case scenario. Figure 18.15 also shows the duty cycle reading through DAC3. The signal read at pin 72 (gray dotted line) needs to be additionally averaged to obtain $d(k)$ (red line). This can be easily done via software by using an average function or a digital filter.

As did in simulation, the closed-loop dynamics are also evaluated in case of variations in the resistance value. This is performed by manually moving the switch S3 mounted on the extRL(C) board after the system reaches a steady-state value. The operation changes the resistance down to $R_{\text{eq}} = R_o/2$. Figure 18.15 shows how the system reacts to such perturbation. The current compensation behaves as explained in the previous Section. In particular, it is shown that the inductor current $i_L(k)$ does not increase instantaneously. Thus, the variation in the load current $i_o(k)$ is compensated by the capacitor through its branch current $i_c(k)$. The $d(k)$ signal is also visible. Both transients and amplitudes are consistent with simulations, validating the stability of the voltage loop to $v_o(k) = v_o^*(k)$ even in case of load variations.

[7]Delay in the converter actuation.

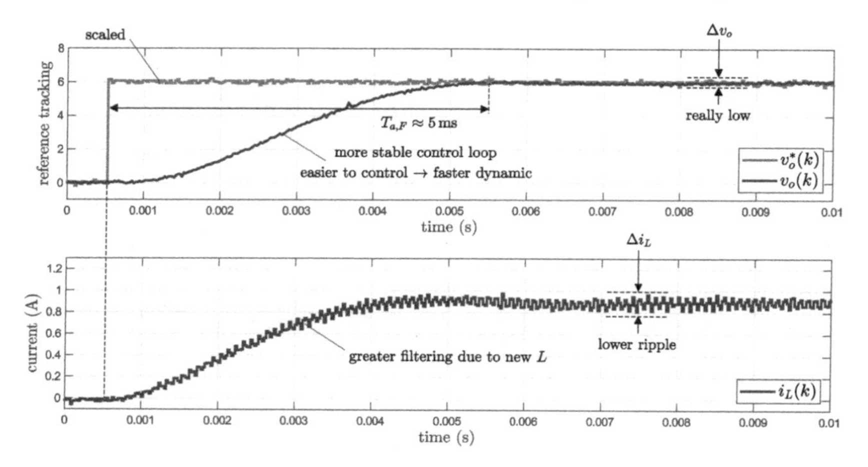

Figure 18.16 Closed-loop voltage and current dynamics measured through an oscilloscope. A large inductor $L = 3.4\,\mathrm{mH}$ is considered, keeping $C = 100\,\mu\mathrm{F}$.

Remark: as already reported in Section 18.2 and 18.3, the parasitic resistances R_L and R_c strongly impact on the damping coefficient ξ. The value of this parameter might be also influenced by snubbers resistors and non-zero impedance of the power supply. Moreover, resistance of the wires and/or the traces of PCBs may impact on ξ too, leading to higher system losses too.

Remark: the firmware can be designed to process the inductor current by using the low-side shunt resistor of leg A. Figure 18.13 already reports such configuration based on the settings provided in the Chapter 17. This measurement can be used in the control loop to (strictly) restrict the current dynamic below $i_{\max} = 2\,\mathrm{A}$, leading to a more advanced control scheme, e.g., a two-degree-of-freedom (2DOF) or a model predictive control-based [40, 38].

18.6 Variations of *LC* Filter Parameters

To evaluate the effectiveness/robustness of the controller, the extRL(C) board allows to change all the *RLC* parameters, see Appendix B. The main purpose of the designed regulator is to guarantee a stable output voltage $v_o(k) = v_o^*(k) < V_{\mathrm{DC}}$ even in case of variations in the filter parameters. Previously, it was demonstrated that the tracking of $v_o^*(k)$ is guaranteed in presence of a variation of R, thanks to the sudden changes of $d(k)$, $i_o(k)$, $i_c(k)$ and $i_L(k)$. Even the *LC* filter parameters can be manually varied. The inductance can be increase from $L = 860\,\mu\mathrm{H}$ up to $L = 4 \cdot 860\,\mu\mathrm{H}$, while the capacitance may decrease from $C = 100\,\mu\mathrm{F}$ to $C = 10\,\mu\mathrm{F}$. The inductors belong to the same

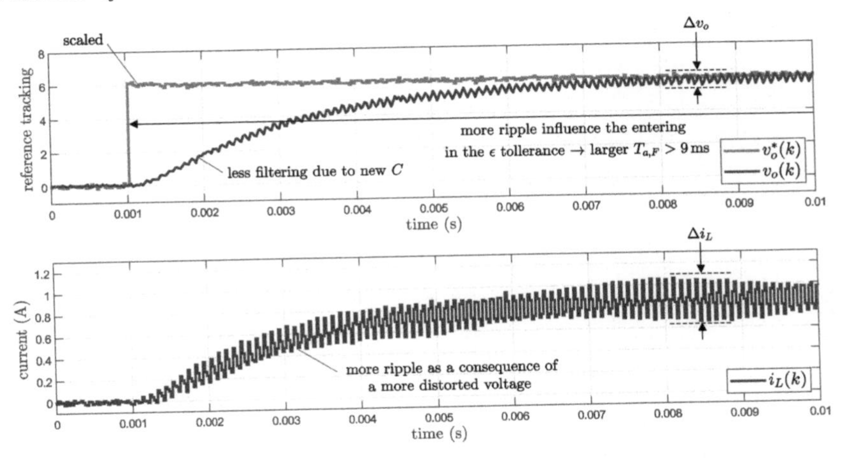

Figure 18.17 Closed-loop voltage and current dynamics measured through an oscilloscope. A small capacitor $C = 10\,\mu\text{F}$ is considered, keeping $L = 860\,\mu\text{H}$.

family (therefore, R_ℓ is the approximately same), while the capacitors shows different R_c (see Appendix B). Since the closed-loop voltage controller is designed on $L = 860\,\mu\text{H}$ and $C = 100\,\mu\text{F}$ (i.e., the reference case), a variation in the LC filter parameters implies different settling time and closed-loop dynamics. Such analysis is useful to understand which parameter variation may lead to potential damage or instability in the system. The firmware architecture is left unchanged.

By manually varying the inductance up to $L = 3.4\,\text{mH}$ (keeping $C = 100\,\mu\text{F}$), the inductor current $i_L(k)$ is largely filtered as well as the load current $i_o(k)$ (see Figure 18.16). A higher inductance value leads to a more stable control loop (i.e., higher stability margin) making it easier to control. This facilitate the controller to regulate the voltage, which is subjected to a lower distortion too. A faster dynamic is achieved as well, that is, $T_{a,F} \approx 5\,\text{ms}$. Moreover, a four times larger inductance implies a four time larger R_ℓ, which is $R_\ell \approx 450\,\text{m}\Omega$. This variation causes an increase in the damping coefficient ξ, reducing the oscillating effects in the voltage/current responses. Moreover, the current ripple is lower than the reference case. Conversely, by manually changing the capacitance value down to $C = 10\,\mu\text{F}$ (keeping $L = 860\,\mu\text{H}$), the capacitor voltage and the output voltage $v_o(k)$ are less filtered than before, as shown in Figure 18.17. This indirectly affects the inductor current $i_L(k)$, which shows a more bouncing response. Nevertheless, the ESR of the new capacitor is $R_c \approx 470\,\text{m}\Omega$, being larger than before. Hence, even if the voltage is less filtered, the damping coefficient is higher, reducing the oscillating effects in both voltage and current waveforms. This is evident comparing Figure 18.17 with 18.14.

Despite both perturbations on the LC parameters, the command tracking is still guaranteed.

19

Cascade Speed Control of a Permanent Magnet DC Motor

After the development of control strategies for static systems such as RL and RLC loads, this chapter shows how to realize a cascade speed control for PMDC motors. The same setup adopted in Chapter 15 is considered for the implementation. In particular, two different arrangements are considered, namely involving a *single motor* and two PMDC machines connected in *back-to-back* (B2B) configuration.

The DecMot Hardware Kit (see Appendix B and Figure 19.1) is adopted to test a single PMDC motor. A Full-Bridge converter configuration is considered (thus, two legs of the TI BOOSTXL-DRV8301 converter are involved) operating in unipolar voltage switching. The DecMot Hardware Kit includes a rotary encoder[1] to measure the rotor speed. This allows to design a closed-loop control scheme based on a speed sensor. The encoder interface is connected to the extPot3 board to provide the speed measurement to the LaunchPad™ F28069M board through the eQEP1/eQEP2 sockets. On the contrary, a sensorless (or encoderless) closed-loop control scheme can be also designed. Among the many observers and estimators formulations available in literature, this chapter presents a model reference adaptive system (MRAS) observer to estimate the rotor speed, commenting its integration/design for a closed-loop scheme and related effects on the closed-loop dynamics. In both cases, aiming to follow a given speed reference $\Omega^*(k)$, the nested PI controllers lead to regulate the dc output voltage of the converter by manipulating the command signal of the ePWM peripherals of the two legs, e.g. $d_\mathrm{A}(k)$ and $d_\mathrm{B}(k)$. The reference speed value can be provided by an external command, which may come from serial communication, or by the extPot3 potentiometers, or internally (e.g., through a pulse generator/constant block). The B2B-PMDC Hardware Kit (see Appendix B and Figure 19.2) is adopted to test two PMDC motors arranged in a B2B configuration. Two Half-Bridge converters are considered to drive each motor (thus, two legs of the TI BOOSTXL-DRV8301 converter are adopted even this time), allowing one direction of rotation for the shaft only. The PMDC motors are mechanically coupled, with one of them behaving as a traction motor, while the other one as a braking one. Such configuration is widely adopted when it is required to characterize or validate

[1]This device is connected to the PMDC motor through a mechanical coupling.

DOI: 10.1201/9781003196938-19

Figure 19.1 DecMot Hardware Kit rearranged for implementations with one PMDC motor only connected to a DC-DC converter (see connection scheme in Figure 19.15). The reader is referenced to Appendix B for further details on this setup.

the operating region/plane of an electrical motor. Indeed, the two motors can be separately controlled. Assuming that one PMDC motor is tracking a speed reference (full cascade structure) while the other is aimed to regulate a friction/braking torque applied to the shaft (i.e., armature current control only), the controlled B2B system is equivalent to a cascade speed control of a PMDC (traction) motor subject to a load torque that can dynamically vary in time. For instances, this allow to fix the speed while moving along the torque axis of the operating region. Sensored (or encoderless) control scheme can be designed as well.

A single motor control behaviur is simulated and analyzed first through Simulink® simulations, being useful even to describe the B2B test case. Then, different firmwares are designed and implemented on the LaunchPad™ F28069M board based on these results. It must be noted that no flux weakening is needed for such motors, since the excitation circuit is replaced by PMs. Moreover, due to the limited size of the PMDC motor, the presence or absence of a back-emf compensation do not particularly impact on the system dynamic. Hence, it is up to the reader to activate (or not) such compensation terms in the designed algorithms.

Figure 19.2 B2B-PMDC Hardware Kit allowing implementations involving two PMDC motors arranged in back-to-back configuration (see connection scheme in Figure 19.23). The reader is referenced to Appendix B for further details.

19.1 Required Hardware

- TI LaunchPad™ Piccolo™ F28069M and BOOSTXL-DRV8301 boards;
- extPot3 custom board (see Appendix B for technical details);
- DecMot Hardware Kit for the control of one PMDC motor (see Appendix B and Figure 19.1);
- B2B-PMDC Hardware Kit for the control of two PMDC motors coupled in back-to-back configuration (see Appendix B and Figure 19.2);
- Power supply set on 10 V and limited to 10 A (i.e. 100 W).

Basically, the setups used in this chapter adopt the same hardware described in Chapter 15. The same recommendations shown there are still valid. Namely, jumpers JP1 and JP2 should be removed to isolate the USB/JTAG port. Moreover, keep in mind that the BOOSTXL-DRV8301 converter requires a dc-link voltage $V_{DC} > 6$ V to properly work, otherwise the `nFAULT` is returned (regardless of the state of the jumpers JP1 and JP2). The reader is referred to the converter details reported in Section 3.2.

In the proposed control schemes, the low-side shunt-resistors are used as current sensor to close the inner current loop (their characterization explained in Chapter 16), whereas the rotary encoder provides the feedback for the speed loop (see Chapter 14 for all details on this device and on the eQEP peripherals).

For the single motor configuration, the LaunchPad™ F28069M board is connected to the BOOSTXL-DRV8301 converter and it is programmed to follow an internally generated square-wave reference speed. This holds also for the B2B configuration, which also includes an additional current loop to manage the dynamic braking acting on the shaft. The application of load torque is related to the rotation of a linear potentiometer on the extPot3 board (ADCB5). As done in Chapter 17 and 18, another potentiometer is used as control enable (ADCA3). The reading of the corresponding ADC channels must be suitably re-scaled. Debugging signals can be sent out of the MCU through SCIA/B (virtual COM port). The DC link voltage is again provided by a power supply set to $V_{\mathrm{DC}} = 10\,\mathrm{V}$, with the output current limited to 10 A. The latter value is higher than in Chapter 15. Indeed, for the B2B configuration, the application of a braking torque on the shaft leads to a current rise, which has to be sustained by the power supply while fulfilling the motor and converter data-sheets (e.g., the rated power of the PMDC motor is 100 W).

In the following, the analytical model of the motor control system is analyzed and a cascade control scheme is designed based on it. Then, some simulations of the controlled system are reported and considered as reference case for the implementation on hardware. Finally, experimental tests are carried out once the simulations returns satisfactory results.

19.2 Linear Model of a PMDC Motor

The model of a PMDC motor was already presented in Section 15.2. The main equations and transfer functions characterizing such machine are reviewed here in the following to introduce the adopted control strategies. The reader is referred to Section 15.2 for all the details on model derivation.

Considering the equivalent circuit reported in Figure 15.1, the equations characterizing the electrical and mechanical behavior of the system are reported in (15.2) and (15.3), that are:

$$
\begin{aligned}
L_{\mathrm{a}} \frac{di_{\mathrm{a}}(t)}{dt} &= v_{\mathrm{a}}(t) - R_{\mathrm{a}} i_{\mathrm{a}}(t) - e\,(t) \\
J \frac{d\Omega(t)}{dt} &= m_e(t) - m_l(t) - \beta \Omega(t)
\end{aligned}
$$

As previously mentioned, the magnetic field and the related flux flowing inside the machine is constant, being generated through permanent magnets. ψ_{PM}.[2] As a matter of fact, a constant K is defined as $K = \psi_{\mathrm{PM}} k_{\mathrm{T}} = \psi_{\mathrm{PM}} k_{\mathrm{e}}$.

[2]No flux weakening can be performed. Thus, the motor speed can not reach values above the rated speed (also called base speed).

Consequently, the electromagnetic torque and back-electromotive force can be computed as reported in equations (15.4) and (15.5), that are:

$$m_e(t) = K i_{\mathrm{a}}(t)$$
$$e(t) = K\Omega(t)$$

For the single motor configuration, the load torque refers to external causes such as additional friction or mechanical load coupled to the PMDC motor. If the configuration does not foresee the application of any load torque, then $m_l(t) = 0$. For the B2B configuration, the load torque is separately managed. Therefore, equation (15.4) is subject to the current control loop implemented for the braking motor. Considering this latter connected to the leg B of the converter, it follows

$$m_l(t) = K i_{\mathrm{B}}(t) \tag{19.1}$$

where $i_{\mathrm{B}}(t)$ is the current flowing in the armature circuit of the braking motor, whereas $m_l(t)$ is the variable load torque.

The state-space representation is derived considering $\boldsymbol{u}(t) = [v_a(t)\ m_l(t)]^T$, $\boldsymbol{x}(t) = [\Omega(t)\ i_{\mathrm{a}}(t)]^T$ and $\boldsymbol{y}(t) = \boldsymbol{x}(t)$. Therefore, the resulting equations in matrix notation are those reported in (15.9) and (15.10), that are:

$$\frac{\mathrm{d}}{\mathrm{dt}}\begin{bmatrix}\Omega(t)\\ i_{\mathrm{a}}(t)\end{bmatrix} = \begin{bmatrix}-\beta/J & K/J\\ -K/L_{\mathrm{a}} & -R_{\mathrm{a}}/L_{\mathrm{a}}\end{bmatrix}\begin{bmatrix}\Omega(t)\\ i_{\mathrm{a}}(t)\end{bmatrix} + \begin{bmatrix}-1/J\\ 0\end{bmatrix} v_{\mathrm{a}}(t)$$
$$\begin{bmatrix}\Omega(t)\\ i_{\mathrm{a}}(t)\end{bmatrix} = \begin{bmatrix}1 & 0\\ 0 & 1\end{bmatrix}\begin{bmatrix}\Omega(t)\\ i_{\mathrm{a}}(t)\end{bmatrix}$$

where the state-space matrices are defined as follows:

$$\boldsymbol{A} = \begin{bmatrix}-\beta/J & K/J\\ -K/L_{\mathrm{a}} & -R_{\mathrm{a}}/L_{\mathrm{a}}\end{bmatrix} \quad \boldsymbol{B} = \begin{bmatrix}-1/J & 0\\ 0 & -1/L_{\mathrm{a}}\end{bmatrix} \quad \boldsymbol{C} = \boldsymbol{I}_2$$

It is important to note that all the parameters of the considered motors (R_{a}, L_{a}, J, β, K) must be estimated through some identification procedures such as the locked rotor test (see Chapter 16) and by observing the open loop mechanical dynamics. These calculations are necessary (and they must be performed as fist task) to follow the controller design procedures presented in the previous chapters. Since the two system configuration are different (i.e., three different motors and two couple of boards and converters are involved), the results of the identification tests will be provided in the following sections for each case.

Finally, the transfer functions $G_i(s)$ and $G_\Omega(s)$, which refer to the electrical and mechanical behavior of the motor, respectively, are derived based on the motor equations in the time domain, as shown in equation (15.11):

$$G_i(s) = \frac{i_{\mathrm{a}}(t)}{v_{\mathrm{a}}(t)} = \frac{1}{R_{\mathrm{a}} + sL_{\mathrm{a}}} \qquad G_\Omega(s) = \frac{\Omega(t)}{m_e(t)} = \frac{1}{\beta + sJ}$$

The PMDC motor in single configuration and the traction motor in B2B

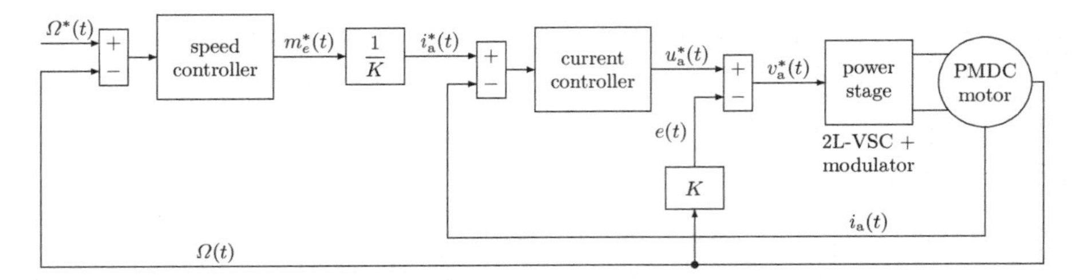

Figure 19.3 Block scheme for a cascade speed control of a PMDC motor with back-emf compensation. This strategy is valid both for Half- or Full-Bridge configurations.

configuration are controlled thanks to a current and a speed control loop which are designed based on $G_i(s)$ and $G_\Omega(s)$, respectively. Moreover, the braking motor in B2B configuration is controlled by a current controller only, which is still design over $G_i(s)$. Further details on the PI-controllers design are reported in the following sections.

19.3 Cascade Control Architecture and Design

The speed regulation of a PMDC motor is achieved by using a so called cascaded control architecture, which is based on two nested loop based on a speed and current controller, respectively. The goal of this scheme is to assure a reference speed tracking for $\Omega^*(t)$ by manipulating the voltage applied to the motor terminals through the power converter. The signal $\Omega^*(t)$ can be specified by changing the position of a potentiometer, by internally generating it or by sending this input via virtual COM port.

Based on to Figure 15.3, which shows the block scheme representation of the cascade speed control for PMDC motors, the back-emf $e(t)$ is the link between the electrical and the mechanical contributions. From a theoretical point of view, a cascade control architecture can be only adopted if the nested loops can be effectively decoupled. In particular, the external loop is required to regulate the shaft speed, as shown in Figure 19.3. The speed controller outputs a reference torque $m_e^*(t)$, which is translated in a reference armature current $i_a^*(t)$ by using the scaling factor $1/K$. Then, the inner current controller outputs a reference voltage $u_a^*(t)$. As a matter of fact, the back-emf $e(t)$ can be considered as an observable disturbance and it can be compensated as shown in Figure 19.3. According to the control theory terminology, this is called a *feed-forward action*, which is computed as:

$$\widehat{e}(t) = \widehat{K}\Omega \quad \text{or} \quad \widehat{e}(t) = \widehat{K}\widehat{\Omega} \tag{19.2}$$

where the superscript $\widehat{\ }$ refers to an estimated variable. Indeed, the feed-forward action is based on an estimation of the back-emf based on the motor parameters (i.e., it is subjected to the accuracy of the results obtained through the identification process) and the speed feedback, which can be returned by a sensor (e.g., an encoder) $\Omega(t)$ or by an observer (e.g., MRAS) $\widehat{\Omega}$.

This feed-forward term allows to move from $u_{\mathrm{a}}^*(t)$ to the reference armature voltage $v_{\mathrm{a}}^*(t)$, which represents the modulating signal to be processed by the modulator stage for generating the PWM signals. This stage is included into the power stage block (not drawn explicitly for the sake of brevity) shown in Figure 19.3, which includes a 2L-VSC (i.e., the DC-DC converter) and the PWM modulator.

The back-emf compensation provides an important contribution, since it represents and additional voltage term (on condition that the dc-link is able to provide the required power). It should be clear that the feed-forward term achieve an effective decoupling only at steady-state, while the cross-coupling persists during transients. Nevertheless, the importance of such compensation term depends on the machine design. Indeed, if the PMDC motor is of small size and low voltage rating, $e(t)$ can be neglected (i.e., no compensation). On the other hand, even the load torque $m_l(t)$ can be estimated and then compensated. However, for the simulation analysis, $m_l(t)$ is considered a controllable disturbance, thus, manipulating its instantaneous value. It can be noted that the current controller is of the same kind of that one designed in Chapter 17, i.e., it is based on an RL circuit (see $G_i(s)$ in (15.11)). In addition, a combined design of both speed and current control loops is needed to ensure that the mechanical dynamic is satisfactory.

Given the feed-forward action, the system is suitable to be controlled with a cascade scheme. This becomes quite evident if the power stage (i.e., DC-DC converter + modulator) and the PMDC motor shown in Figure 19.3 are substituted with the equivalent linear average representation of the motor dynamics shown in Figure 15.3. This require to approximate $v_{\mathrm{a}}^*(t)$ with $v_{\mathrm{a}}(t)$ (see Section 17.3.1).

In particular, it is important to remind that this architecture has to fulfill cascade constraints in terms of bandwidth between the nested loops, i.e., aiming to decouple the electrical and mechanical dynamics. The outer (speed) loop must be consistently slower than the inner (current) one. This means that, from the outer loop point of view, a current reference $i_{\mathrm{a}}^*(t)$ sent as input of the current loop is immediately provided as output $i_{\mathrm{a}}(t) = i_{\mathrm{a}}^*(t)$, i.e., allowing to approximate the inner closed-loop transfer function as a unitary gain. This requirement translates in a constraint on the bandwidths. Indeed, this behavior is guaranteed by imposing a cut-off frequency for the inner current controller at least ten times smaller than the outer speed one.

$$\omega_\Omega \leq \omega_\Omega/10 \qquad \text{or} \qquad \tau_{G_\Omega} \geq 10\tau_{G_i} \tag{19.3}$$

where ω_i and ω_Ω are the bandwidths of the current and speed controller, respectively. If equation (19.3) is satisfied, the two controllers can be designed separately, i.e., each one is tuned based on its specific transfer function. This means that the control action of the speed controller is not influenced by the dynamics of the inner loop, since the current regulator is much faster than the outer one.

The PI-based speed controller is defined as $R_\Omega(s) = k_{p,\Omega} + k_{i,\Omega}/s$ and the current one as $R_i(s) = k_{p,i} + k_{i,i}/s$, where the integrator time constants are $T_{i,\Omega} = k_{p,\Omega}/k_{i,\Omega}$ and $T_{i,i} = k_{p,i}/k_{i,i}$, respectively. By imposing equation (19.3), $R_\Omega(s)$ would be designed based on $G_\Omega(s)$ only, while $R_i(s)$ based on $G_i(s)$. It is worth noting that this simplifies the controllers design. Indeed, since both $G_\Omega(s)$ and $G_i(s)$ are 1st order transfer functions, $R_\Omega(s)$ and $R_i(s)$ can be explicitly designed using the pole/zero cancellation (i.e., still within linear control theory) as shown in Chapter 5. This leads to the following open-loop transfer functions:

$$L_i(s) = \frac{i(t)}{e_i(t)} = R_i(s)G_i(s) = \frac{k_{p,i}/R_\mathrm{a}}{sT_{i,i}} \tag{19.4}$$

$$L_\Omega(s) = \frac{\Omega(t)}{e_\Omega(t)} = R_\Omega(s)G_\Omega(s) = \frac{k_{p,\Omega}/\beta}{sT_{i,\Omega}} \tag{19.5}$$

where $T_{i,\Omega} = J/\beta$ and $T_{i,i} = L_\mathrm{a}/R_\mathrm{a}$. The open loop transfer functions behave as a pure integrators ($\phi_m \approx 90°$) and their cut off frequencies $\omega_{c,\Omega}$, $\omega_{c,i}$ depends explicitly on the system parameters. As shown in Chapter 5, it follows that:

$$k_{p,\Omega} = \omega_{c,\Omega} J \qquad k_{p,i} = \omega_{c,i} L_\mathrm{a} \tag{19.6}$$

$$k_{i,\Omega} = \omega_{c,\Omega} \beta \qquad k_{i,i} = \omega_{c,i} R_\mathrm{a} \tag{19.7}$$

Then, the closed loop transfer functions are:

$$F_\Omega(s) = \frac{\Omega(t)}{\Omega^*(t)} = \frac{L_\Omega(s)}{1 + L_\Omega(s)} \qquad F_i(s) = \frac{i(t)}{i^*(t)} = \frac{L_i(s)}{1 + L_i(s)} \tag{19.8}$$

where L_Ω, F_Ω and L_i, F_i are subject to the time constants τ_{F_Ω} and τ_{F_i}, respectively.

The cascade constraints assure that the control action of the speed controller is not influenced by the dynamics of the inner loop, which is approximated as a unitary gain $F_i(s) \approx 1$, thus, describing steady state operations only. As an example, knowing that the time constant of the electrical transfer function $G_i(s)$ is $\tau_{G_i} = L_\mathrm{a}/R_\mathrm{a}$ and the corresponding settling time is $T_{a,G_i} = 5\tau_{G_i}$, a final settling time T_{a,F_i} can be defined as $T_{a,F_i} = 1/10 \times T_{a,G_i}$ in order to speed up the closed-loop system response. Therefore, since $\tau_{F_i} = T_{a,F_i}/5$, it follows that $\omega_{c,i} = 1/\tau_{F_i}$. The speed controller bandwidth has to satisfy equation (19.3), that is $\omega_{c,\Omega} = \omega_{c,i}/10$.

In case the cascade constraint is not respected, the open loop transfer function $L_\Omega(s)$ becomes:

$$L_\Omega(s) = R_\Omega(s)G_\Omega(s)F_i(s) \tag{19.9}$$

which is an higher order transfer function compared to the previous L_Ω and, therefore, it is not suitable for pole/zero cancellation.

Reminding that the power supply is set on $V_{DC} = 10\,V$, the output of both the PI controllers must be saturated. Namely, the current controller has an upper limit equal to 10 V, whereas the speed controller must be saturated at a certain $\tilde{m}_e$, which depends on the specific system parameters. Therefore, the two controllers adopt an anti-windup structure like that one shown in Chapter 6. It is important to remind that the coefficient k_{aw} for a back-calculation scheme is computed as $k_{aw} = k_i/k_p$.

Moving the attention to the B2B configuration, the braking motor is driven by a current loop only, as shown in 19.3, where the reference torque can be provided through an ADC channel, internally generated or sent through serial COM port. The associated current controller is still designed with a pole/zero cancellation as described previously. In principle, both motors mounted on the B2B-PMDC Hardware Kit are independent, but they show quite similar rated parameters. Therefore, the current control design for the traction motor is be used even for the braking one.

Discrete-time PI controllers $R_\Omega(z)$ and $R_i(z)$ are derived based on the considerations reported so far for the continuous-time counterpart. They can be integrated in simulations using Rate Transition blocks to mimic the discretization process, as shown for the previous exercises. The sampling time of the controller is set equal to $T_s = T_{sw} = 1/f_{sw}$, using Forward Euler as integration method.

It is important to note that this cascade control scheme holds for both the setups considered in this chapter, since it can be equally used for Half- or Full-Bridge implementations (provided that the reference voltage signal is suitably conditioned). Thus, this approach is used to control both the single PMDC motor and the traction motor in the B2B hardware kit. Further details are given in the following sections about the implementations into MCU.

19.4 System Simulations

The speed control of the single PMDC motor is simulated in Simulink® to analyze its design and check the resulting performance of the controller. This step is carried out for the Full-Bridge configuration of the DC-DC converter working in unipolar voltage switching. To this purpose, the power electronics and the PMDC motor load are modeled through Simscape™ Electrical elements, thanks to the Specialized Power Systems ™ library (*Simscape/Electrical/Specialized Power Systems/*) as shown in Chapters 6, 17 and 18.

In particular, the **Mosfet** blocks available in the subset */Fundamental Blocks/Power Electronics* are used to represent the switches mounted on the BOOSTXL-DRV8301 converter board. The semiconductor characteristics taken from the converter data-sheet, e.g., R_{on} and R_{d}, are set into the Simulink® blocks as shown in Chapters 15, 17 and 18. The two-level Full-Bridge configuration is obtained by rearranging these blocks based on the equivalent circuit and Simulink® scheme reported in Section 15.5. The switches are driven by a PWM modulator with a carrier frequency $f_{\mathrm{sw}} = 30\,\mathrm{kHz}$. The PMDC motor is modeled through the **DC Machine** block which can implement both a wound-field and permanent magnet DC machine by selecting the motor model in the *Field type* frame. The armature circuit (A+, A-) consists in an inductor L_{a} and resistor R_{a} connected in series with the back-emf. In the permanent magnet DC machine, there is no field current creating the excitation flux, which is established by the magnets. Thus, k_{e} and k_{T} are constant as expected. It is important to underline that *Torque constant* has to be selected in the *Specify* frame. The DC Machine block already include a lumped (first-order) mechanical model based on the inertia J, the viscous friction coefficient β (or B_m), and, as a further option, the Coulomb friction torque (which is not used here, i.e., = 0). The load torque $m_l(t)$ can be provided as an external input by selecting *Torque TL* in the *Mechanical input* frame. By selecting -1 as sample time, this block uses the largest integration step-size allowed by the **Powergui** block. This latter is essential to manage the numerical integration of the Simscape™ elements, as explained in Section 6.1.2.

19.5 Full-Bridge Configuration

A Full-Bridge configuration (or H-bridge) foresees to use two legs of the adopted boosterpack (legs A and B), for which two ePWM modules are required, e.g., ePWM1 and ePWM2. The terminals of the PMDC motor are connected to the central point of legs A and B, as shown in Figure 15.12. Hence, the switches 1_{A}, 1_{B} and 2_{A}, 2_{B} are considered. As detailed in Chapter 15, the H-bridge configuration allows to spin the motor in both directions (i.e., clockwise and counter-clockwise).

In the following simulations, it is assumed that some identification routines have been already applied to the single-motor configuration to estimate the system parameters. This procedure will be detailed in the following sections. The PMDC motor is then characterized by: $R_{\mathrm{a}} = 0.529\,\Omega$, $L_{\mathrm{a}} = 0.865\,\mathrm{mH}$, $K = 0.0232\,\mathrm{Nm/A}$, $\beta = 90.13\,\mu\mathrm{N\,m\,s}$ and $J = 7.12\,\mu\mathrm{g\,m}^2$, which are slightly different in comparison to the ones adopted in Chapter 15. It can be noted that the mechanical coupling with the rotary encoder increases both equivalent inertia and friction coefficient. The supply voltage is $V_{\mathrm{DC}} \approx 10\,\mathrm{V}$ and the switching frequency is set equal to $f_{\mathrm{sw}} = 30\,\mathrm{kHz}$.These parameters can be initialized in *Model Properites/Callbacks/InitFcn* or in a separate `m`-file as shown here in the following::

```
%% power supply
Vdc      = 10;
%% carrier frequency definition
fsw      = 30e3;
Tsw      = 1/fsw;
%% sampling time definition
Ts       = Tsw;
%% step-size definition
Tsim     = Tsw/400;
%% motor parameters
Ra       = 0.529;
La       = 8.651e-4;
K        = 0.0232;
B        = 9.013e-5;
J        = 7.118e-6;
%% current controller
tauGc    = La/Ra;
TaGc     = 5*tauGc;
TaI      = TaGc/10;
wcI      = 5/TaI;
kpI      = wcI*La;
kiI      = wcI*Ra;
%% speed controller
tauGm    = J/B;
TaGm     = 5*tauGm;
wcW      = wcI/100;
TaW      = 5/wcW;
kpW      = wcW*J;
kiW      = wcW*B;
%% encoder settings
Tw       = 5*Ts;
cpr      = 2400;
minRPM   = 1/(cpr*Tw)*60;
minRAD   = minRPM*pi/30;
%% MRAS settings
Tsm      = 5*Ts;
kpMR     = kpI;
kiMR     = kiI;
```

Based on this data, the electrical transfer function is characterized by a small time constant $\tau_{G_i} = L/R = 1.6\,\text{ms}$, which leads to a settling time of $T_{a,G_i} = 8.2\,\text{ms}$. In order to improve such uncontrolled dynamic, the settling time of the current control loop is set to $T_{a,F_i} = T_{a,G_i}/10 = 818\,\mu\text{s}$. This choice implies $\tau_{F_i} = T_{a,F_i}/5 = 164\,\mu\text{s}$ and a controller bandwidth equal to $\omega_{c,i} = 1/\tau_{F_i} =$

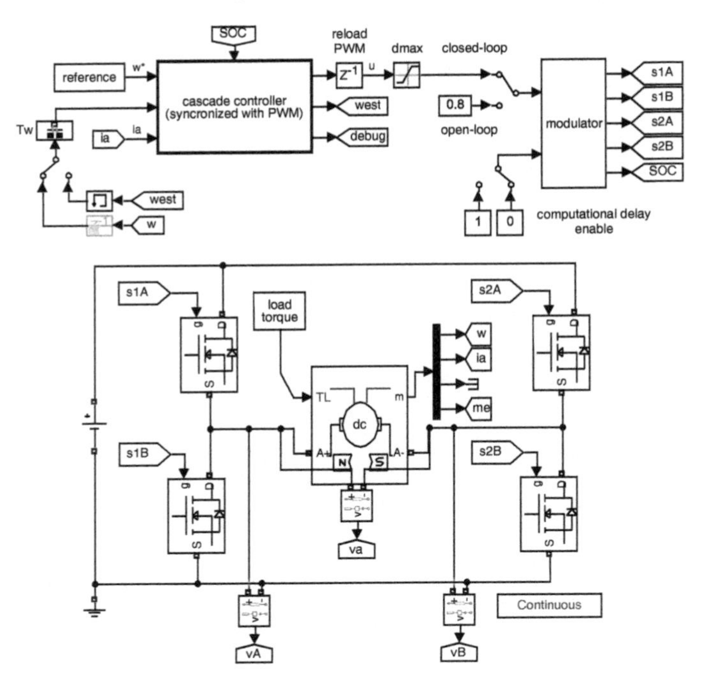

Figure 19.4 Simulink® scheme to simulate the closed-loop speed control of a PMDC motor driven in Full-Bridge configuration unipolar voltage switching.

6115 rad/s. The values of $k_{p,i}$ and $k_{i,i}$ are explicitly determined as:

$$k_{p,i} = \omega_{c,i} L_{\mathrm{a}} = 5.29\,\Omega \qquad k_{i,i} = \omega_{c,i} R_{\mathrm{a}} = 3235\,\Omega/\mathrm{s} \tag{19.10}$$

The speed control dynamic is designed to fulfill the cascade constraint. Namely, the speed controller bandwidth is set to $\omega_{c,\Omega} = \omega_{c,i}/100 = 61.15\,\mathrm{rad/s}$. Therefore the outer loop is significantly slower than the inner one. Therefore, the resulting settling time is $T_{a,F_\Omega} = 81.8\,\mathrm{ms}$, while the corresponding values of $k_{p,\Omega}$ and $k_{i,\Omega}$ are:

$$k_{p,\Omega} = \omega_{c,\Omega} J = 4.35 \times 10^{-4}\mathrm{N\,m\,s} \qquad k_{i,\Omega} = \omega_{c,\Omega}\beta = 0.0055\,\mathrm{N\,m} \tag{19.11}$$

Discrete-time PI controllers $R_\Omega(z)$ and $R_i(z)$ are derived based on these considerations. The sampling time of the controller is set equal to $T_s = T_{\mathrm{sw}} = 1/f_{\mathrm{sw}}$, using Forward Euler as integration method. Getting closer to the MCU implementation, the system variables are referred to the discrete-time domain.

The equivalent circuit shown in Figure 15.12 is then realized in Simulink® including the control scheme discussed previously (see Figure 19.3), as reported in Figure 19.4. This scheme is available in the file:

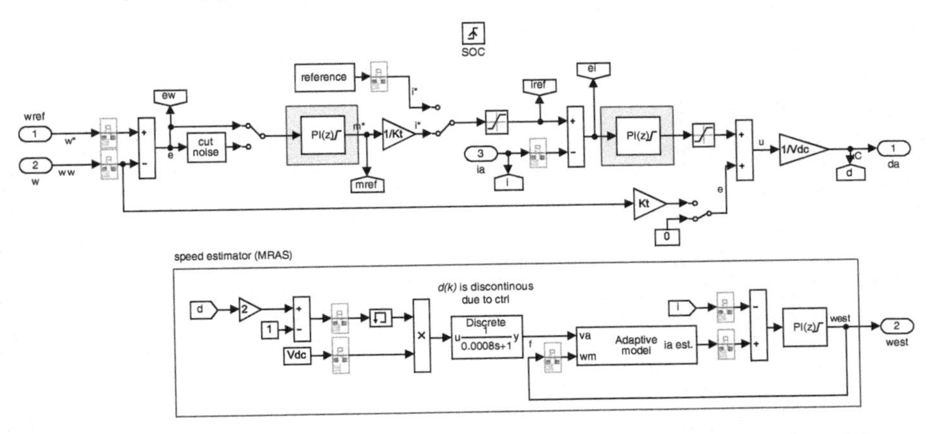

Figure 19.5 Detailed view of the cascade controller subsystem (see Figure 19.4). The top part of the scheme reports the two nested control loops realized based to the scheme shown in Figure 19.3, whereas the bottom shows a possible realization of the MRAS observer.

FBsimDC_closed.slx (solver: fixed step - ODE4, step size: T_{sim})
(powergui: simulation type *Continuous*)

Similarly to the previous chapters, the simulation time step T_{sim} is set quite small. Moreover, the **reference** subsystem includes different kinds of inputs to operate the system differently.

The actuation of the two legs (i.e., two duty cycles for two modulator stages) is done by manipulating one modulation signal $d(k)$ only, from which the gating signals of the four switches are generated. This duty cycle comes from a processing done on the desired reference armature voltage $v_a^*(t)$, which has to be divided by V_{DC} provided by the power supply, i.e., 10 V in this case. A detailed description of the unipolar voltage switching operation is reported in Section 15.5.2. However, it is of great importance to recall that the PMDC motor is subjected to the difference of the leg voltages v_{A} and v_{B}, which is defined as:

$$v_{\text{a}}(k) = v_{\text{A}}(k) - v_{\text{B}}(k)$$

The adopted working operation foresees that the duty-cycles of the two legs d_{A} and d_{B} are linked as follows:

$$d_{\text{B}}(k) = 1 - d_{\text{A}}(k)$$

Therefore, the armature voltage can be rewritten as:

$$v_{\text{a}}(k) = (2d_{\text{A}}(k) - 1)\, V_{\text{DC}}$$

As a result, $v_{\text{a}}(k)$ jumps between 0 and -10 V or from 0 and 10 V depending

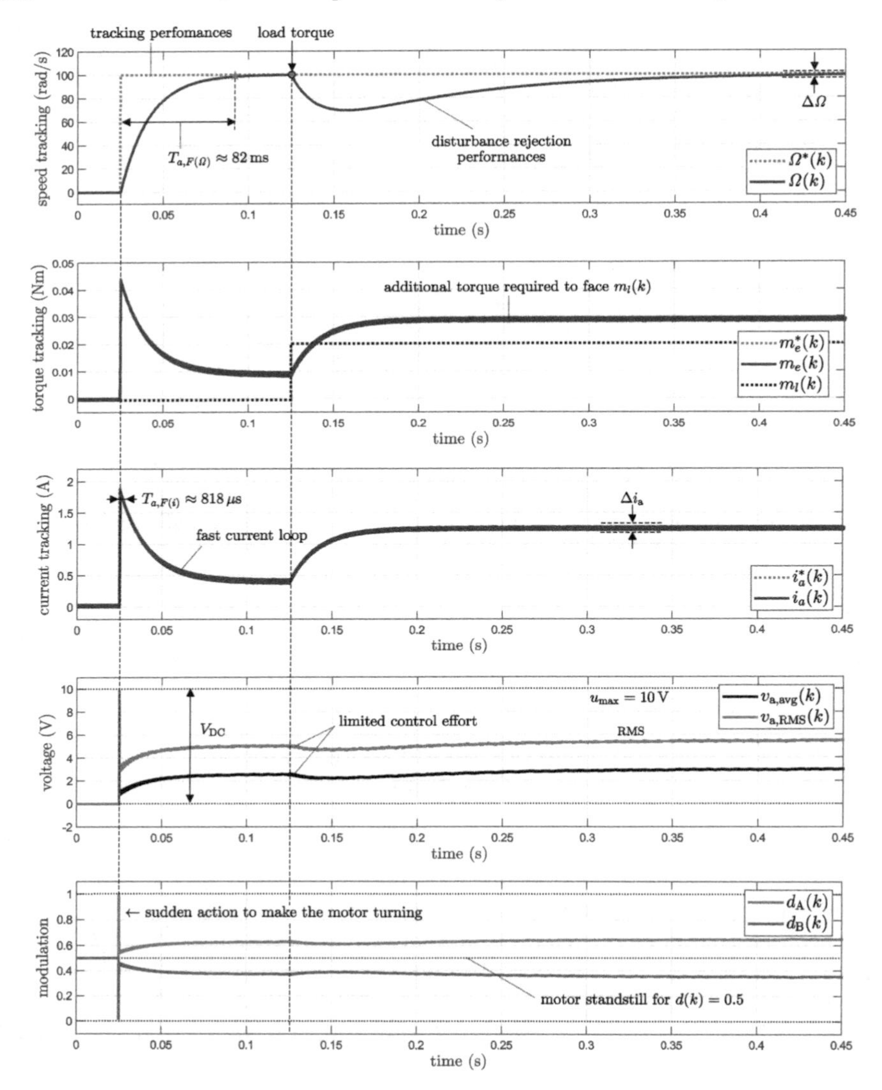

Figure 19.6 Closed-loop dynamics of the cascade speed control of a single PMDC motor.

on the desired direction of rotation (that is, on the value of the duty cycle, as shown in Section 15.5.2). Indeed, it must be remembered that Full-Bridge converters allow to run DC motors in both clockwise and anti-clockwise directions of rotations without modifying the physical connections of the setup. Speed and current feedbacks are taken from the measurement port of the DC machine block. The aim of this exercise is to design a *cascade control strategy*.

Namely, this means to control and stabilize the motor speed $\Omega^*(k)$ despite the presence of a load torque, while achieving a settling time $T_{a,F_\Omega} = 81.8\,\text{ms}$. To check these performances, a step-wise speed reference $\Omega^*(k)$ jumping from $-100\,\text{rad/s}$ up to $100\,\text{rad/s}$ is applied. Figure 19.6 shows the closed-loop dynamics of the cascade speed control of a single PMDC motor for a step of $100\,\text{rad/s}$ in the reference speed. Both speed $\Omega(k)$, torque $m_e(k)$ and, consequently, armature current $i_\text{a}(k)$ are correctly following their related references. It can be noted that the settling times are respected both for the current and the speed, being $\approx 818\,\mu\text{s}$ and $\approx 82\,\text{ms}$, respectively. The current control loop is quite fast compared to the speed control loop. The references $m_\text{e}^*(k)$ and $i_\text{a}^*(k)$ are almost coincident with the actual variables $m_e(k)$ and $i_\text{a}(k)$, while the transient of $\Omega(k)$ is visible. It follows an exponential behavior. Current and torque responses are scaled by the factor K. Due to the different system parameters and controller settings, the plots reported in Figure 19.6 show that the current behavior is pretty different with respect to those reported in Chapters 15 and 17.

Since the reference $\Omega^*(k)$ consists in a positive speed only (i.e., clockwise rotating direction), the voltage is varying between 0 and V_DC. At the time instance in which the speed reference changes from 0 to $100\,\text{rad/s}$, the current $i_\text{a}(k)$ grows up quickly, starting the rotation of the motor shaft. Therefore, the back-emf starts acting on the armature circuit after less than 1 ms. Then, the current drops down to values close to 0.5 A thanks to the $R_i(z)$ controller. The current peak (which translates in torque peak as well) is due to a sudden variation in the duty cycles $d(k) = d_\text{A}(k)$ and $d_\text{B}(k)$, after which they both approach the related steady-state values. A pulse in the average and RMS value (which is always greater than zero) of the armature voltage $v_\text{a}(k)$ are measured as well.

It can be noted that both speed and the current ripples, $\Delta\Omega$ and Δi_a, are quite small, as a consequence of the unipolar voltage switching operation.

After the steady state is reached, a load torque $m_l(k) = 0.02\,\text{N m}$ is applied to the motor to test the robustness of the cascade control scheme. The load torque is defined into the subsystem **load torque** (see Figure 19.4). The command-tracking performances are quite different from the disturbance-rejection ones, with the latter characterized by a longer time constant. Indeed, in order to compensate the additional load torque applied to the system, the cascade control requires a time interval which is larger than the settling time $T_{a,\Omega}$. This effect can be changed by setting a higher bandwidth $\omega_{c,\Omega}$. It must be noted that the control effort to compensate $m_l(k)$ is quite small, since the duty cycles and the armature voltage are slightly higher compared to their values before the load torque transients. Besides the verification of the settling time of the closed-loop system, the choice of the sampling time to avoid issues related to peripheral and task scheduling, e.g., controller calculation and actuation updates, is investigated as discussed in Section 13.10 and 17.4. To this purpose, the elements involved in the closed-loop control scheme loaded into the MCU, i.e., current reading (ADC module), encoder reading (eQEP),

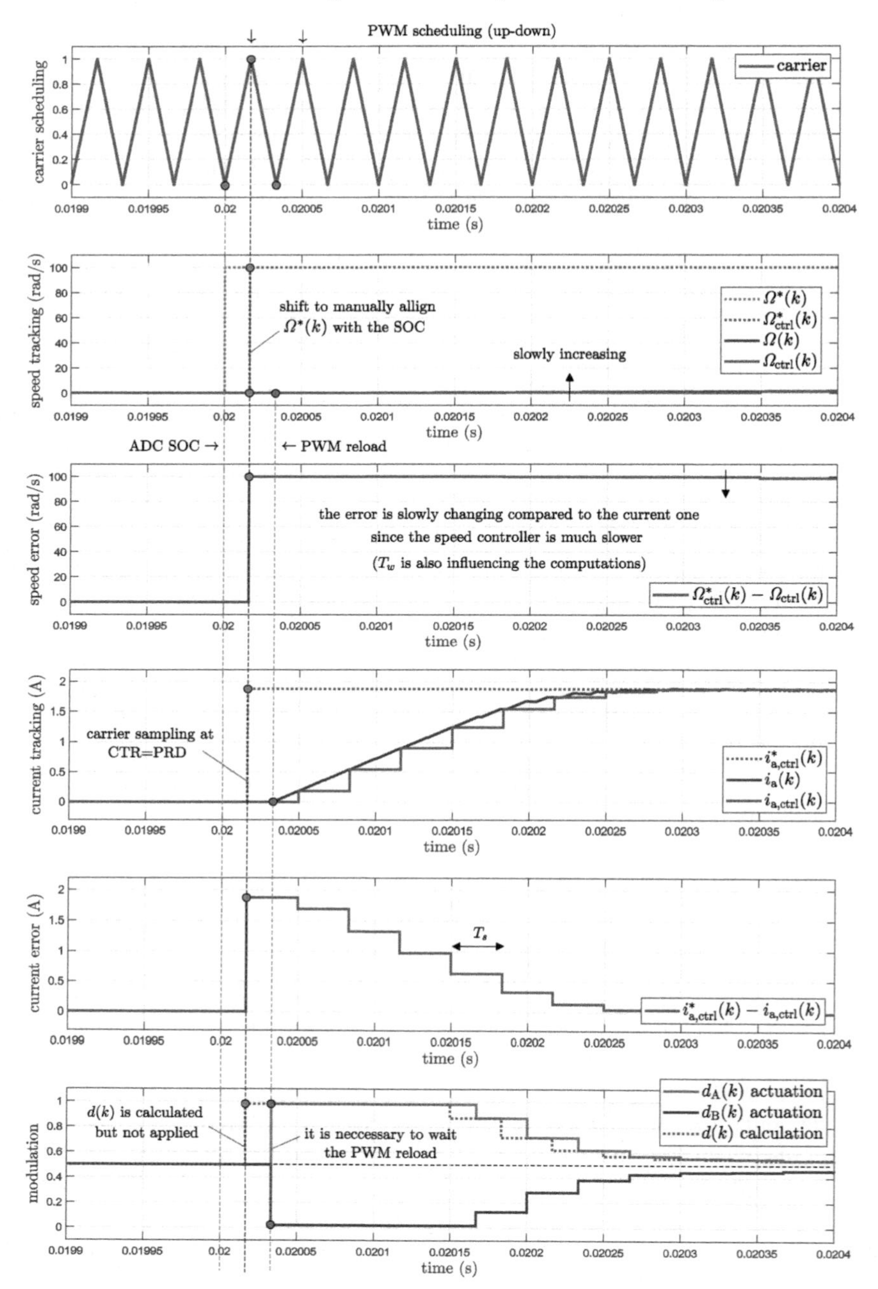

Figure 19.7 Part I of Figure 19.8.

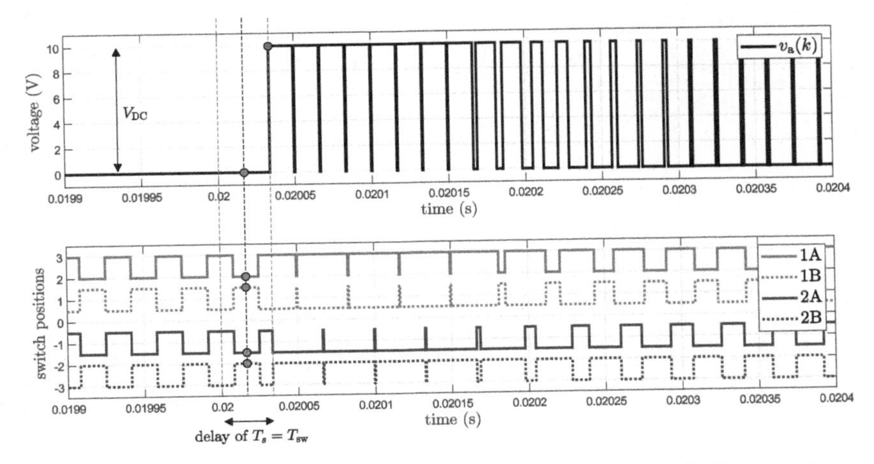

Figure 19.8 Part II of Figure 19.7. Analysis of the MCU scheduling and latency within a sampling period. Due to the different bandwidth between the loops, the command-tracking performances of the speed and the related speed error are slowly changing compared to the current dynamics and the related error. The duty cycles actuation happens at every PWM reload action.

2L-PWM actuation (ePWM module) and a PI-based controllers, are scheduled to be executed within the sampling period $T_s = T_{sw} = 1/f_{sw}$. The target hardware is subjected to a similar *sequential* events execution as that one observed for the current control of an *RL* load (Section 17.4): It is summarized in Figure 19.7 and 19.8.

Therefore, the controller adopted in the Simulink® simulation is consistent with such event sequence.

In particular, a trigger signal SOC (i.e., a boolean square-wave) is generated inside the modulator subsystem every time the carrier equals 1. The SOC is used as trigger source for a triggered subsystem which realizes a cascaded controller synchronized with the carrier peaks. Inside this block, The $R_{\Omega}(z)$, $R_i(z)$ controllers are modeled by using the PID controller blocks available in the Simulink® library, as shown in Figure 19.5. A *manual switch* is used to include or exclude the outer speed loop.[3] Therefore, the closed-loop performances of both traction and braking motors can be tested by using the same simulation file (i.e., valid both for single-motor and B2B configurations). Similarly, the back-emf can be compensated or not.

A speed reference $\Omega^*(k)$ is internally generated by using a Pulse Generator block (synchronous with the zero value of the triangular carrier). Since the ADC SOC trigger is generated at the carrier peak, $\Omega^*(k)$ is delayed by $T_s/2$ to provide a correct voltage error at the speed controller input. The resulting

[3]If necessary, the subsystem **cut noise** can be used to bring the speed error equal to zero every time it is below a certain threshold.

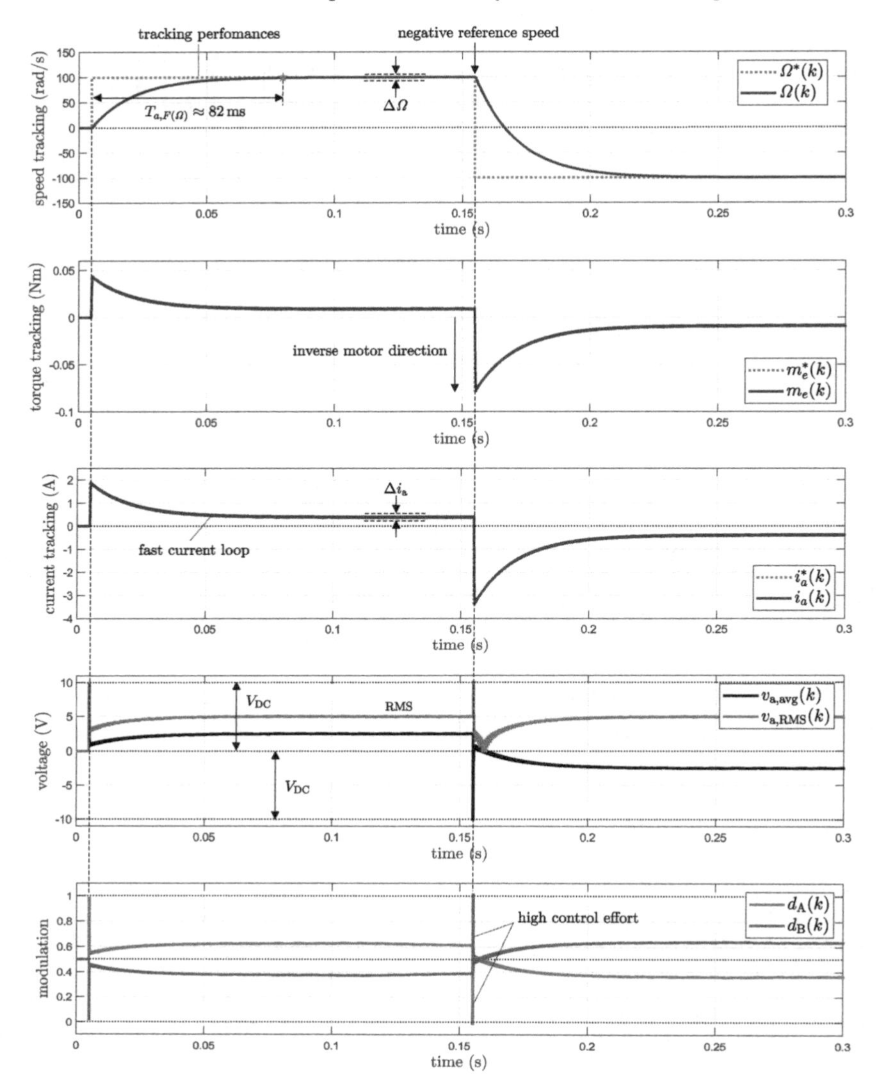

Figure 19.9 Closed-loop dynamics of the cascade speed control of a single PMDC motor rotatign in clockwise and anti-clockwise direction.

reference speed is called $\Omega^*_{\mathrm{ctrl}}(k)$, where the subscript ctrl refers to the quantity that are processed by the controller. The same holds even for the other variables. The closed-loop performances for a different speed profile are shown in Figure 19.9. The system responses are similar to those reported in Figure 19.6, despite higher peaks during the speed transition from 100 rad/s to −100 rad/s (a greater control effort is required). The speed measurements emulates the

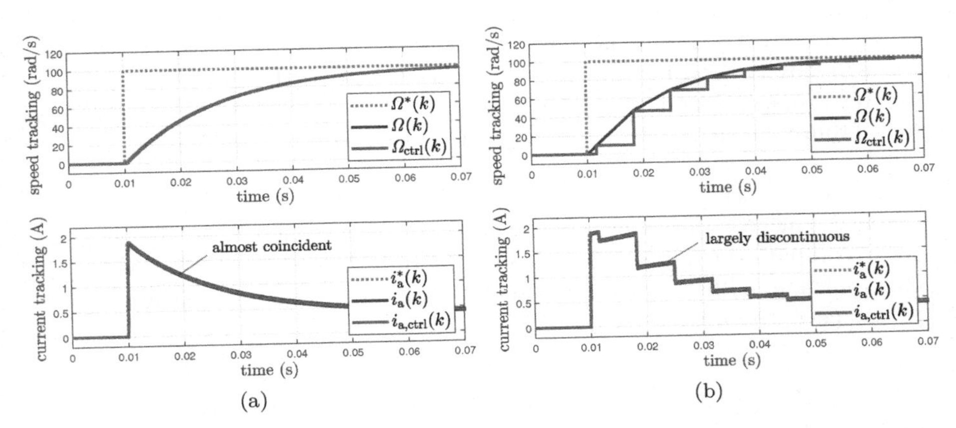

Figure 19.10 Comparison closed-loop dynamics (in terms of speed and armature current transients) for (a) $T_w = 5T_s$ and (b) $T_w = 200T_s$

presence of a rotary encoder. By considering the standard approach proposed in Section 14.4 it follows that:

$$\Omega(k) \approx \frac{x_p(k) - x_p(k-1)}{T_w} = \frac{\Delta x(k)}{T_w} \tag{19.12}$$

where $\Omega(k)$ is the speed computed at time step k, $x_p(k)$ and $x_p(k-1)$ are the position counters at time step k and $k-1$ respectively, T_w is the acquisition time window, which is the inverse of speed calculation rate. The value of T_w is of paramount importance since it is directly related to the speed computation:

$$\Omega_{\min} = \frac{\text{round}}{\min} \text{ or } \frac{\text{revolution}}{\min} = \underbrace{\frac{1}{\text{cpr} \cdot T_w}}_{\frac{\text{revolution}}{\text{sec}}} 60 \tag{19.13}$$

Considering a constant cpr, a large value of T_w implies an higher speed measurement accuracy, being able to measure down to small value of $\Omega(k)$. Nevertheless, in order to keep the synchronized event sequence in the embedded target, the acquisition time must be a multiple of T_s. Hence, a large value of T_s may lead to a quite slow speed computation task adding a relevant delay in the whole cascaded control action (e.g., the duty cycles tend to saturate for long period). In the worst case scenario this may lead to system instability.

Based on this fact, even if a sufficiently large value of T_w is necessary to well approximate the actual speed, it is quite often to keep T_w in a $[2T_s,\ 10T_s]$ range for switching frequencies below 30 kHz.[4] In order to analyze the effect of

[4]For larger switching frequencies this range is extended.

the acquisition time T_w, Figure 19.10 (a) and (b) compare the closed-loop dynamics in terms of speed and current transients for $T_w = 5T_s$ and $T_w = 200T_s$, respectively. It is quite evident from Figure 19.10(b) that large values of T_w creates a discontinuous behavior even in a fast control loop like the current one. From now on, $T_w = 5T_s$ is set, which leads to $\Omega_{\min} = 15.71\,\text{rad/s}$.

Remark: in case the cascade control loop is required to operate within particularly low speed (e.g., even lower than the $\Omega_{\min}$) it is better to change the speed computation approach, as described in 14.4. Moreover, by decreasing the bandwidth $\omega_{c,\Omega}$ the acquisition time might be increased without introducing too much delay in the control loop.

19.5.1 Model reference adaptive system (MRAS) observer

One of the evergreen research areas in the field of electrical drives is the study of estimators and observers aimed to describe the state variables of a motor. In order to achieve a robust cascade architecture, the closed-loop performances strongly depends on the quality of those feedbacks.

In general, current and voltage sensors are low cost devices and simple to measure. Therefore, it is widely common to assume that those sensors are always present in a motor control application. Combining them with the mathematical description of the motor dynamics, it si possible to estimate many variables, e.g., electromagnetic torques, speeds and fluxes. In particular, there are different kinds of speed sensors available on the market. Their costs and characteristics are relatively variable, and they are strongly affecting the final cost and performances of a product, especially for small/medium size low voltage drives. On the other hand, from medium to high voltage drives, the problem of getting a safe and clean transmissions of the speed (or position) measurement from the motor site to the control cabinet is of great importance and mostly solved by adopting expensive wiring harness. Due to that, even for two different reason, the possibility to replace speed sensors with estimators/observers could be a wise option despite the power rating of the drive.

In the electrical drives field, control schemes which are not based on direct measurements of speed or position in favor of estimators/observers are called encoderless schemes or, more in general, sensorless schemes.

The aim of this Section is to built an encoderless cascade speed control of a PMDC motor by using the so called model reference adaptive system (MRAS) speed observer to estimate the rotor speed. The MRAS observer is only one candidate among the large family of observers, which are not fully detailed here.

Remark: from the terminology point of view, it is important to underline that the *estimators* are based on open-loop methods which benefits of a easy implementation although they are quite sensitive to parameter uncertainty and

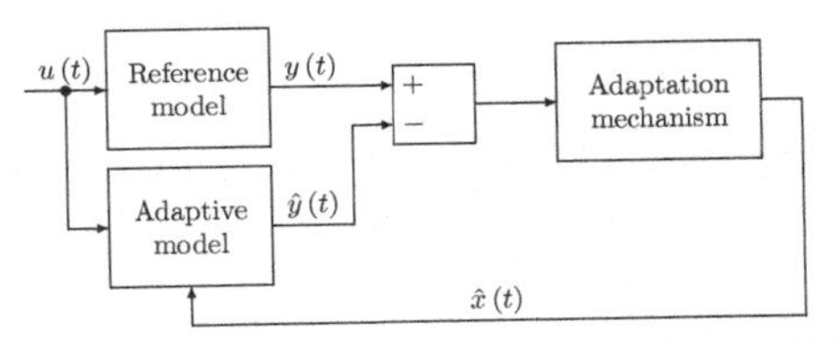

Figure 19.11 Block scheme showing the working principle of a generic MRAS observer.

measurement errors. Instead, the *observers* are based on closed-loop methods which provide better accuracy than estimators, at the expense of a greater complexity of the algorithms.

The MRAS is defined as an observer because it compensates simultaneously the armature resistance and rotor time constant variations (subjected to large changes during operation) through a PI controller, which is used as *adaptation mechanism* (i.e., in a closed-loop manner). This approach can be both applied to the single-motor or B2B configuration.

The operating principle of the aforementioned MRAS observer is based on an *adaptive model*, which is a function of the variable to be estimated. Referring to the general scheme reported in Figure 19.11, for the simplicity of the PMDC motor model, the implementation proposed in the following does not foresee any reference model. The measurement $y(t) = i_{\mathrm{a}}(t)$ is directly compared to its estimation $\hat{y}(t) = \hat{i}_{\mathrm{a}}(t)$ (see Figure 19.5). A PI-based controller is used as adaptation mechanism to provide the (output) speed estimation $\hat{x}(t) = \widehat{\Omega}(t)$. The armature voltage is considered as input variable $u(t) = v_{\mathrm{a}}(t)$. The loop iteration refines $\hat{x}(t)$ until $\hat{y}(t) \approx y(t)$. To this purpose, the electrical contribution in a PMDC motor model is reported here:

$$v_{\mathrm{a}}(t) = L_{\mathrm{a}}\frac{di_{\mathrm{a}}(t)}{dt} + R_{\mathrm{a}}i_{\mathrm{a}}(t) + K\Omega(t) \tag{19.14}$$

This expression is used to predict the deviation of the current at time instant t (reference model). Thus, $i_{\mathrm{a}}(t)$ can be computed by solving this differential equation. In view of an implementation on MCUs, the adaptive model is discretized considering a forward Euler method:

$$\frac{di_{\mathrm{a}}(t)}{dt} \approx \frac{i_{\mathrm{a}}(k+1) - i_{\mathrm{a}}(k)}{T_{sm}} \tag{19.15}$$

where T_{sm} is a generic discretization step. Then, the discrete-time adaptive model follows as:

$$i_{\mathrm{a}}(k+1) = \left(1 - \frac{T_{sm}}{L_{\mathrm{a}}}R_{\mathrm{a}}\right) i_{\mathrm{a}}(k) + \frac{T_{sm}}{L_{\mathrm{a}}}\left[v_{\mathrm{a}}(k) - K\Omega(k)\right] \tag{19.16}$$

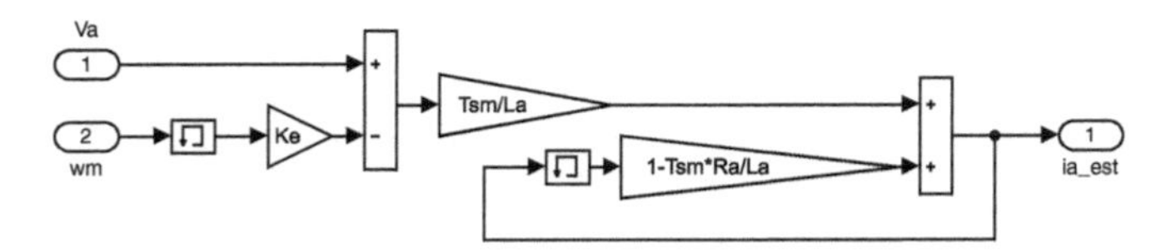

Figure 19.12 Detailed view of the adaptive model subsystem (see Figure 19.5).

The corresponding Simulink® implementation of this equation is reported in Figure 19.12, which details the adaptive model subsystem. Indeed, the MRAS observer is included inside the triggered subsystem to process the involved variables based on the right event sequence scheduling.

A careful sensitivity analysis of this equation shows that there is the need to invert the signs of the sum reported in Figure 19.11. Indeed, by applying a virtual displacement on the mechanical speed, it can be found out that:

$$\frac{\Delta i_{\mathrm{a}}}{\Delta \Omega} = -\frac{T_{sm} K}{L_{\mathrm{a}}} \tag{19.17}$$

This equality suggests that an increase in the speed causes a drop in the armature current. Indeed, inverting the signs before the PI regulator leads to the following scenarios:

$$\begin{aligned} \hat{i}_{\mathrm{a}}(k+1) > i_{\mathrm{a}}(k+1) \quad &\rightarrow \quad \Omega(k+1) \uparrow \Rightarrow \hat{i}_{\mathrm{a}}(k+2) \downarrow \\ \hat{i}_{\mathrm{a}}(k+1) < i_{\mathrm{a}}(k+1) \quad &\rightarrow \quad \Omega(k+1) \downarrow \Rightarrow \hat{i}_{\mathrm{a}}(k+2) \uparrow \end{aligned}$$

Besides the accuracy of the system parameters, the stability of the proposed observer is related to T_{sm}. Similarly to the encoder trade-off between sensing accuracy and latency, the MRAS observer has to be designed to well approximate the speed measurements (i.e., a small T_{sm} to increase the iterative refinements). However, if the speed controller is relatively slow and the actual speed value does not need to be update every T_s, the discretization time can be set as $T_{sm} = 5T_s$ as for the encoder. This is helpful to reduce the computations which has to be scheduled within T_s (which may enlarge the overall execution time), while keeping the synchronization between computations since T_{sm} is a multiple of T_s. Therefore this would be the preferable choice.

Since the Adaptive Model is based on the armature current dynamic, the $k_{p,MR}$ and $k_{i,MR}$ parameters of the the PI controller can be set equal to those computed for the current control loop, i.e., $k_{p,MR} = k_{p,i}$ and $k_{i,MR} = k_{i,i}$. These value define the expected time of convergence of the estimated value $\hat{\Omega}(k)$. As previously said, this approach is suitable for both Half- and Full-Bridge configurations (i.e., feasible for single-motor and B2B test cases). The only difference would be in the definition of the armature voltage, which is related to the specific adopted topology. The MRAS observer is assumed to be used in the Full-Bridge configuration scheme adopted to drive the single PMDC motor. The armature voltage is $v_{\mathrm{a}}(k) = (2d_{\mathrm{a}}(k) - 1)V_{\mathrm{DC}}$, as shown in

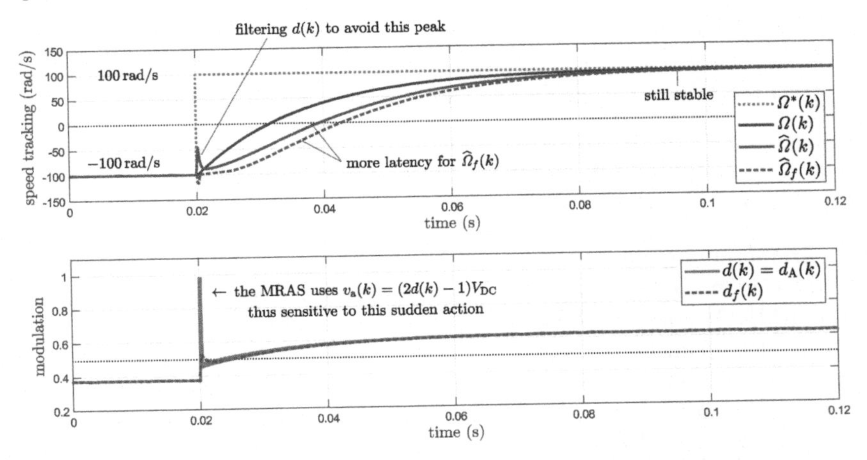

Figure 19.13 Comparison between command-tracking performances of the speed loop when an encoder, a MRAS observer, and a MRAS observer with a low pass filter are used. The control scheme is used to drive a single PMDC motor in Full-Bridge configuration.

the implementation scheme shown in Figure 19.5. The command-tracking performances of the speed loop are reported in Figure 19.13. The speed dynamics given by using the encoder (blue line) is kept as reference case. The MRAS observer (solid purple) is well approximating the measured speed, apart from the introduction of a processing latency. In particular, the speed reference is changed from $-100\,\text{rad/s}$ to $100\,\text{rad/s}$, requiring a sudden control action to meet the desired settling time T_{a,F_Ω}. The duty cycle $d(k) = d_A(k)$ is used to compute $u(t) \approx u(k) = v_a(k)$. Therefore, a peak in $d(k)$ is reflected in the $\hat{\Omega}(k)$ estimation. Based on the application requirements, if this peak is an issue, a low pass filter can be used to smooth the speed dynamic, denoted as $\hat{\Omega}_f(k)$, by filtering the duty cycle $d_f(k)$. Figure 19.13 shows that this countermeasure effectively reduces the peak in $\hat{\Omega}_f(k)$ at the cost of an additional latency. In both cases, the use of a MRAS observer lead to a stable cascade control scheme.

19.6 Single Motor Configuration

The implementation of the cascade speed loop for a single PMDC motor corresponding to the simulation proposed in Section 19.5 is reported below. First, the procedure required for the identification of the system parameters is described. Then, a step-by-step procedure for the realization of the firmware for the speed control is proposed.

Table 19.1 Estimated electrical parameters for PMDC motor control.

τ_{G_i} (ms)	R_a (Ω)	L_a (mH)
1.635	0.529	0.865

19.6.1 Parameter identification

The parameters involved in the speed control loop (namely, R_a, L_a, J and β) should be identified through suitable tests. Indeed, parameter identification of the PMDC motor is suggested to design the PI controller properly. In this case, they can be estimated separately through the measurement of the dynamics of the electrical and mechanical parts of the system, respectively. Indeed, assuming to connect one PMDC motor to a Half-Bridge converter and an encoder (see Figure 19.1), its armature resistance and inductance can be identified through a locked-rotor test, as reported in Section 16.2. Thus, R_a can be evaluated measuring the voltage and the current at steady-state, whereas the identification of L_a requires the study of the transient of the current. In particular, the time constant τ_{G_i} is essential to this aim. Indeed, an exponential interpolation of the measured data should be performed in the MATLAB® editor through functions like `fittype` or `fminsearch`. Using the same firmware proposed in Section 16.2 (this time considering the calibrated gain and offset for the considered ADC channel) and current/voltage probe connected to an oscilloscope, the electrical parameters can be evaluated. The results are reported in Table 19.1.

Actually, still one electrical parameter is missing, that is, the back-emf constant $K = \psi_\mathrm{PM} k_\mathrm{e}$ which is accounting for the flux of the permanent magnets. Moreover, this parameter is numerically equal to the torque constant $\psi_\mathrm{PM} k_\mathrm{T}$, that allows to compute $m_e(k)$ knowing $i_\mathrm{a}(k)$. This K can be evaluated by supplying the PMDC motor in open loop configuration with the same voltage step adopted for the locked rotor test. Indeed, at steady state it holds that:

$$V_\mathrm{a} = R_\mathrm{a} I_\mathrm{a} + K\Omega \tag{19.18}$$

Hence, knowing the armature resistance R_a and measuring the voltage V_a, the current I_a and the mechanical speed of the rotor Ω at steady state, it is possible to compute K as:

$$K = \frac{V_\mathrm{a} - R_\mathrm{a} I_\mathrm{a}}{\Omega} \tag{19.19}$$

Alternatively, this parameter can be estimated by mechanically connecting the motor under study with another machine, building up a setup similar to the B2B one. This second motor turns the first one, causing an open circuit voltage drop on its terminals. Thus, K can be computed measuring this voltage and the rotational speed of the two machines. $\Omega(k)$ can be still measured through the rotary encoder. Moreover, its steady-state profile can be visualized

Table 19.2 Estimated mechanical parameters for PMDC control.

$\tau_{G\Omega}$ (ms)	K (Nm/A)	β (µN m s)	J (µg m^2)
79	0.0232	90.13	7.12

thanks to the virtual COM port, even if some delay is introduced. However, it is assumed to be sufficiently low not to compromise the characterization of mechanical transient, which is typically much slower than the electrical one. To this purpose, a firmware exploiting the same block chain shown in Section 14.6 for speed measurement is built. The blocks presented in Section 16 for running the motor in open loop are included as well. Indeed, this time the rotor is not locked any more. If some delay issues arise, the DAC peripherals can be used for data acquisition.

The mechanical dynamics is related to the following equation:

$$m_e(k) - m_l(k) = \beta\Omega(k) + J\frac{d\Omega(k)}{dt} \tag{19.20}$$

The friction coefficient and the inertia of the motor can be computed similarly to the electrical parameters. Indeed, β can be evaluated at steady state without applying any load torque $m_l(k)$ on the shaft, whereas the inertia needs a measurement of the mechanical dynamic. Therefore, the motor can be still supplied with $V_{\text{DC}} = 10\,\text{V}$ with a train of pulses with duty-cycle equal to, e.g., $d(k) = 0.2$. Then, the electromagnetic torque can be computed as:

$$m_e(k) = K_t i_a(k) = K i_a(k) \tag{19.21}$$

The resulting mechanical parameters are reported in Table 19.2. It is important to underline that J includes the effect of both the inertia of the motor and of the rotary encoder.

It can be noted that the values adopted for the simulation proposed in Section 19.5 are those obtained from such characterization procedure, which, in principle, must be done as first task.

Firmware Environment

c28069_DCchar_hbF.slx (solver: fixed step - ODE4, step size: $T_s = 50\,\mu\text{s}$)

This firmware is obtained by joining the schemes realized in Section 14.6 and 16.2 (f_{sw} can be set equal to 20 kHz or 30 kHz). Indeed, the motor is supplied with a $d(k) = 0.2$ pulse duty cycle once again and the speed measurement through the encoder is still valid: this signal is processed so that the angular speed in rad/s is obtained. In addition, the GPIO block is replaced with a SCI Transmit block set as reported in Section 10.5. A Rate Transition block with sample time equal to 0.01 s is added and the Data Type Conversion block is set on single values. Finally, the resulting Simulink® scheme is shown in Figure

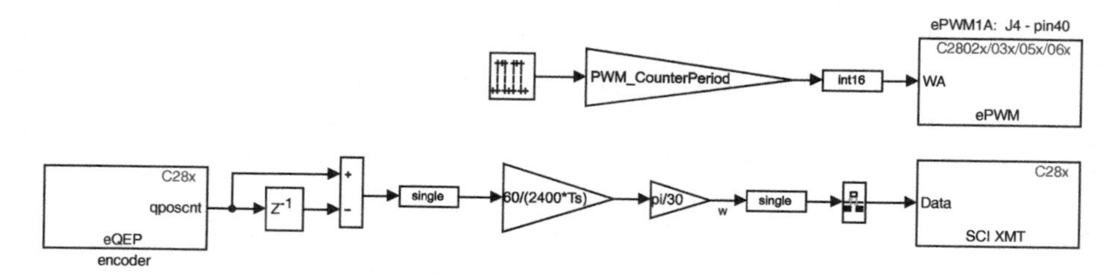

Figure 19.14 Simulink® scheme included in c28069_DCchar_hbF.slx.

19.14. It is important to note that voltage and current measurements adopted for the proposed computations are performed with voltage and current probes connected to an oscilloscope. The testing environment for the acquisition of the mechanical speed is the same as the one proposed in Section 16.3.

19.6.2 Control implementation

The closed-loop dynamics of a cascade speed control of a single PMDC motor are tested through the DecMot Hardware Kit (see Figure 19.1).

Both legs (phase) A and B are used for this implementation (H-bridge): the connection scheme is reported in Figure 19.15.
Therefore, two ePMW modules are used and they are operated in up-down counting mode at $f_{\mathrm{sw}} = \mathrm{PWM}_{\mathrm{frequency}} = 30\,\mathrm{kHz}$. Reminding that for the F28069M board $f_{\mathrm{ck}} = \mathrm{CPU}_{\mathrm{frequency}} = 90\,\mathrm{MHz}$, it follows:

$$\mathrm{TBPRD} = \mathrm{PWM}_{\mathrm{counter-period}} = \frac{\mathrm{CPU}_{\mathrm{frequency}}}{2 \cdot \mathrm{PWM}_{\mathrm{frequency}}} = 1500 \qquad (19.22)$$

where the sampling time is $T_s = T_{\mathrm{sw}} = 1/\mathrm{PWM}_{\mathrm{frequency}}$.

The reference speed can be internally generated in the firmware or externally provided both using the available potentiometers on the extPot3 board or through COM port (e.g., SCIA). The analog solution is adopted to add versatility to the control scheme in the proposed solution (e.g. use one of the linear potentiometer as control enable). The needed parameters can be initialized in *Model Properites/Callbacks/InitFcn* or in a separate `m`-file as done for simulation.

A step-by-step procedure for programming the microcontroller for this task is reported here in the following.

Firmware Environment

c28069_DCcontrol_fbF_uni (solver: fixed step - ODE4, step size: $T_s = 1/2f_{sw} = 16.7\,\mu\mathrm{s}$)

Open a new blank Simulink® project and configure the environment as shown

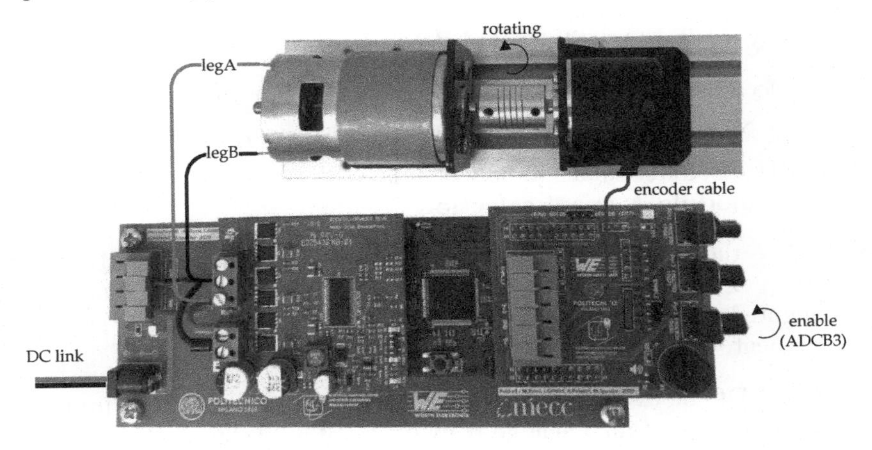

Figure 19.15 Connections of the DecMot Hardware Kit with the LaunchPad F28069M through the BOOSTXL-DRV8301 converter and the extPot3 board. A Full-Bridge configuration is realized and the encoder wires are interfaced with module 2 of the eQEP peripheral.

in Section 9. Remember that the BOOSTXL-DRV8301 is enabled by settling **GPIO50** high (see Section 3.2).

- Insert an **ePWM** block and use the same settings shown in Section 13.5:
  - Select **ePWM1** module with TBPRD corresponding to $f_c = 30\,\text{kHz}$, i.e., $\text{PWM}_{\text{counter-period}} = 1500$;
  - Remember to unflag Enable ePWM1B and tick Inverted version of ePWMxA. This option allows to drive 1_{B} as the complement of 1_{A};
  - Keep all the other settings as reported in Section 13.5.
- Insert a second **ePWM** block: select **ePWM2** module and use the same settings adopted for the previous block.
- Add two **gain** blocks to normalize the modulation signals over the peak values of the carrier $\text{PWM}_{\text{counter-period}}$. Include two **data type conversion** blocks set on `int16` to represent the duty cycle within a 16-bit base.

These two ePWM modules are arranged to operate the converter in Full-Bridge configuration, i.e., $d(k) = d_{\text{A}}(k)$ and $d_{\text{B}}(k) = 1 - d_{\text{A}}(k)$. This relationships can be achieved by using simple Add and Constant blocks. The reader is referenced to Section 15.5.4 to see their connections. For up-down counting mode, the measurement of an average current $i_{\text{a}}(k)$ is feasible thanks to the ePWM module which sends a synchronization signal that triggers the SOC of the related ADC modules when the carrier equals TBCTR=TBPRD; further details on this topic are provided in the previous Chapters. To this aim, the following settings are summarized here below:

- In the **Event Trigger** (Tab) of ePWM1:
  - Select **Enable ADC start of conversion for module A**;
  - Set **Number of event for SOCA to be generated** to First event;
  - Select Counter equals to period (CTR=PRD) as **Start of conversion for module A event selection**.

Every time the carrier signal reaches the maximum value TBPRD, a SOC input (i.e., ePWM1_ADCSOCA) is sent to the ADC. This last block has to be configured as well. The onboard shunt on leg 1 is exploited for the current measurement.

- Add an **ADC** block and select **ADCINA0** as **Conversion channel** in the **Input Channels** tab. Then, go back in the **SOC Trigger** tab:
  - **Sampling mode**: Single sampling mode;
  - **SOC trigger number**: SOC0;
  - **SOCx acquisition window**: 7;
  - **SOCx trigger source**: ePWM1_ADCSOCA;
  - **ADC will trigger SOCx**: No ADCINT;
  - **Sample time**: T_s;
  - **Data type**: uint16.
- Add a **Data Type Conversion** block set to int16 (since positive signals only are used) or `single` to exploit 32 bit.
- Place a **Gain** block for scaling the digital signal over $[0, 1]$, since a 12-bit reading returns values ranging from 0 (0 V) up to $2^{12} - 1 = 4095$ (3.3 V).
- Build a proper conversion chain for the current sensing by removing the estimated offset (e.g., 2064) as well as using the scaling factor g_{adc} to obtain Ampere values after moving to `single` data type. See the characterization procedure in Chapter 16.
- Regarding the speed and current controllers, use the same discrete **PID controller** blocks adopted in simulation; set the type of the controllers as PI (Parallel) with sample time T_s. For the *speed controller* P=$k_{p,\Omega}$ and I=$k_{i,\Omega}$ are set, with the controller output $m_e^*(k)$ saturated within the range $[-Ki_{\text{a,max}},\ Ki_{\text{a,max}}]$, where $i_{\text{a,max}} = 10\,\text{A}$. For the *current controller*, P=$k_{p,i}$ and I=$k_{i,i}$ are set, with the controller output $v_{\text{a}}^*(k)$ saturated within the range $[0,\ V_{\text{DC}}]$. A more conservative choice might be to consider $0.98 \cdot V_{\text{DC}}$ as upper bound. In both blocks, select an external source for the **reset** with **rising** condition, and set back-calculation method as anti-windup strategy for both of them with Kb=k_{aw}.

- Redundant saturation blocks can be used to protect the system by limiting the required voltage and torque.
- Add a **Gain** block to normalize the speed controller output (which varies in the $[-Ki_{\mathrm{a,max}},\ Ki_{\mathrm{a,max}}]$ range) through a scaling factor $1/K$. Then, the signal spans a $[-i_{\mathrm{a,max}},\ i_{\mathrm{a,max}}]$ range, becoming a reference current. Depending on the value of k_{aw}, dynamics close to the upper/lower bounds are damped by the back-calculation strategy.
- Add a **Gain** block to normalize the current controller output (which varies in the $[0,\ V_{\mathrm{DC}}]$ range) trough a scaling factor $1/V_{\mathrm{DC}}$. Then, the signal spans a $[0,\ 1]$ range, being consistent with the definition of $d(k) = d_{\mathrm{A}}(k)$ (and, then, $d_{\mathrm{B}}(k)$ is computed).
- Add a **Pulse Generator** (sample based) block to create the reference $\Omega^*(k)$. Set the amplitude equal to $A = 200$, number of samples $n_s = 60000$, pulse width 30000 samples, phase delay 0, and sample time T_s. It is equivalent to have a square-wave $\Omega^*(k)$ of period $60000T_s = 2\,\mathrm{s}$ with $T_{\mathrm{on}} = 1\,\mathrm{s}$. It can be noted that $\Omega^*(k)$ is synchronized with the zero value of the triangular carrier CTR=0. Then, a Add and a Constant blocks are placed to bias $\Omega^*(k)$ and to generate a reference signal ranging from $[-100,\ 100]$ rad/s.
- Add a Gain set on K, a constant block, a Manual Switch and an Add block to allow the inclusion of the back-emf compensation in the control scheme.
- Regarding the speed measurement, add an eQEP block and build the related conversion chain to process `qposcnt` in order to get $\Omega(k)$; the reader is referred to Chapter 14, Section 16.3 and to the file c28069_DCchar_hbF.slx. It can be noted that these eQEP blocks are scheduled through the sample time T_w. To this purpose, it must be noted that T_w must be set large enough to provide a good approximation of the measured speed. Indeed, a too low value fo T_w may badly compute the speed value within the acquisition time, while a too large T_w introduces further actuation delay that may lead the controllers to saturate. As previously explained, a trade-off must be found. IN this implementation, $T_w = 5T_s$ is considered.

Arrange these aforementioned blocks as shown in Figure 19.16. In order to manually manage the activation of the PI controller, the external reset (rising) is connected to a potentiometer (ADCINB3/pin68). Without moving it, the control algorithm is forced not to provide any actuation signal. To this aim:

- Add an **ADC** block and select ADCINB3 as **Conversion channel** in the **Input Channels** tab. Then, go back in the **SOC Trigger** tab:
  - **Sampling mode**: Single sampling mode;
  - **SOC trigger number**: SOC1;
  - **SOCx acqusition window**: 7;

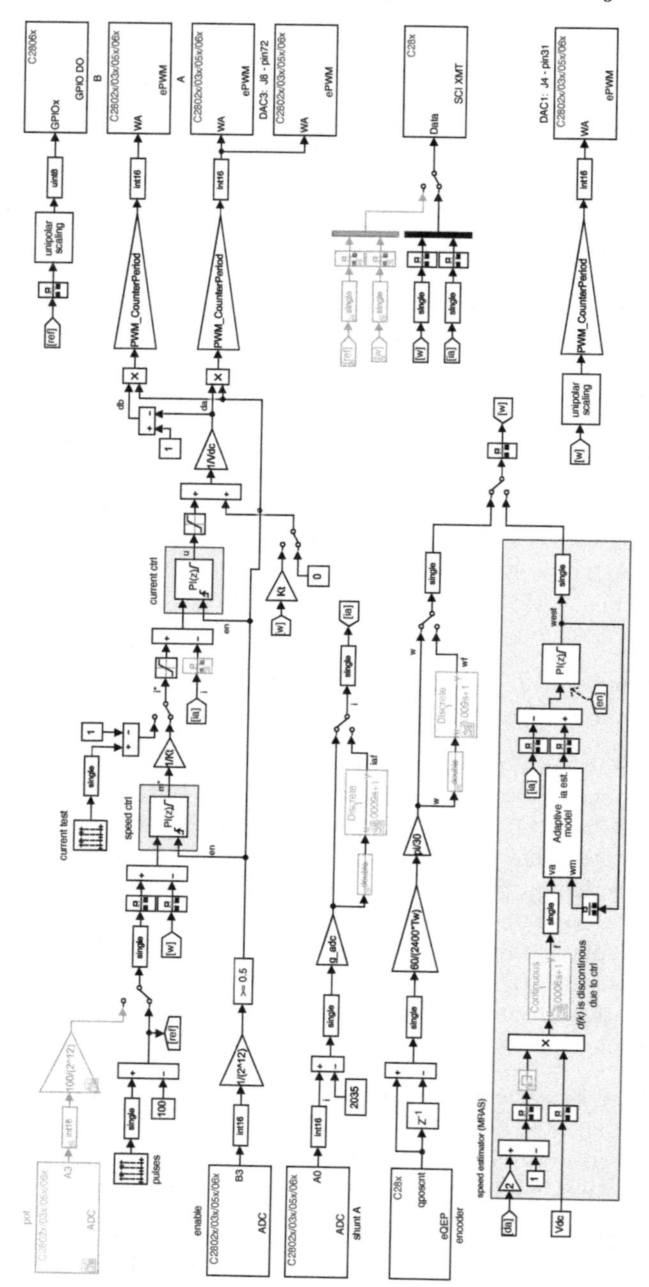

Figure 19.16 Simulink® scheme included in c28069_DCcontrol_fbF_uni.

  - **SOCx trigger source**: Software;
  - **ADC will trigger SOCx**: No ADCINT;
  - **Sample time**: $T_{sig} = 0.001\,\text{s}$;
  - **Data type**: uint16.

- Add a **Data Type Conversion** block set to int16 (since positive signals are used only) and the 12-bit scaling to convert the interval from 0 (0 V) to $2^{12} - 1 = 4095$ (3.3 V) in a range of variation $[0, 1]$.

- Add a **Compare to Constant** block, which is used to evaluate if signals are greater or equal than a Constant value 0.5. If this condition is true, the output switches from zero to one and it resets the integrator of the PI controller. The enable signal is defined as $\text{en}(k)$.

- Additionally, $\text{en}(k)$ can be used to add a further protection redundancy. Indeed, by using a Product block, the enable signal is multiplied by the actuation variable. As long as $\text{en}(k) = 0$, the duty cycle is forced to stay equal to zero $d(k) = 0$, i.e., no switching is performed.

- To visually evaluate the step-wise reference $\Omega^*(k)$, the Pulse Generator can be connected to a **GPIO** block (e.g., GPIO7) and read through a voltage probe. A proper bias of 100 and a scale factor of 1/200 are used to re-normalize the signal in a [0, 1] range. This processing allows to visualize the reference signal ranging in [0, 3.3] V at pin 79 through an oscilloscope.

- An MRAS observer can be realized by copying the corresponding block scheme adopted in the simulation file FBsimDC_closed.slx. Those blocks are suitable for implementation since they were arranged for discrete-time domain executions already.

It must be noted that some Memory blocks might be required to avoid algebraic loops. This is quite common when recursive loop are present. A manual switch is then used to select if the outer speed controller is feed by the speed measurements coming from the encoder or by the speed estimation computed by the MRAS observer, as shown in Figure 19.16. An initialization script either in *Model Properties/Callbacks/InitFcn* or as a separate `m`-file should be edited to set the system parameters. The latter approach is used in this exercise (see `data_init_DC.m`). The script is quite similar to the one used for simulation. The additional settings are:

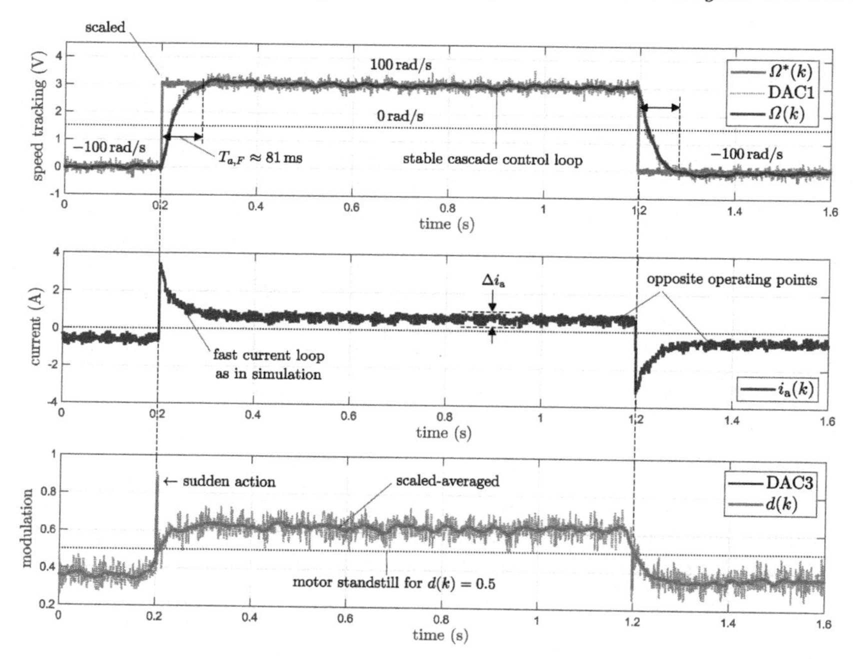

Figure 19.17 Closed-loop dynamics of the PMDC subjected to a step variation in the speed reference from −100 up to 100 rad/s.

```
%% F28069M clock
CPU_frequency        = 90e6;
%% carrier frequency definition
PWM_frequency        = 30e3;
%% TBPRD (counting mode) definition
PWM_counterperiod    = CPU_frequency/(2*PWM_frequency);
%% sampling time definition
Ts                   = 1/PWM_frequency ;
%% ADC theoretical gain
g_adc                = (530e-3)/62;
```

Optional: similarly to the principle adopted to link $\Omega^*(k)$ to GPIO7, the firmware could be also prepared to visualize the duty cycle $d(k)$ and the rotor speed through the DAC peripherals. In this case:

- Copy the one of the previous **ePWM** block, select ePWM8 (which is connected to DAC3 in J8, pin 72), and keep the other settings. Connect this block in parallel to ePWM1 (i.e., connect $d(k)$ just after the Data type conversion block). Hence, $d(k)$ can be observed by connecting a voltage probe to pin 72.

Figure 19.18 Example of load torque application on the shaft by locking it with a hand.

Similarly, the measured speed can be sent to an other channel of the DAC peripheral:

- Copy the previous **ePWM** block, select ePWM7 (which is connected to DAC1 in J4, pin 31), and keep the other settings. Copy and paste the Gain block set on PWM_Counter_Period and the data type conversion set on int16. The measured speed is first scaled through a 1/400 gain and, then, it is biased by constant which can be defined by the reader depending on specific needs. For instance, a 0.65 bias cause a speed variation up to $100\,\mathrm{rad\,s}/400 + 0.65 \rightarrow 0.9 \times 3.3 = 3\,\mathrm{V}$. This might be not satisfactory in presence of a speed overshoot, because the DAC saturates at 3.3 V. Hence, a more conservative choice could be to use a bias of 0.45. It must be remembered that this value is application dependent. Finally, $\Omega(k)$ can be observed through an oscilloscope by connecting a voltage probe to pin 31.

Figure 19.17 reports the closed-loop behavior of the cascade speed control scheme obtained by executing the firmware and observing the speed $\Omega(k)$, the armature current $i_{\mathrm{a}}(k)$ and the duty cycle applied on leg 1 $d(k) = d_{\mathrm{A}}(k)$ through an oscilloscope. In particular, $\Omega(k)$ and $d(k)$ were measured with voltage probes connected to DAC1 and DAC3, respectively, while $i_a(k)$ was acquired through a current probe.

As done in simulation, Figure 19.17 shows the command tracking performances of the controlled system in terms of speed and current dynamics. The resulting $\Omega(k)$ (blue line) is following the reference signal $\Omega^*(k)$ (red line). It must be noted that those two signals are expressed in a $0 - 3.3\,\mathrm{V}$ range. In particular, the speed lies in a shorter $0 - 3\,\mathrm{V}$ interval thanks to the 0.65 bias in the DAC1 processing. This is aimed to still have voltage margin to detect

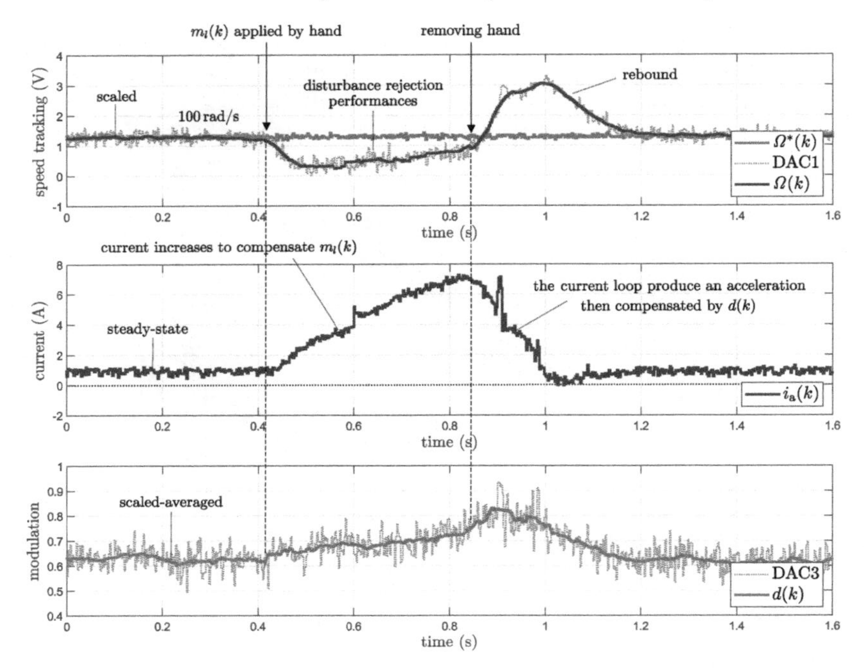

Figure 19.19 Closed-loop dynamics of the PMDC subjected to a load torque acting on the shaft.

overshoots. Therefore, it must be noted that 0 V corresponds to −100 rad/s, whereas 3 V stands for 100 rad/s. Consequently, zero speed is equivalent to 1.5 V. DAC1 results (gray dotted line) are averaged to get a good approximation of $\Omega(k)$ dynamic. The settling time is quite similar to the expected one ($T_{a,F} = 82\,\text{ms}$), showing a response which is comparable to the simulated one. It can be noted that the speed shows some oscillations which are a consequence of the encoder measurement. Those variation in $\Omega(k)$ are reflected in the $i_a(k)$ measurement as well, resulting in a greater current ripple Δi_a.

The Pulse generator which provides the internal reference $\Omega^*(k)$ is synchronized with the zero value of the triangular carrier (TBPRD=0). Nevertheless, such synchronization is not as strict as other tasks, e.g., counter comparison in ePWM, which is based on hard synchronization (hard-sync) event through interrupts. Therefore, it might happens that the rising edge of the step-wise $\Omega^*(k)$ is not perfectly synchronized with the ADC SOC (i.e., carrier peak). As already observed in simulation and in the previous implementations of closed-loop controls for *RL* and *RLC* loads, an actuation delay up to T_s might be present as a consequence of this displacement.

On the other hand, the duty cycle processed by DAC3 is fully scaled among a [0, 3.3 V] range, where 3.3 V corresponds to $d(k) = 1$. The signal read at pin 72 (gray dotted line) needs to be additionally averaged to obtain $d(k)$ (red

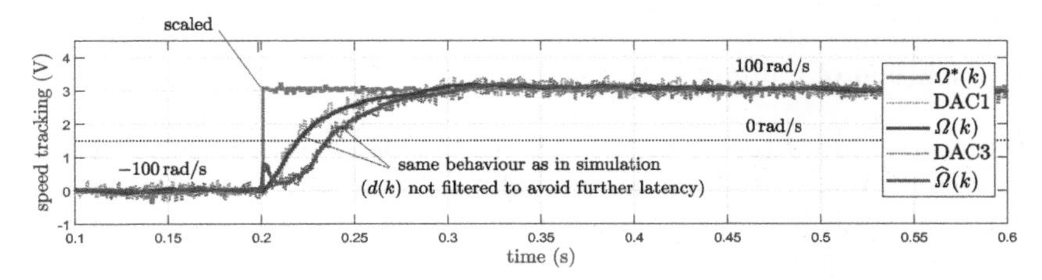

Figure 19.20 Comparison between command-tracking performances of the speed loop when an encoder and a MRAS observer are used.

line). This can be easily done via software by using an average function or a digital filter. Despite this processing, $d(k)$ still show spikes in the correspondence of the step variations of the reference signals, being consistent with the behavior observed in simulation. Indeed, this is a consequence of the sudden control action provided by the PI-based current controller.

The steady-state values of $\Omega(k)$, $i_a(k)$ and $d(k)$ are in line with the final value theorem, as shown in Chapter 5.

Furthermore, in analogy with simulations, the robustness of the closed-loop dynamics are evaluated in case a load torque is applied to the shaft. This can be carefully performed by simply hand-locking the rotor, as shown in Figure 19.18. This operation is equivalent to applying a variable an not-a-priori known $m_l(k)$ on the rotor shaft. Figure 19.19 shows how the system reacts to such variable perturbation. As expected, the disturbance rejection performances differ from the command-tracking ones, with the former being much slower. In particular, the first plot shows that $\Omega(k)$ tends to decrease when the hand is holding the shaft, while the current $i_a(k)$ increases, trying to compensate $m_l(k)$. Even if slow, the load disturbance is rejected by the cascade architecture until $\Omega(k)$ tracks $\Omega^*(k)$ again.

Nevertheless, the rotor accelerates when $m_l(k)$ is removed, producing a rebound effect. This behavior is due to the large rising of $i_a(k)$ during $m_l(k) \neq 0$, which has to rapidly decrease when $m_l(k) = 0$ producing an acceleration in the rotor speed. The latter is, then, compensated by reducing $d(k)$. However, it requires some time as a consequence of the slow dynamic of the speed loop.

It must be noted that the scaling of $\Omega^*(k)$ and $\Omega(k)$ for the GPIO7 and DAC1 reading is different with respect to that one adopted Figure 19.17. Indeed, in this case 100 rad/s correspond to $\approx$ 1.2 V to keep a sufficiently large voltage margin to observe the speed overshoot without saturating the peripheral.

Finally, Figure 19.20 reports a comparison of the closed-loop dynamics for a speed loop which processes the speed measured by the encoder LPD3806-600BM-G5-24C (blue line) and the one obtained by using an MRAS observer. The measured curves are in accordance with the simulated results even if the MRAS observer shows a larger latency in comparison to simulations. Moreover, the implemented MRAS do not include a low pass filter on the duty cycle

$d(k)$. Indeed, it is left up to the reader test possible behavioral differences due to the inclusion of a filter.

Implementation hints: even if the realization of the firmware follows a step-by-step procedure, some issue might arise anyway. Therefore, a hands-on experience creates a practical know-how which is useful to solve potential issues in a limited amount of time. Some examples are:

- **Noise**—if the setup shows a relevant noise of electrical nature, it must be remembered that the MOSFET are always switching ones per period due to the presence of a dead time. Therefore, if f_{sw} is in an audible range, this may create a sort of chattering sound that can be avoided by increasing the switching frequency. In case f_{sw} is fine, the noise might be related to a continuous computational error of the speed and/or current control. Indeed, the feedback signals might be really noisy (and noisy speed measurements/estimations are reflected in the current control action) or not accurate due to a bad characterization of the the ADC conversion chain (i.e., the experimental characterization should be repeated). To this end, keep in mind that discrete low pass filters may be added in the feedback path, at the cost of additional latency.

- **Execution overload**—every time many blocks or computations are included in the firmware, execution overload effects might happen due to the fact that all computations can not be performed within T_s. This issue is dependent on MCU communication (e.g., serial COM) in particular, which requires a lot of resources from the target. Therefore, if the processor has to solve a large number of operations, the communication stage may lead to overloads. In general, this issue depends on many key factors, e.g, value of T_s, size of data types, etc. A simple way to verify whether the execution is overrunning or not, is to use a free GPIO channel, implement a unitary pulse generator with a slow period (e.g., 1 s) in fixed-point unit (i.e., reduced computational resources and latency) and connect the corresponding output pin to an oscilloscope through a voltage probe. When the signal period starts to differ from the expected one, the MCU is close to overload without performing all the required computations in one clock cycle.

- **Limit the resources**—if additional blocks are added to the firmware during the developing stage, but they are not strictly necessary for the required task, they should be commented out to avoid to allocate useful resources for them and avoid overruns.

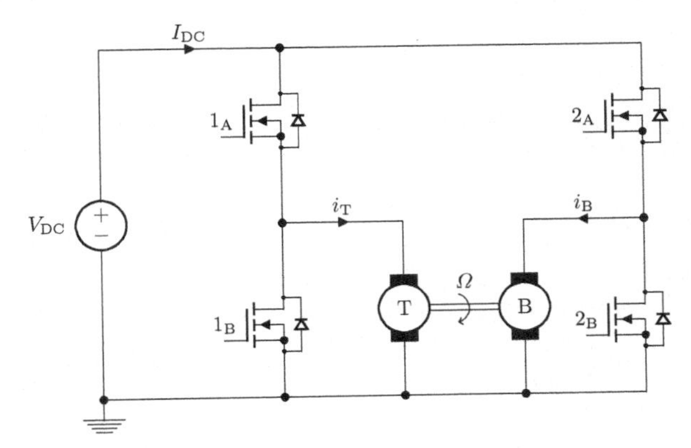

Figure 19.21 Equivalent circuit of the test setup involving the two PMDC motors in back-to-back configuration.

19.7 Back-to-Back (B2B) Configuration

The development of a cascaded speed control for a single PMDC motor both using a rotary encoder or a MRAS observer is useful to understand and analyze in details all the design aspects of cascade schemes and resulting dynamic responses for different operating points. A further point which can be investigated is the dynamic regulation of a load (braking) torque to emulate specific load conditions. This condition can be realized considering a B2B configuration with two PMDC motors coupled through an elastic joint. By referring to the B2B-PMDC hardware kit (see Appendix B), the aim of the exercise proposed here in the following is to achieve a test condition for which the traction motor is driven by a cascade speed control loop able to track a reference speed $\Omega^*(k)$ (by using a rotary encoder or a MRAS observer), whereas the braking motor is driven by a current control loop able to dynamically regulate the value of friction torque applied to the shaft.

Each PMDC motor is connected to one leg only (i.e., Half-Bridge configuration) of the BOOSTXL-DRV8301 converter, thus, allowing one rotating direction only for the whole setup. The single motor exercise already covered all the design and implementation aspects which are useful even for this exercise. Therefore, a simple review of the most important concepts is provided in the following.

The equivalent circuit of the B2B setup is reported in Figure 19.21. In this representation, the electrical parameters of the motor are not explicitly shown to allow the visualization of the mechanical coupling of the two machines. The traction motor is connected to leg 1 (phase A), whereas the braking one to leg 2 (phase B), as shown in Figure 15.1. The traction machine is operated as a motor, since the absorbed power is positive (see port current and voltage

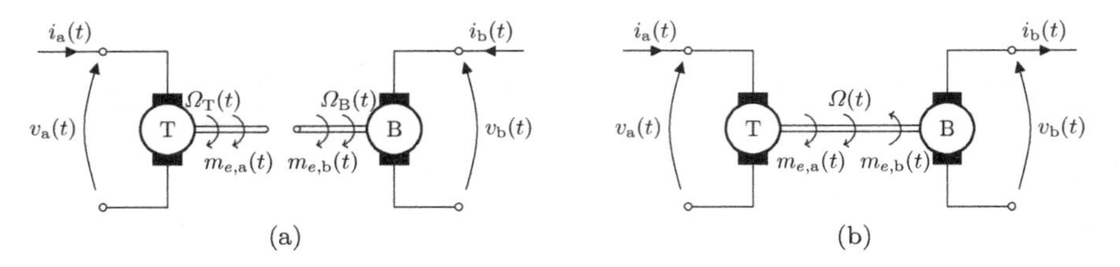

Figure 19.22 Mechanical and electrical port conventions for the test setup involving the two PMDC motors in back-to-back configuration: (a) no mechanical coupling (both machines spin their shaft in the same direction when operated as motors) and (b) with mechanical coupling (the braking motor works as a generator).

in Figure 19.22), while the braking machine is working as a generator, being $i_B < 0$ and, consequently, absorbing negative power (see Figure 19.22).

Indeed, this machine is subjected to the back-emf due to the motion of the shaft caused by the traction motor. It must be noted that it is recommended to physically connect the two motor terminals in such a way to keep the same sign of the back emfs for both of them. Namely, this means that positive currents in both motors would cause the rotation of the shaft in the same direction. This holds even for the torque direction.

Moreover, based on the equivalent circuit reported in Figure 19.21, if the leg B connected to the braking motor is operated at $d_B(k) = 0$, it follows that $\boldsymbol{S}(k) = [0\ 1]^T$. This implies a short circuit across the motor terminals (i.e., the armature circuit). Therefore, an open-loop control for the braking motor is not enough to limit its armature current. In the latter case, when no load torque is applied, the braking motor is equivalently connected to an open circuit, since its current is forced to be 0 (which does not necessarily mean $d(k) = 0$). On the other hand, when the reference (braking) current is greater than 0 in absolute value, the PI-based controller regulates the current flow through the motor, causing a load torque $m_l(k)$ applied on the shaft and a net current flow toward the low-side shunt on leg B as well. This underlines that the electrical connections of the two motors are of paramount importance. An opposite application of the back-emf of the braking motor would create a high loop current flowing between the two machines, causing an increase both in the electrical and mechanical stress, other than an overheating of the windings.

In principle, high-power motors require some equipment to dissipate losses associated to the application of load torque on the shaft to avoid overheating of the traction machine. This is not the case for the considered test setup, due to the limited size of the PMDC motors and the reduced amount of energy that is flowing in the system.

Due to the connections of both motors to the same dc link, the current that is flowing into the dc side equals the sum of the two currents flowing in

Table 19.3 Identified parameters of the BSB-PMDC Hardware Kit.

R_{a} (Ω)	L_{a} (mH)	K (Nm/A)	β_{eq} (N m s)	J_{eq} (kg m^2)
0.578	0.920	0.0245	$2.10 \cdot 10^{-4}$	$1.192 \cdot 10^{-5}$

each motor (i.e., in each leg):

$$I_{\mathrm{DC}} = i_{\mathrm{T}}(k) + i_{\mathrm{B}}(k) = i_{\mathrm{a}}(k) + i_{\mathrm{b}}(k) \tag{19.23}$$

where the labels T and B stand for traction and braking, respectively. It is quite evident that the traction motor is related to the current $i_{\mathrm{a}}(k)$ (from leg A) and the braking motor to i_{b} (from leg B). From now on these two currents are called $i_{\mathrm{a}}(k)$ and $i_{\mathrm{b}}(k)$. It must be noted that the proposed implementation foresees that the two involved legs are independently controlled. Thus, the braking motor may make the shaft rotate if activated before the traction one.

The model of this system now includes two electrical and one mechanical equations, which describe the behavior of the coupled PMDC motors (rigid shaft and coupling elements are assumed). In the continuous-time domain, it follows that:

$$L_{\mathrm{a}}\frac{di_{\mathrm{a}}(t)}{dt} = v_{\mathrm{a}}(t) - R_{\mathrm{a}}i_{\mathrm{a}}(t) - K\Omega(t) \tag{19.24}$$

$$L_{\mathrm{b}}\frac{di_{\mathrm{b}}(t)}{dt} = v_{\mathrm{b}}(t) - R_{\mathrm{b}}i_{\mathrm{b}}(t) - K\Omega(t) \tag{19.25}$$

$$J_{\mathrm{eq}}\frac{d\Omega(t)}{dt} = m_{e,\mathrm{a}}(t) - m_{e,\mathrm{b}}(t) - \beta_{\mathrm{eq}}\Omega(t) \tag{19.26}$$

where J_{eq} and β_{eq} are the equivalent inertia and friction coefficient associated to the whole system (i.e., the two PMDC motors and the mechanical joint), whereas $m_{e,\mathrm{b}}(t)$ and $m_{e,\mathrm{a}}(t)$ correspond to $m_e(t)$, $m_l(t)$ (see equation (15.3) and Figure 19.22), respectively. Moreover, it can be noted that the subscripts a and b now denote the two motors for all the system parameters. Given the sample time T_s and an integration method, the B2B model and the related variable can be discretized.

19.7.1 Parameter identification

Since a new couple of PMDC motors is installed in the B2B-PMDC Hardware Kit, the characterization of the electrical and mechanical parameters of the system should be repeated. Thus, referring to the same procedure reported in Section 19.6.1, the parameters shown in Table 19.3 are identified. The electrical parameters can be determined independently for the two motors, whereas it is convenient to keep the two electrical machine mechanically coupled to determine the overall J_{eq} and β_{eq}. It is important to note that, due to the similarity of the motors, average electrical parameters can be considered. Therefore, the electrical identification in terms of resistance R and inductance L is performed, leading to $R_\mathrm{b} = R_\mathrm{a}$ and $L_\mathrm{b} = L_\mathrm{a}$. In addition, the identification of the back-emf constants can be performed by measuring the open circuit voltage of the non-controlled motor at steady state.

Based on these values, the parameters of the speed and current controllers are computed again using a pole-zero cancellation. Both the current PI are identical. For what regards the current loop, starting from the time constant $\tau_{G_i} = L_\mathrm{a}/R_\mathrm{a} = 1.6\,\mathrm{ms}$ and the corresponding settling time $T_{a,G_i} = 5\tau_{G_i} = 8\,\mathrm{ms}$, the controlled system is set to have $T_{a,F_i} = T_{a,G_i}/10 = 796\,\mu\mathrm{s}$. Therefore, knowing that $\tau_{F_i} = T_{a,F_i}/5$, it follows that $\omega_{c,i} = 1/\tau_{F_i} = 6283\,\mathrm{rad\,s}$. Thus, the values of $k_{p,i}$, $k_{i,i}$ are explicitly determined as:

$$k_{p,i} = \omega_{c,i} L_\mathrm{a} = 5.78\,\Omega \qquad k_{i,i} = \omega_{c,i} R_\mathrm{a} = 3631\,\Omega/\mathrm{s} \tag{19.27}$$

The speed control dynamic is designed to fulfill the cascade constraint. Namely, the speed controller bandwidth is set equal to $\omega_{c,\Omega} = \omega_{c,i}/100 = 62.83\,\mathrm{rad/s}$. Therefore, the outer loop is significantly slower than the inner one. Therefore, the resulting settling time is $T_{a,F_\Omega} = 79.6\,\mathrm{ms}$, while the corresponding values of $k_{p,\Omega}$ and $k_{i,\Omega}$ are:

$$k_{p,\Omega} = \omega_{c,\Omega} J = 0.749\,\mathrm{mN\,m\,s} \qquad k_{i,\Omega} = \omega_{c,\Omega} \beta = 0.0132\,\mathrm{N\,m} \tag{19.28}$$

Discrete-time PI controllers $R_\Omega(z)$ and $R_i(z)$ are derived based on these considerations. The sampling time of the controller is set equal to $T_s = T_\mathrm{sw} = 1/f_\mathrm{sw}$, using Forward Euler as integration method.

19.7.2 Control implementation

The closed-loop dynamics are tested for both the sensored (encoder) and sensorless (MRAS observer) solution on the B2B-PMDC Hardware Kit (see Figure 19.23). Even in this implementation, legs (phase) A and B are used. However, each motor is driven through one leg only, i.e., through a Half-Bridge converter. The connection scheme for this setup is reported in Figure 19.23. These two ePMW modules are operated in up-down counting mode at $f_\mathrm{sw} = \mathrm{PWM_{frequency}} = 30\,\mathrm{kHz}$. Reminding that for the F28069M board $f_\mathrm{ck} = \mathrm{CPU_{frequency}} = 90\,\mathrm{MHz}$, it follows that:

$$\mathrm{TBPRD} = \mathrm{PWM_{counter-period}} = \frac{\mathrm{CPU_{frequency}}}{2 \cdot \mathrm{PWM_{frequency}}} = 1500$$

as already seen for the control of the single PMDC motor. Moreover, the sampling time is $T_s = T_\mathrm{sw} = 1/\mathrm{PWM_{frequency}}$. Once again, reference speed can be internally generated in the firmware or externally provided both using the available potentiometers on the extPot3 board or through COM port (e.g., SCIA). The analog solution is adopted to add versatility to the control scheme in the proposed solution. However, the reference braking torque is obtain by varying a second potentiometer on the extPot3 board. The required

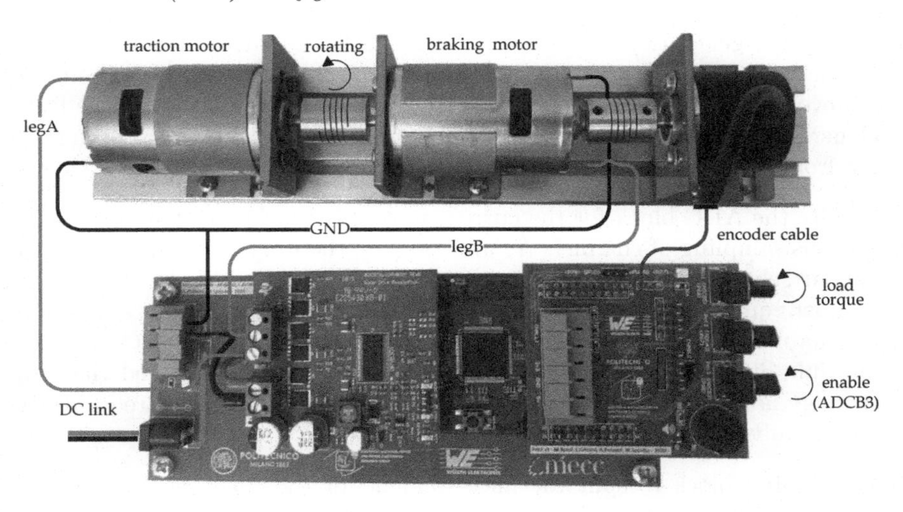

Figure 19.23 Connections of the B2B-PMDC Hardware Kit with the LaunchPad F28069M through the BOOSTXL-DRV8301 converter and the extPot3 board. A Half-Bridge configuration is realized for each motor and the encoder wires are interfaced with module 2 of the eQEP peripheral.

parameters can be initialized in *Model Properites/Callbacks/InitFcn* or in a separate `m`-file.

A step-by-step procedure for programming the microcontroller for this task is reported here in the following.

Firmware Environment

c28069_B2Bcontrol_hbF (solver: fixed step - ODE4, step size: $T_s = 1/f_{\mathrm{sw}}$)

Open a new blank Simulink® project and configure the environment as shown in Section 9. Remember that the BOOSTXL-DRV8301 is enabled by settling **GPIO50** high (see Section 3.2). This firmware can be realized starting from the file c28069_DCcontrol_fbF_uni built previously. Indeed, this implementation requires a speed and a current loop for the traction motor, as well as speed and current measurements. Moreover, the braking motor requires the same current control scheme adopted previously and a MRAS observer is needed to test sensorless runs of the code. The control (and actuation) enable is managed again through the potentiometer connected to ADCB3, whereas DAC modules can be used for measuring speed and duty cycle with an oscilloscope. Therefore, it is suggested to copy and paste the whole scheme included in c28069_DCcontrol_fbF_uni in this new blank project and, then, to slightly modify the ePWM blocks and the processing of the modulating signals in order to move from a Full-Bridge to an Half-Bridge configuration. Indeed, referring to the previous file:

- Modify the generation of the speed reference. As a matter of fact, the back-to-back setup is allowed to spin the shaft in one direction only. Therefore, remove the Constant and the Add blocks and set the amplitude of the Pulse Generator equal to 100. Therefore, the system tracks a speed reference ranging from 0 up to 100 rad/s.

- Modify the ADC block for the current measurement by allowing the Simultaneous sampling of channels A0 and B0, i.e., the modules connected to the onboard shunts on legs 1 and 2, respectively. To this aim, set the SOC trigger number equal to SOC0-SOC1 and modify the same entry in the ADC block for enabling the PI blocks (channel B3) setting it equal to SOC4. Indeed, multiple SOC trigger numbers are not allowed. It can be noted that this strategy is preferred to the use of two separated ADCs since it requires less use of embedded resources.

- Add a Mux block to split the measurement for the currents on both legs. Then, copy and paste the same conversion chain adopted for channel A0. It is important to modify the offset and the gain based on the identification process for the second leg.

- Copy and paste the current control loop from the traction motor. It must be remembered that the two current PI controllers share the same parameters. Then, move the gain block set equal to PWM_Counter_Period, the Data Type Conversion block set on int16 and the ePWM2 close to this second control loop for the braking motor. The reference braking torque is generated through the potentiometer connected to channel B5 of the ADC peripheral. Set the parameters of this block as follows:

  - Select **ADCINB5** as **Conversion channel** in the **Input Channels** tab. Then, go back in the **SOC Trigger** tab;
  - **Sampling mode**: Single sampling mode;
  - **SOC trigger number**: SOC3;
  - **SOCx acquisition window**: 7;
  - **SOCx trigger source**: Software;
  - **ADC will trigger SOCx**: No ADCINT;
  - **Sample time**: 0.001;
  - **Data type**: uint16.

  Then, use a Data Type Conversion block set on int16 and a Gain block with parameter $1/2^{12}$ to re-scale the reading of the ADC channel into a [0, 1] range. After that, this signal is translated into a braking torque through a Gain block set on `-maxBr`, that is, $-0.05\,\mathrm{N\,m}$. A redundant saturation block can be used to limit the load torque up to this value.

- Leave all the other blocks unchanged.

Rearrange the aforementioned blocks as shown in Figure 19.24.

Remark: it must be noted that the overall firmware scheme is pretty large and complex compared to the previous exercises. Hence, it may be possible to saturate the computational resources of the MCU during the execution of the firmware. To avoid this issue, some parts of the scheme which are not strictly necessary for the execution can be commented. For example, the blocks implementing the MRAS observer during the sensored control. Moreover, Data Type Conversion blocks can be inserted and/or edited to reduce the processing effort during computations as well as DAC blocks can be excluded from the firmware.

An initialization script both in *Model Properites/Callbacks/InitFcn* or as a separate m-file should be edited to set the system parameters. The latter approach is used in this exercise (data_init_B2B.m) and the code is reported here in the following:

```
%% F28069M clock
CPU_frequency         = 90e6;
%% carrier frequency definition
PWM_frequency         = 30e3;
Tsw                   = 1/PWM_frequency
%% TBPRD (counting mode) definition
PWM_counterperiod     = CPU_frequency/(2*PWM_frequency)
%% sampling time definition
Ts                    = 1/PWM_frequency;
%% ADC theoretical gain
g_adcA                = 4.65/550;
g_adcB                = 4.65/450;
%% maximum braking torque definition
maxBr                 = 0.05;
%% encoder sample time definition
Tw                    = Ts*5;
```

were the parameters of the two motors and of the controllers are not reported for the sake of brevity. It can be noted that the encoder acquisition window is set to $T_w = 5T_s$. However, the same consideration reported previously on this parameter are still valid and a new trade-off value can be determined if necessary. As done for all the implementations proposed in this book, Figure 19.25 shows the command tracking performances of the controlled system. The resulting $\Omega(k)$ (blue line) is following the reference signal $\Omega^*(k)$ (red line). It must be noted that those two signals are expressed in a [0, 3.3 V] range as consequence of the measurement performed via GPIO7 and DAC1,

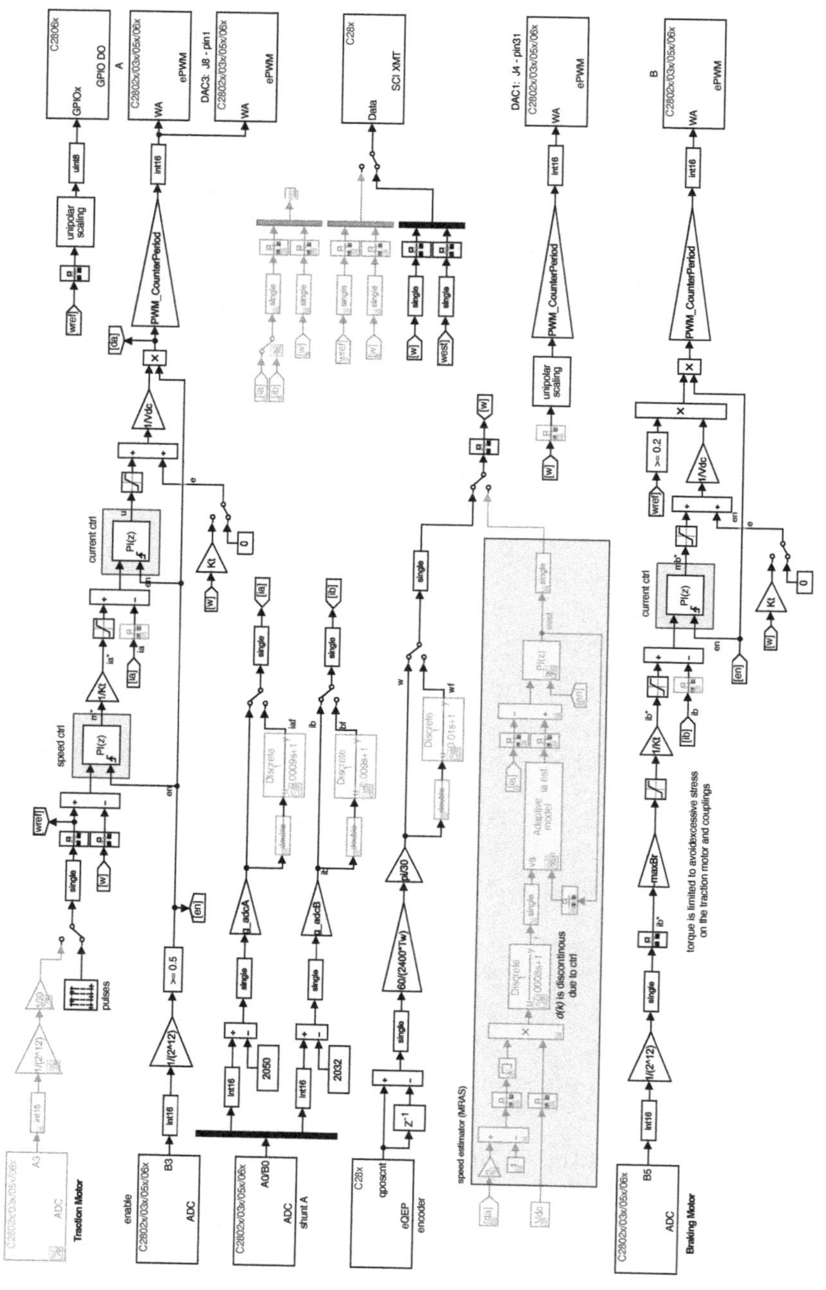

Figure 19.24 Simulink® scheme included in c28069_B2Bcontrol_hbF.

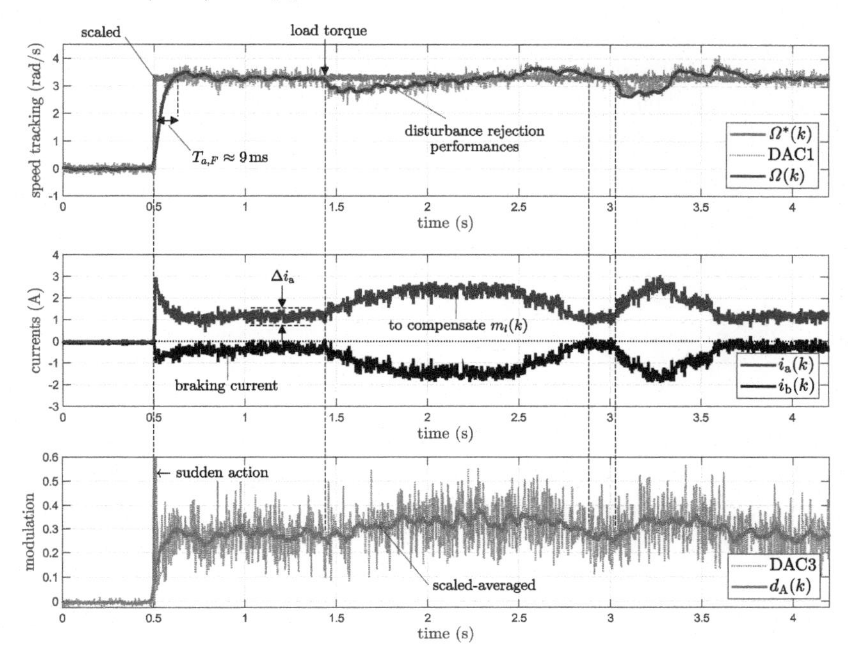

Figure 19.25 Closed-loop dynamics of the back-to-back setup subjected to a step variation in the speed reference and to the application of some braking torque by the braking PMDC machine.

respectively. In particular, it must be noted that 0 V corresponds to 0 rad/s, whereas 3 V stands for 100 rad/s. Actually, 3.3 V represents the highest value which is possible to visualize in case of speed overshoots (after which the peripheral saturates).

The settling time is pretty similar to the expected one $T_{a,F_\Omega} \approx 79.6$ ms. The same holds for the current control loop. Even in this case, the speed shows some oscillations which are a consequence of the encoder measurement. Those variation in $\Omega(k)$ are reflected in the current measurement as well, resulting in an increase of the current ripple.

Figure 19.25 also shows the duty cycle measurement through DAC3. The signal read at pin 72 (gray dotted line) needs to be additionally averaged to obtain $d(k)$ (red line). This can be easily done via software by using an average function or a digital filter. Despite this processing, $d(k)$ still shows spikes in the correspondence of the step variation of the reference signals, similar to the single motor case. This is a consequence of sudden control action of the PI-based current controller.

In addition, these plots show the response of the system to the application of a dynamical load torque on the shaft through the braking motor. Indeed, current $i_b(k)$ is proportional to $m_l(k)$ and it is almost symmetrical with respect to $i_\mathrm{a}(k)$. The evolution in time of $m_l(k)$ depends on the dynamical variation of the potentiometer connected to ADCB5. The plots show a rejection to disturbances which is similar to that one observed in the single PMDC motor implementation.

Part V

Real-Time Control in Power Electronics: Load Emulation

20

Debugging Tools and Firmware Profiling

In the previous chapters, firmware design has been described in a step-by step manner, helping the reader to understand how to make an embedded control scheme out of system simulations, which is consistent with linear control theory. Nevertheless, there are a many debugging techniques that can be adopted both to check possible mistakes or to verify the performances of the generated code. Among them, the processor-in the-loop (PIL) testing technique and the external mode execution are adopted in this book and described in this Chapter. Both approaches lead to test numerical equivalence between model computations, produced code and expected behavior (predicted from simulations). As the system is running, information on block states, inputs, outputs, and time execution can be displayed within the Simulink® Editor. This allows the designer to investigate specific problems and/or potential improvements in blocks, parameters, or interconnections.

20.1 Processor-in-the-loop with Simulink®

The Simulink® code generation workflow allows to perform processor-in-the-loop (PIL) simulations of a given setup. In the previous chapters, simulation schemes have been derived to design and analyze the closed-loop control dynamics. PIL simulations extends this good practice by considering the control part running online in a dedicated processor (i.e., target specifications are required) while still emulating the system/load offline in a simulation platform (e.g., host PC). To this aim, the control algorithm is translated/compiled into C-code and downloaded into a development board (e.g., MCU). This latter performs a data exchange with the host PC through a serial communication, which has to be set (e.g., type, speed). Hence, PIL is effectively a testing technique which should be performed after simple system simulations and before working on real firmware implementation, i.e., prior to include peripheral blocks and their hardware constraints. This operation allows to:

- *Test whether the model and generated code are numerically equivalent*: PIL tests are designed to highlight possible problems with algorithm execution in the embedded environment. For instance, it can be used to check whether

DOI: 10.1201/9781003196938-20

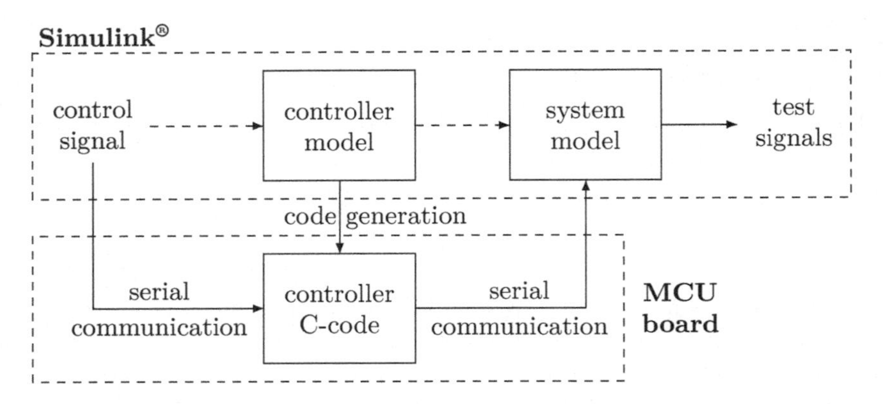

Figure 20.1 Block diagram showing the main steps to be carried out during PIL simulations.

the considered control loop fits within the execution time available on the embedded processor.

- *Observe code coverage*: PIL allows an evaluation of the automatically generated code.
- *Perform code execution profiling*: this testing technique is useful to identify the most demanding operations inside the control loop.

Therefore, the PIL simulations are aimed to evaluate the performances of a controller running on a target board before its implementation on a real system. From the hardware point of view, a control platform such as the LaunchPad™ F28069M board is required only. Instead, the load is still emulated.

As an example, by considering the closed-loop speed control of a PMDC motor as shown in Chapter 19, PIL simulations may be performed to gain confidence and to check that the control algorithm performs as expected once it is physically running on the MCU, without connecting the TI BOOSTXL-DRV8301 BoosterPack or the DC machine as well. A block diagram showing the main steps of this procedure is shown in Figure 20.1. In particular, a simulation environment is used to model and verify the behavior of a speed-loop control for a DC motor. Basically, starting from the simulation environment that represents the whole controlled system, the control task which have to be executed in the MCU are selected (e.g., PI controllers and sensing path) and translate in C-code. This latter is then downloaded into the MCU, while the PMDC motor model is kept in the scheme, but it is emulated only (i.e., code is not generated for this part of the model, see Figure 20.2).

Through a communication channel, the Simulink® environment (on the host PC) exchanges stimulus signals (e.g., test vectors) with the MCU board for each sample interval of the simulation. Every time the target processor

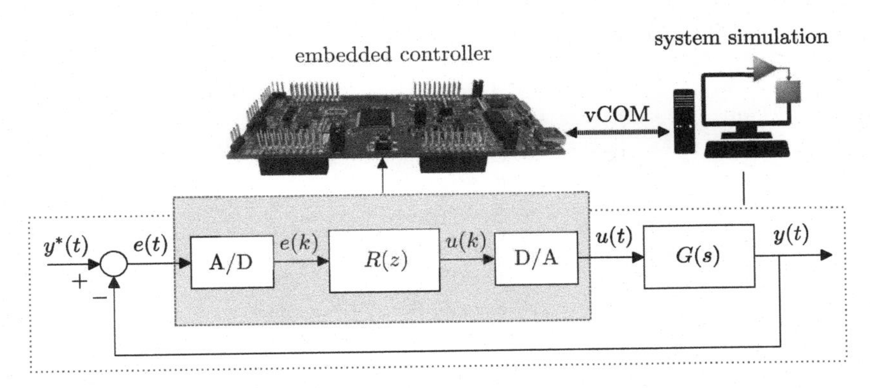

Figure 20.2 PIL workflow between the LaunchPad™ F28069M and the host PC developed in Simulink®.

receives signals from Simulink®, it executes the on-board code for one sample step. The results, e.g., a control action, are returned to the Simulink® scheme through a communication channel. Then, one sample cycle of the simulation is complete, and the process keeps repeating for the next sample interval. It can be noted that in each sample period, PIL simulations do not run in real time. Indeed, the Simulink® environment and the object code exchange I/O data depending on the serial communication settings. Although the workflow reported here in the following describes how the performances of a motor control application are assessed using PIL testing, it is possible to generalize this procedure for any power electronic-based application, framework and processor/MCU supported by the Simulink® environment.

20.1.1 PMDC motor control implementation through PIL

The closed-loop speed control of a PMDC motor presented in Chapter 19 is considered. The control architecture is based on two nested loops, i.e., the speed and current ones, with a feedforward compensation of the back-emf. The design rules for both controllers are already discussed in the previuous Chapter and, thus, they are not reported here.

For the sake of simplicity, the PMDC motor equivalent circuit is created in Simulink® based on transfer function blocks instead of using Simscape™ Electrical elements. This implies no ripple on the armature current $i_a(k)$ and the rotor speed $\Omega(k)$, since a steady-state analysis is performed (see Chapter 6). The control scheme is included in the file:

c28069_PMDC_PIL.slx (solver: fixed step - ODE4, step size: T_{sim})

The system parameters reported here in the following can be initialized in *Model Properites/Callbacks/InitFcn* or in a separate m-file.

```
%% motor data
Vdc    = 10;
imax   = 5;
Ra     = 0.529;
La     = 8.651e-4;
Kt     = 0.0232;
B      = 1.803e-4;
J      = 3.56e-6;
%% modulation and sampling
fsw    = 20e3;
Tsw    = 1/fsw;
Ts     = Tsw;  or Ts = Tsw/2
Tsim   = Ts/100;
%% current controller
tauGc  = La/Ra;
TaGc   = 5*tauGc;
TaI    = TaGc/10;
wcI    = 5/TaI;
kpI    = wcI*La;
kiI    = wcI*Ra;
%% speed controller
tauGm  = J/B;
TaGm   = 5*tauGm;
wcW    = wcI/10;
TaW    = 5/wcW;
kpW    = wcW*J;
kiW    = wcW*B;
```

The generation of the reference signal to operate the system in several working conditions as well as the cascaded control loops are shown in the schemes reported in Figures 20.3 and 20.4. Before going on with the proper PIL simulation, it is necessary to setup the Simulink® environment. While connecting the LaunchPad™ F28069M board to the host computer, it must be ensured that the Serial Communication is set properly on a COM port (see Section 8 for reference). Due to the embedded nature of the PIL simulations, the solver type has to be necessarily **fixed step** (i.e., discrete-time domain). For instance, the proposed exercise uses fixed step - ODE4 (Runge-Kutta) solver with step size $T_{\text{sim}} = T_s/100$. These settings can be edited by opening the **Solver** panel in the **Model Configuration Parameters** window. Moreover, PIL simulations requires the enable for PIL blocks creation. To this aim:

- Click on **Hardware Implementation** panel, verify that **TI Piccolo**

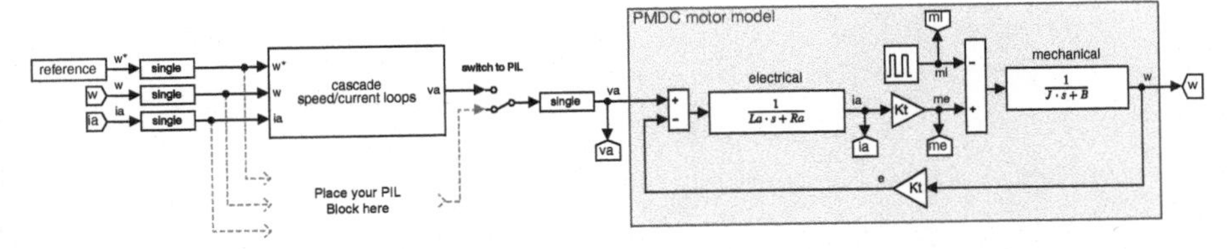

Figure 20.3 Simulink® scheme included in c28069_PMDC_PIL.slx.

F28069M LaunchPad is set as **Hardware board**. Then, keep the settings as reported in Chapter 3 and 4.

- Click on **Code Generation** panel and then on **Verification**. Open the **Advanced parameters** section and set the **Create block** frame by selecting **PIL** in the drop-down menu.

As an alternative, the last setting can be found by using a shortcut. Namely, open the **Model Configuration Parameters** and type `CreateSILPILBlock` in the upper search box. This operation directly links to the *Code Generation/Verification/Advanced parameters* section.

Considering the scheme shown in Figure 20.3, the control subsystem (which includes both speed/current controllers and back-emf compensation) is the object of the verification through PIL testing. Only this part of the Simulink® scheme has to be implemented and run online into the LaunchPad™ F28069M board.

To this aim, enable the settings for running PIL over Serial by typing the instructions reported here below in the Command Window:

```
setpref('MathWorks_Embedded_IDE_Link_PIL_Preferences','COMPort','COM10');
setpref('MathWorks_Embedded_IDE_Link_PIL_Preferences','BaudRate',115200);
setpref('MathWorks_Embedded_IDE_Link_PIL_Preferences','enableserial',true);
```

where the user has to set the COM port which has been defined in the PC device manager, e.g., COM10 in this example.

Then, **right-click on the control subsystem** and select **C/C++**

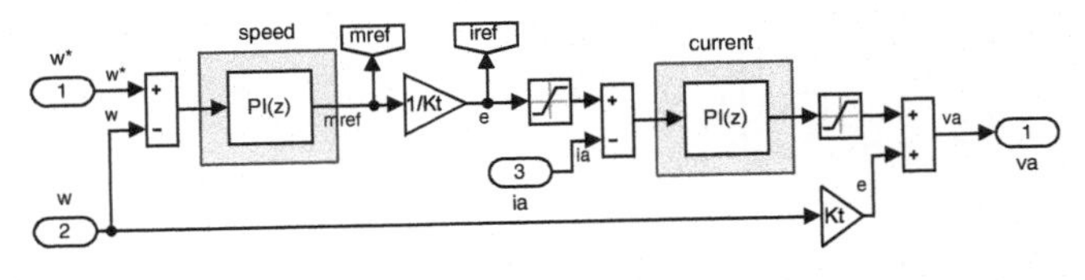

Figure 20.4 Inner structure of the cascade speed/current loops subsystem.

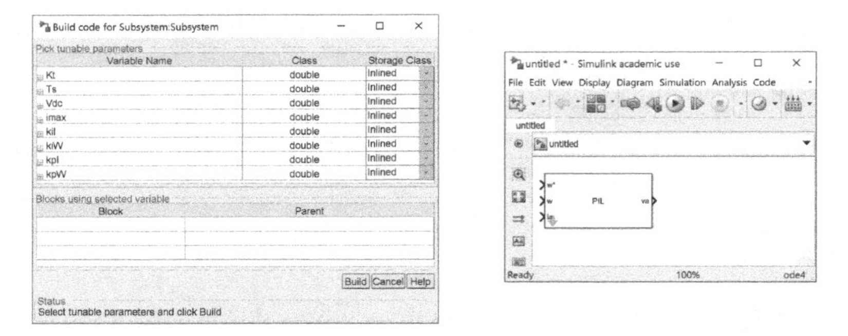

Figure 20.5 Subsystem parameters and the resulting PIL block.

Code/Deploy this Subsystem to Hardware from the list. This command Deploy this Subsystem to Hardware opens a window which summarizes the control parameters defined inside the subsystem. Select **inlined** for all of them as reported in Figure 20.5 and then click on the **Build** button. This starts the translation of this piece of scheme in C-code, which ends with the opening of a new `untitled.slx` file containing a PIL block only. This is the translated control subsystem. Copy such block and paste it in the highlighted area in the initial Simulink® model, as shown in Figure 20.6. Now, the file c28069_PMDC_PIL.slx is enabled to perform PIL simulations, besides the standard ones. Indeed, by clicking on Run, the PIL block of the control subsystem is downloaded into the MCU board and available for executions. Therefore, the control actions are computed online on the MCU and sent as data arrays to the PMDC motor model input. This latter simulates the motor behavior and the required feedback signals are transmitted back to the MCU. The manual switch allows to test the consistency of the results and compare the closed-loop dynamics when the system is driven by the control scheme running in Simulink® or implemented on-board the MCU.

In this example, a step-wise speed reference $\Omega^*(k)$ is applied to analyze the command-tracking performance of speed and current loops, respectively.

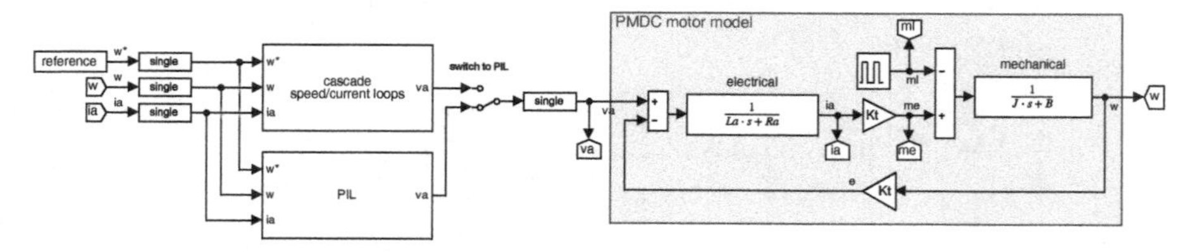

Figure 20.6 Simulink® scheme included in c28069_PMDC_PIL.slx, which involves the translated block for PIL simulations.

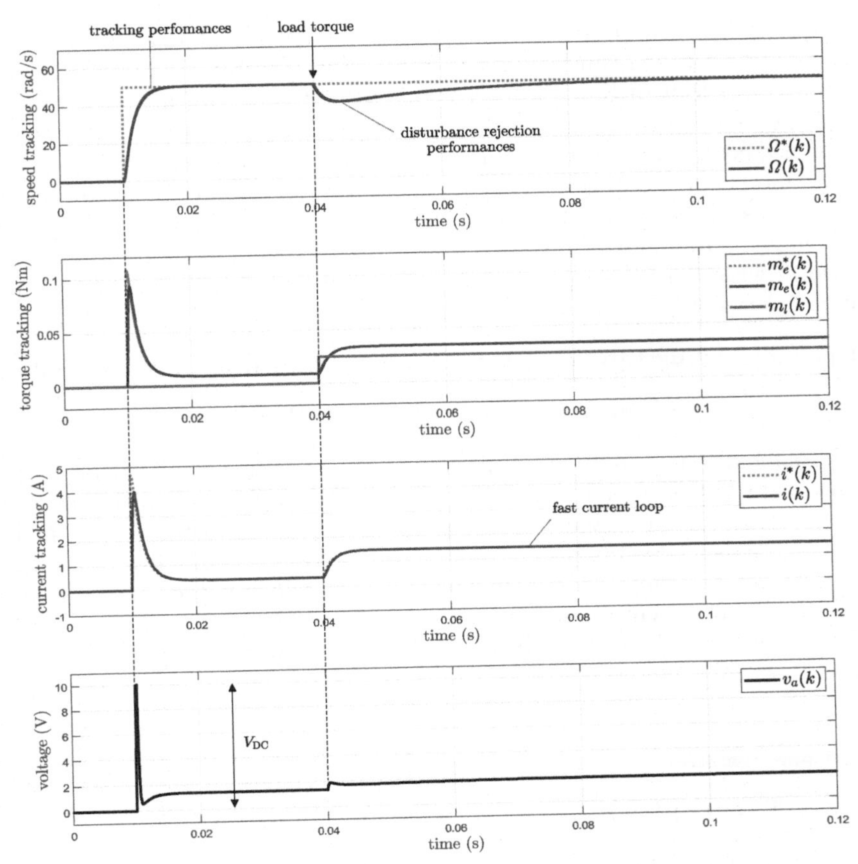

Figure 20.7 Closed-loop dynamics resulting from standard and PIL simulations.

A load torque $m_l(k)$ is also applied to test the disturbance rejection and to verify the robustness of the proposed control scheme. Both strategies return the same dynamics, which are reported in Figure 20.7. It is possible to note that the PIL simulations are subjected to a slower processing, since the computations of the control action are performed on-board the MCU (serial communication is required),[1] while in standard simulations those are computed through the host PC CPU. The PIL framework enables another widely used

[1]Note that, from MATLAB® version 2019b it might be neccessary to specify the desired behavior for denormal results for arithmetic operations. Denormal numbers are any non-zero numbers whose magnitude is smaller than the smallest normalized floating-point number. Some hardware targets as the LaunchPad™ F28069M board flush denormal results from arithmetic operations to zero. You can set this behavior in **Model Configuration Parameters/Math and Data Types**, then change the parameter `Denormal Behavior` to `Flush To Zero (FTZ)` (instead of `Gradual Underflow` which is given by default).

feature which is the profiling of the task execution time and function execution time of the real-time algorithm running on the MCU board. This tool is named **Profiler** or Simulink® Profiling. The Profiler captures performance data while the software is running the simulation, e.g., it identifies which parts of a control algorithm require most time or most iterations to be computed and executed. Such information is useful to evaluate the need for code refinements and where to focus the optimization efforts in the algorithm. To this end, profiling is particularly useful to analyze the pipelining in real-time platform which may benefit of multi-core processors on the target hardware. There are two ways to enable the Profiler:

- Open the **Debug tab**, select **Performance Advisor/Simulink Profiler**.
- Select **Analysis/Performance Tools/Show Profiler Report**

When simulation is complete, Simulink® generates and displays the performance data in the simulation profile report in the form of a panel. This data shows the amount of simulation time spent executing each function in the model.[2] The *block hierarchy view* of the report shows the profiling information in a nested tree form. The first row of the table, which is the top of the tree, corresponds to the entire model. Subsequent rows at the first level of nesting correspond to single or group blocks which can be individually assessed.

20.2 External Mode Execution with Simulink®

In the framework of firmware debugging, an alternative to PIL simulations is the real-time *external mode* execution. In this mode, Simulink® is still used to generate and download the firmware code into the MCU board, but the resulting executable is able exchanges data with Simulink® via a shared memory interface.

Compared to the PIL technique, this kind of execution is useful to debug the firmware inner computations during the design stages. As a matter of fact, the external mode allows to run the whole code on the microcontroller and to visualize some variables of interest using scope and display blocks in Simulink®. The variables are recorded from hardware through the serial communication. The real-time code stores contiguous response data in memory accessible to Simulink® until a data buffer is filled (i.e., data captured within one buffer are contiguous). Then, the firmware carries on running while Simulink® plots data. After model computations are finished, data transfer runs at a lower priority while the process waits for another interrupt to trigger the

[2]In particular, the Profiler measures the time required to execute each invocation of these functions.

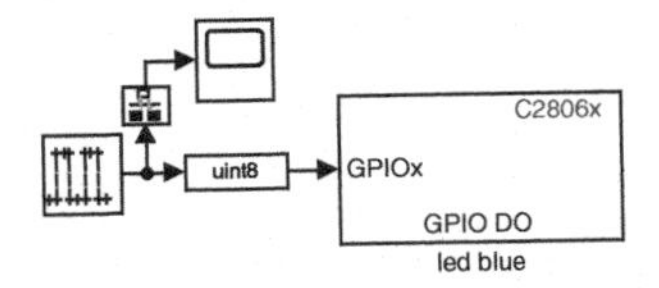

Figure 20.8 Simulink® scheme included in c28069_ledB_EXT. This scheme can be obtained from c28069_ledB_F.slx proposed in Example 1, Section 11.3. In this case, the signal source is a Pulse Generator.

next model update. Depending on the firmware complexity, such data transfer could be less critical than maintaining deterministic real-time updates at the required sample interval (i.e., not a priori).

The serial communication between PC and board is automatically settled without any block like Serial Send and SCI Receive. Hence, differently from what is described in Part III, this time only one Simulink® file is needed both for programming and debugging purposes, since the external mode automatically manages the data flow through the virtual COM. Scopes and displays can be inserted directly in the firmware scheme. It is important to note that there is a maximum number of plots that can be used depending on the performances and capabilities of both the serial communication and the host PC. Indeed, code size and memory consumption are increased when the External Mode is enabled.

20.2.1 Simulink® setup for external mode execution

The same options set so far are to be ensured even in this case. Thus, the proper settings for the Serial Communication as shown in Section 8 must be verified. Moreover, the reader is referred to Chapter 9 for preparing the environment for implementations on the MCU board (e.g., settings for jumpers, switch positions and virtual COM port).

Considering Example 1 in Section 11.3, which consists in driving the blue led (GPIO39) in toogle mode. A Constant block is used to produce an internal command signal which drives the GPIO39 with 0 or 1 values. The sample time of the constant block is set equal to $T_{sig} = 1000T_s = 0.1\,\text{s}$. The led is turned on/off accordingly. As an alternative, the constant block can be replaced by a Pulse Generator with gain $A = 1$, number of samples per period 1000, pulse width 50% (that is, 500 samples) and sample time T_s. The reader is referred to Section 11.3 for the remaining settings. Then, place a Simulink® Scope block connected to the Constant or Pulse Generator block output, as shown in Figure 20.8 to evaluate the produced values. Use a Rate Transition block to improve the accuracy of data sampling and set the sample time equal to $T_s/2$. Then, once the firmware design is ready, the external mode execution can be set and enabled as follows:

- Open the **Model Configuration Parameters** window and go in the **Hardware Implementation** menu. Check if the TI Piccolo LaunchPad™ F28069M board is set as **Hardware board**. Among the list of menus in

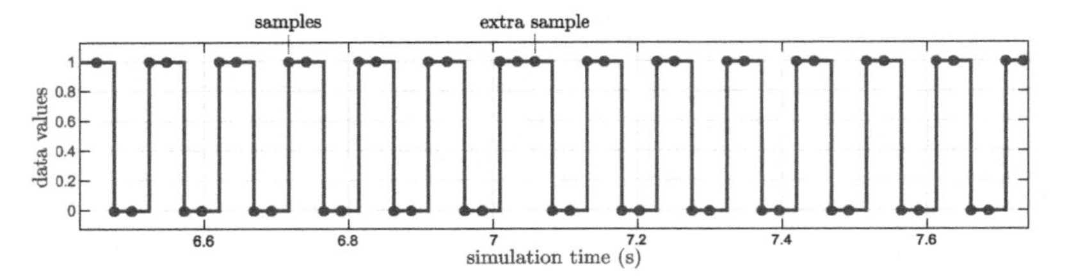

Figure 20.9 Plot from the scope block in External Mode execution.

Target Hardware Resources, look for **External mode**. There, enter the serial **COM port** set in the PC device manager, e.g., COM10 in this example. Keep the other settings as they are.

- Coming back to the Simulink® editor, set the desired simulation time, e.g., 20 s, and click on the monitor & tune icon to run external mode execution.

These settings are necessary to program the MCU and to run the External Mode execution for the supported MCU devices. Starting from MATLAB® release R2018a, this procedure is easier due to the introduction of a specific "TI Piccolo F28069M LaunchPad" target as Hardware board, which has default settings matching the LaunchPad™ SCI pins and oscillator frequency.[3] In previous releases, it was required to manually adjust the oscillator frequency and the SCI pins to match the LaunchPad™ board layout. This settings already include a target buffer size which specifies how much memory should be allocated to store signals for the external mode.[4]

In addition, some export settings for the variables of interest can be found in external mode control panel, by clicking on the icon Control Panel  .
Once the external model execution is launched, the model is built, compiled and deployed in the platform. The simulation automatically starts and the data values can be assessed through Scopes. In this case, the pulses used to make the led blinking can be visualized as shown in Figure 20.9. An other important feature of the external mode is that constant blocks can be edited during the execution, allowing several real-time verifications for different steady-state working conditions.

Unfortunately, the external mode suffers from a communication bottleneck. It may happen that the data exchange via COM fails, causing an abrupt stop of the Simulink® simulation. This scenario may be potentially dangerous,

[3]In particular, it is possible to specify which GPIO channels are used for the External Mode SCI connection. The corresponding pins cannot be used by other peripherals.

[4]The number of words N_w required by the external mode can be computed as $N_w = N_{signals} \cdot 2 \cdot (N_{samples} + 1)$. In case more samples than those allowed by the memory allocation are required, Simulink® automatically truncates the scope traces every time $N_{samples}$ points are plotted. It can be noted that cases of insufficient memory on the target result in a build error.

since the MCU may keep executing the C-code with the last memorized state. Based on Figure 20.9, it is particularly evident that the data samples are not equally distributed everywhere. Extra samples may appear, implying that the data exchange is affected by latencies. Therefore, this issue explains why this kind of execution is suitable for debugging purposes, but it is recommended to double check the results even with other evaluation tool.

Remark: it must be remembered that the XDS interface typically has two serial interface channels, one for debugging and the other one for auxiliary communication, including UART. If the External Mode connection to the device is unsuccessful, a possible explanation is that the debug channel was selected instead of the auxiliary communication channel during the COM port setup.

21

Electric Propulsion Case Studies

This last Chapter proposes two realistic cases study that deal with electric propulsion on vehicles. No solution is reported in the following pages, since the aim of those two exercises is to leave the reader free to organize the control logic to check its understanding on the topics and concepts explained in the previous Sections. Since those two exercises involve a tramway vehicle and an electric car, no actual implementation on hardware is feasible for both of them. However, the reader is invited to perform PIL simulations to test the performance of the designed control strategy. The first case study deals with a DC motor control. The reader is referred to Chapter 19 for all the details of this control. Instead, the electric car involved in the second exercise is controlled through an AC brushless drive. Such strategy is not explained in the previous chapters of this book. However, this second exercise reports some guidelines on how to deal with a synchronous motor and how to simplify its modeling. Then, the control logic of such machines is not so different with respect to that one adopted for PMDC motors.

21.1 Urban Tramway

Tramway vehicles "Carelli 1928" by ATM company (still in operation in Milan and San Francisco, see Figure 21.1) are moved by four DC motors with series excitation. Their main characteristics are reported here in the following:

- Line voltage: $V_{\mathrm{DC}} = 600\,\mathrm{V}$;
- Rated power of each motor: $P_{\mathrm{r}} = 21\,\mathrm{kW}$;
- Maximum speed of the vehicle: $v_{\mathrm{max}} = 42\,\mathrm{km/h}$;
- Rated speed of the motor: $\Omega_{\mathrm{r}} = \Omega_{\mathrm{b}} = 970\,\mathrm{rpm}$
- Efficiency: $\eta = 0.9$;
- Armature circuit time constant: $\tau_{\mathrm{a}} = 10\,\mathrm{ms}$;
- Mass of the vehicle at no load: $m_{\mathrm{T}} = 15\,\mathrm{t}$;

DOI: 10.1201/9781003196938-21

Figure 21.1 A tramway vehicle "Carelli 1928" running in Milan.

- Maximum loading capacity: 130 passengers;
- Diameter of the wheel: $d = 680\,\text{mm}$;
- Gearbox ratio (motor-to-wheels): $\rho = 13/74$;
- The equivalent friction force on the shaft is proportional to Ω and, at rated speed, it is 1/10 of the traction force.

For the sake of simplicity, it is possible to assume that the four motors of the tramway behave like one equivalent DC motor with separate excitation having excitation rated voltage equal to $V_{\mathrm{er}} = 60\,\mathrm{V}$, excitation rated current of $I_{\mathrm{er}} = 5\,\mathrm{A}$ and excitation time constant $\tau_{\mathrm{e}} = 0.1\,\mathrm{s}$. This kind of machine differs from the PMDC motor in the generation of the flux on the rotor. Indeed, this flux is not constant any more and it can be regulated depending on the working point of the machine. Namely, field weakening is regulated through a dedicated current loop to exceed the base speed Ω_{b}. The equation that should be considered to tune the PI controller for the excitation is:

$$v_{\mathrm{e}}(t) = L_{\mathrm{e}}\frac{\mathrm{d}i_{\mathrm{e}}(t)}{\mathrm{d}t} + R_{\mathrm{e}}i_{\mathrm{e}}(t) \tag{21.1}$$

Thus, the back emf can be computed through the following equation:

$$e(t) = k_{\mathrm{e}}i_{\mathrm{e}}(t)\,\Omega(t) \tag{21.2}$$

and the equation for torque computations becomes:

$$m_{\mathrm{e}}(t) = k_{\mathrm{T}}i_{\mathrm{e}}(t)\,i_{\mathrm{a}}(t) \tag{21.3}$$

Even for a separately excited DC machine, it can be assumed that $k_{\mathrm{e}} = k_{\mathrm{T}} = k$.

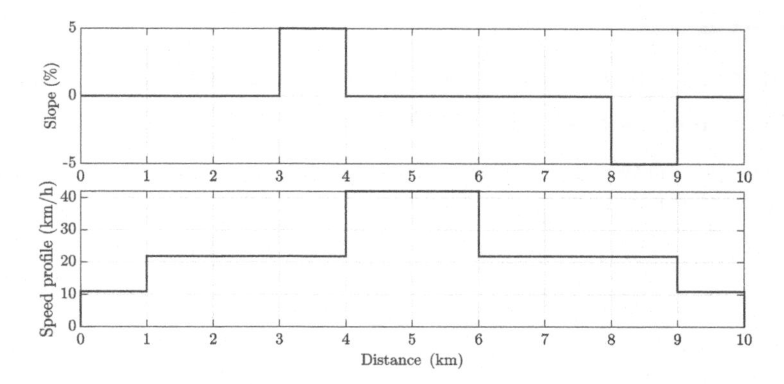

Figure 21.2 Plots of the slope and speed profile of the track.

Based on this data:

- Design and simulate speed and current control loops in order to cover a 10 km track considering the speed and slope profiles reported in Table 21.1 and in Figure 21.2.[1] The slope can be computed as $100 \tan(\vartheta)$, where ϑ is the angle between the rising path and the ground, i.e., $\vartheta = \arctan(\text{slope } \%/100)$;
- Test the developed control logic on the microcontroller performing a PIL simulation.

Table 21.1 Track profile for the tramway "Carrelli 1928." The vehicle starts from 0 speed and it stops at the end of the track.

Track (km)	Slope %	Speed
$0-1$	0	$v_r/2$
$1-3$	0	v_r
$3-4$	5	v_r
$4-6$	0	v_{max}
$6-8$	0	v_r
$8-9$	-5	v_r
$9-10$	0	$v_r/2$

[1] In the reality, tramway vehicles go forward in coasting mode once they reach a steady state speed. For the sake of simplicity, in this exercise it is possible to assume that the traction torque is always applied on the motor to follow the speed profile with no error.

Example of the system parameters computations

Rated parameters of the motor

1. Overall mass of the vehicle loaded with all the passengers (standard weight $m_{\mathrm{P}} = 80\,\mathrm{kg}$):

$$m_{\mathrm{tot}} = m_{\mathrm{T}} + m_{\mathrm{P}} N_{\mathrm{P}} = 25.4\,\mathrm{t}$$

2. Rated speed of the equivalent motor in rad/s:

$$\Omega_{\mathrm{r,\ rad/s}} = \Omega_{\mathrm{r}} \frac{2\pi}{60} = 101.58\,\mathrm{rad/s}$$

3. Mechanical power of the equivalent motor:

$$P_{\mathrm{tot}} = 4P_{\mathrm{r}} = 84\,\mathrm{kW}$$

4. Electrical power absorbed by the equivalent DC motor:

$$P_{\mathrm{tot,\ el}} = \frac{P_{\mathrm{tot}}}{\eta} = 93.33\,\mathrm{kW}$$

5. Rated speed of the vehicle in m/s:

$$v_{\mathrm{r,\ m/s}} = \rho \frac{d}{2} \Omega_{\mathrm{r,\ rad/s}} = 6.07\,\mathrm{m/s}$$

6. Friction power acting on the vehicle at rated speed:

$$P_{\mathrm{fr,r}} = \frac{1}{10} P_{\mathrm{tot}} = 8.40\,\mathrm{kW}$$

7. Friction torque acting on the shaft of the equivalent motor at rated speed:

$$m_{\mathrm{fr,r}} = \frac{P_{\mathrm{fr,r}}}{\Omega_{\mathrm{r,\ rad/s}}} = 82.69\,\mathrm{N\,m}$$

8. Rated torque of the equivalent motor:

$$m_{\mathrm{r}} = \frac{P_{\mathrm{tot}}}{\Omega_{\mathrm{r,\ rad/s}}} = 826.95\,\mathrm{N\,m}$$

9. Rated armature current of the equivalent motor:

$$I_{\mathrm{ra}} = \frac{P_{\mathrm{tot,\ el}}}{V_{\mathrm{DC}}} = 155.56\,\mathrm{A}$$

10. Rated back-emf acting on the armature circuit of the equivalent motor:

$$E_r = \frac{P_{tot}}{I_{ra}} = 540\,\mathrm{V}$$

11. Torque constant of the equivalent motor:

$$K = \frac{m_r}{I_{ra} I_{re}} = 1.06\,\mathrm{Nm/A^2}$$

12. Power loss associated to the equivalent armature winding:

$$P_l = P_{tot,\ el} - P_{tot} = 9.33\,\mathrm{kW}$$

Armature and excitation circuit parameters

1. Armature resistance of the equivalent motor:

$$R_a = \frac{P_l}{I_{ra}^2} = 0.39\,\Omega$$

2. Armature inductance of the equivalent motor:

$$L_a = \tau_a R_a = 3.86\,\mathrm{mH}$$

3. Excitation resistance of the equivalent motor:

$$R_e = \frac{V_{re}}{I_{re}} = 12\,\Omega$$

4. Excitation inductance of the equivalent motor:

$$L_e = \tau_e R_e = 1.2\,\mathrm{H}$$

Mechanical parameters

1. Equivalent inertia of the vehicle seen on the motor shaft:

$$J_{eq} = m_{tot} \frac{v_{r,\ m/s}^2}{\Omega_r^2} = 90.62\,\mathrm{kg\,m^2}$$

2. Friction coefficient[2]:

$$\beta = \frac{m_{fr,r}}{\Omega_r} = 0.81\,\mathrm{N\,m\,s}$$

[2]it can be assumed that the friction torque acting on the whole vehicle can be computed as $m_{fr}(t) = \beta \Omega_r(t)$

Figure 21.3 Example of an electric car during battery recharging [8].

21.2 Electric Racing Car

A full electric racing car[3] like that one shown in Figure 21.3 is equipped with a permanent-magnet synchronous motor (PMSM). It has a rated mechanical power of 180 kW and it is fed by a battery stack working at 400 V. As an example, see the car denoted in Figure 21.3. Its main characteristics are reported here in the following:

- Efficiency $\eta = 0.95$;
- Stator resistance $R_\mathrm{s} = 0.1\,\Omega$;
- Stator inductance $L_\mathrm{s} = 0.8\,\mathrm{m}\Omega$;
- Nominal speed $\Omega_\mathrm{r} = 400\,\mathrm{rad/s}$;
- Pole pairs $p = 2$;
- Permanent-magnet flux $\phi_\mathrm{PM} = 0.5\,\mathrm{Wb}$.

Moreover, assume that:

- The motor control is able to keep $\cos\varphi = 1$;
- The converter is operated through a PWM modulation with third harmonic injection, allowing to reach a phase voltage with first harmonic peak value equal to $V_\mathrm{peak} = \dfrac{2}{\sqrt{3}}\dfrac{V_\mathrm{DC}}{2}$;
- The car weights 1500 kg, whereas the driver 80 kg;
- The maximum achievable acceleration of the vehicle is $a_\mathrm{max} = 4\,\mathrm{m/s^2}$;

[3]The vehicle considered in this exercise is the electric counterpart of the Alfa Romeo Giulietta Quadrifoglio Verde. Please, note that such electric car is does not exist and, therefore, it is not available on the market.

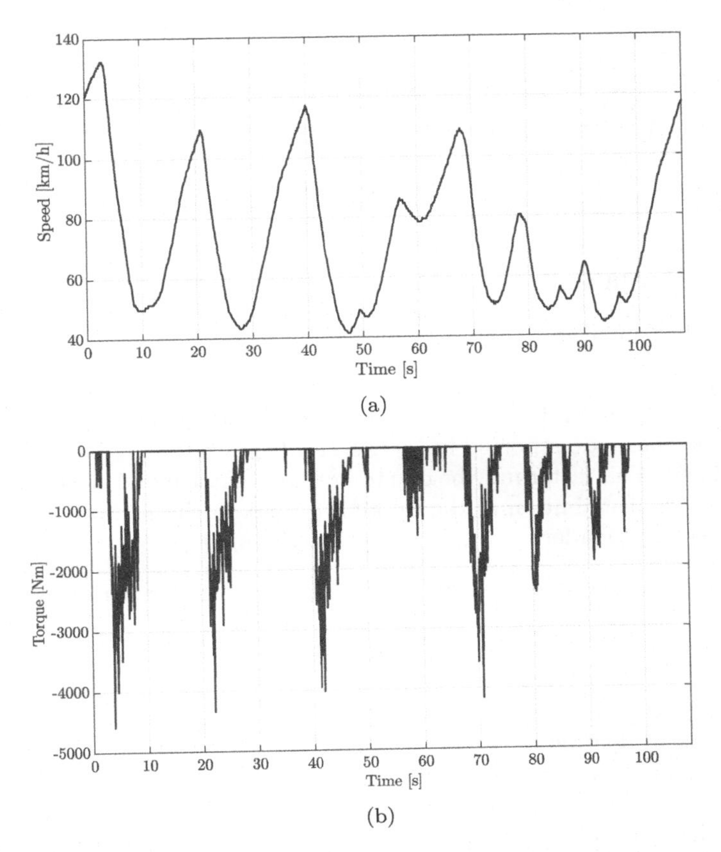

Figure 21.4 Speed profile of the car running on the Varano circuit (also available in Ex_E_car.mat file, variable: speed_kmh) (a) and corresponding equivalent braking torque applied on the motor shaft before the gearbox (also available in Ex_E_car.mat file, variable: T_braking) (b).

- The friction coefficient is $\beta = 0.78\,\mathrm{N\,m\,s}$, where the equivalent friction torque acting on the motor shaft can be computed as $m_{\mathrm{fr}} = \beta\Omega$;
- The radius of each tyre is $r = 30\,\mathrm{cm}$;
- The motor is connected to each axle through a gearbox with an equivalent gear ratio equal to $\rho_{\mathrm{eq}} = 1/3$;
- The car is running on the Varano circuit. Its speed profile is reported in Figure 21.4 (a), whereas the corresponding mechanical braking torque is shown in Figure 21.4 (b).

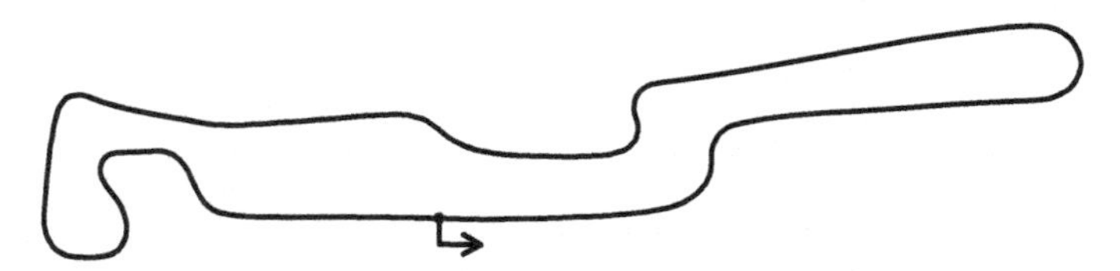

Figure 21.5 Map of the track in Varano.

Knowing this data:

- Design a field-weakening strategy for the PMSM;
- Design and simulate a speed control of the motor to follow the speed profile reported in Figure 21.4 (a) and considering the corresponding braking torque applied on each axle reported in Figure 21.4 (b) (see the map of the track in Figure 21.5). This latter may be due to both mechanical and electromagnetic braking. The distribution of the braking torque on those two contributions is left up to the reader;
- Test the developed control logic on the microcontroller performing a PIL simulation.

Hint#1: linear model of a PMSM

Before briefly presenting the dynamical model of the PMSM, it is fundamental to point out some assumptions to simplify this study. Namely, the machine to be considered is perfectly symmetric (isotropic machine) with infinite iron permeability. In addition, the fundamental harmonic for each electrical variables is considered only. In view of these two assumptions, it can be stated that the electrical equations of the three-phase isotropic PMSM are the following[4]:

$$
\begin{aligned}
v_{\mathrm{a}} &= R_{\mathrm{s}} i_{\mathrm{a}} + p\phi_{\mathrm{a}} \\
v_{\mathrm{b}} &= R_{\mathrm{s}} i_{\mathrm{b}} + p\phi_{\mathrm{b}} \\
v_{\mathrm{c}} &= R_{\mathrm{s}} i_{\mathrm{c}} + p\phi_{\mathrm{c}} \\
\phi_{\mathrm{a}} &= L_{\mathrm{ss}}(\theta_m) i_{\mathrm{a}} + M_{\mathrm{ss}}(\theta_m) i_{\mathrm{b}} + M_{\mathrm{ss}}(\theta_m - \tfrac{2}{3}\pi) i_{\mathrm{c}} + \phi_{\mathrm{PM}}(\theta_m) \\
\phi_{\mathrm{b}} &= M_{\mathrm{ss}}(\theta_m) i_{\mathrm{a}} + L_{\mathrm{ss}}(\theta_m - \tfrac{2}{3}\pi) i_{\mathrm{b}} + M_{\mathrm{ss}}(\theta_m + \tfrac{2}{3}\pi) i_{\mathrm{c}} + \phi_{\mathrm{PM}}(\theta_m - \tfrac{2}{3}\pi) \\
\phi_{\mathrm{c}} &= M_{\mathrm{ss}}(\theta_m - \tfrac{2}{3}\pi) i_{\mathrm{a}} + M_{\mathrm{ss}}(\theta_m + \tfrac{2}{3}\pi) i_{\mathrm{b}} + L_{\mathrm{ss}}(\theta_m + \tfrac{2}{3}\pi) i_{\mathrm{c}} + \phi_{\mathrm{PM}}(\theta_m + \tfrac{2}{3}\pi)
\end{aligned}
$$

where p is the derivative operator, i.e., d/dt, R_{s} is the phase resistance, $v_{\mathrm{a,b,c}}$ are the voltages applied on each phase of the motor, $i_{\mathrm{a,b,c}}$ are the currents flowing in each phase of the motor, $\phi_{\mathrm{a,b,c}}$ stand for the magnetic flux linked

[4]All the currents $i_{\mathrm{a,\,b,\,c,\,d,\,q}}$, voltages $v_{\mathrm{a,\,b,\,c,\,d,\,q}}$, fluxes $\phi_{\mathrm{a,\,b,\,c,\,d,\,q}}$, angles θ_m, speeds ω_m and Ω, torque m_e and m_l are in general function of time. In this paragraph, this dependency on t is not made explicit for the sake of brevity.

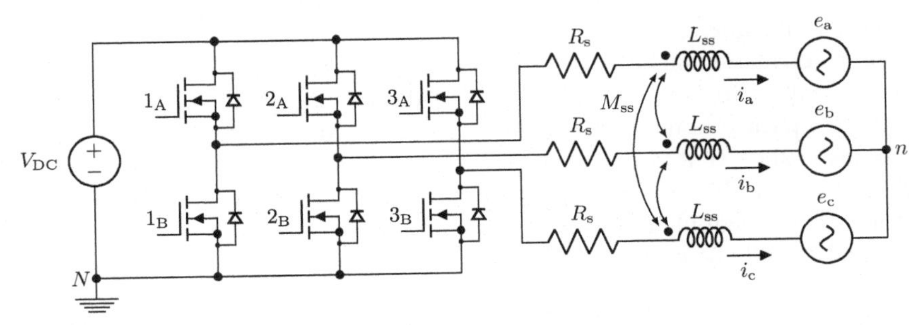

Figure 21.6 Equivalent circuit of a three-phase PMSM connected with a three-phase inverter.

with each winding of the motor, L_{ss} is the phase inductance (equal for all the windings), M_{ss} is the mutual inductance (equal for all the magnetic couplings between windings), ϕ_{PM} is the magnetic flux generated by the permanent magnets and θ_m is the electrical position of the rotor. This angle differs from the mechanical position of the rotor by a factor equal to n_p, that is the number of pole pairs of the machine. Indeed, the periodicity of the electrical variables is $1/n_p$ of the mechanical one. In addition, it is important to note that, in general, all the phase voltages, currents, fluxes and the rotor position θ_m are functions of time. The previous set of equations can be simplified reminding that the motor adopted in this chapter is isotropic. Thus, the dependence of the self and mutual inductances on the rotor position disappears. In addition, the stator windings are connected in such a way that $i_a + i_b + i_c = 0$ for every considered time instant. Consequently, the new set of equations becomes:

$$
\begin{aligned}
v_a &= R_s i_a + p\phi_a \\
v_b &= R_s i_b + p\phi_b \\
v_c &= R_s i_c + p\phi_c \\
\phi_a &= L_s i_a + \phi_{PM}(\theta_m) \\
\phi_b &= L_s i_b + \phi_{PM}(\theta_m - \tfrac{2}{3}\pi) \\
\phi_c &= L_s i_c + \phi_{PM}(\theta_m + \tfrac{2}{3}\pi)
\end{aligned}
$$

where $L_s = L_{ss} - M_{ss}$ is the synchronous inductance. The corresponding equivalent circuit of a PMSM connected to an inverter is shown in Figure 21.6. From this sketch and the previous equations, it is possible to note that the voltage generators e_a, e_b and e_c represent back emfs and they are due the presence of the permanent magnets which spin together with the rotor.

Despite the fact that these new equations are far simpler than the previous ones, they still do not allow the development of a general control strategy in the phase variables, since they show a dependence on the rotor position. Therefore, the Park transformation is introduced to obtain a new equivalent, decoupled machine. This operation introduces a change in the reference frame for the electrical equations, moving from a phase-based reference frame to a new one with two orthogonal axis called direct and quadrature axis (d and q) and a separate homopolar (0) axis. The typical structure of the electrical machines is such that the homopolar components are nil for all the electrical variables during their normal operation. Therefore, starting from a three-phase motor, a new decoupled two-phase machine is obtained. This transformation is defined for all the electrical variables as follows:

$$\begin{bmatrix} v_{\mathrm{d}} \\ v_{\mathrm{q}} \\ v_{0} \end{bmatrix} = \mathbf{T}(\theta_m) \begin{bmatrix} v_{\mathrm{a}} \\ v_{\mathrm{b}} \\ v_{\mathrm{c}} \end{bmatrix}$$
$$\begin{bmatrix} i_{\mathrm{d}} \\ i_{\mathrm{q}} \\ i_{0} \end{bmatrix} = \mathbf{T}(\theta_m) \begin{bmatrix} i_{\mathrm{a}} \\ i_{\mathrm{b}} \\ i_{\mathrm{c}} \end{bmatrix}$$
$$\begin{bmatrix} \phi_{\mathrm{d}} \\ \phi_{\mathrm{q}} \\ \phi_{0} \end{bmatrix} = \mathbf{T}(\theta_m) \begin{bmatrix} \phi_{\mathrm{a}} \\ \phi_{\mathrm{b}} \\ \phi_{\mathrm{c}} \end{bmatrix}$$

where the transformation matrix $\mathbf{T}(\theta_m)$ is defined as:

$$\mathbf{T}(\theta_m) = \sqrt{\frac{2}{3}} \begin{bmatrix} \cos(\theta_m) & \cos\left(\theta_m - \frac{2}{3}\pi\right) & \cos\left(\theta_m + \frac{2}{3}\pi\right) \\ -\sin(\theta_m) & -\sin\left(\theta_m - \frac{2}{3}\pi\right) & -\sin\left(\theta_m + \frac{2}{3}\pi\right) \\ 1/\sqrt{2} & 1/\sqrt{2} & 1/\sqrt{2} \end{bmatrix} \tag{21.4}$$

$\sqrt{(2/3)}$ is a coefficient that ensures the orthogonality of the transformation. Then, it can be noted that this operation allows to define a number of new reference frames, since angle θ_m can be chosen arbitrarily. For convenience, a set of transformed axes which rotates synchronously with the rotor of the machine under study is adopted commonly. In this way, all the AC electrical variables appear as constants at steady state. Namely, the new transformed electrical equations of a PMSM are:

$$\begin{aligned} v_{\mathrm{d}} &= R_{\mathrm{s}} i_{\mathrm{d}} + p\phi_{\mathrm{d}} - n_p \Omega \phi_{\mathrm{q}} \\ v_{\mathrm{q}} &= R_{\mathrm{s}} i_{\mathrm{q}} + p\phi_{\mathrm{q}} + n_p \Omega \phi_{\mathrm{d}} \\ \phi_{\mathrm{d}} &= L_{\mathrm{s}} i_{\mathrm{d}} + \phi_{\mathrm{PM}} \\ \phi_{\mathrm{q}} &= L_{\mathrm{s}} i_{\mathrm{q}} \end{aligned}$$

where Ω is the mechanical speed of the rotor and the equations for the hompolar components of the voltage and the flux are not reported since they are both nil. It can be noted that the electrical speed of the rotor and of all the electrical variables can be defined as $\omega_m = n_p \Omega$. Moreover, space vectors can be defined using complex numbers. The d-axis component is considered as the real part of the space vector, whereas the projection on the q-axis becomes its imaginary part. Namely, the electrical model can be rewritten in a more compact form as follows:

$$\overline{v_{\mathrm{s}}} = v_{\mathrm{d}} + jv_{\mathrm{q}} = R_{\mathrm{s}}\overline{i_{\mathrm{s}}} + p\overline{\phi_{\mathrm{s}}} + jn_p\Omega\overline{\phi_{\mathrm{s}}}$$
$$\overline{\phi_{\mathrm{s}}} = \phi_{\mathrm{d}} + j\phi_{\mathrm{q}} = L_{\mathrm{s}}\overline{i_{\mathrm{s}}} + \phi_{\mathrm{PM}}$$

From this compact notation, the term $jn_p\Omega\overline{\phi_{\mathrm{s}}}$ can be highlighted. This is a motional term which is a consequence of the adoption of a moving reference frame and it is responsible for the electromechanical conversion of the machine. Moreover, combining the voltage and flux equations, the following dynamical expressions can be obtained:

$$v_{\mathrm{d}} = R_{\mathrm{s}}i_{\mathrm{d}} + L_{\mathrm{s}}pi_{\mathrm{d}} - n_p\Omega L_{\mathrm{s}}i_{\mathrm{q}}$$
$$v_{\mathrm{q}} = R_{\mathrm{s}}i_{\mathrm{q}} + L_{\mathrm{s}}pi_{\mathrm{q}} + n_p\Omega L_{\mathrm{s}}i_{\mathrm{d}} + n_p\Omega\phi_{\mathrm{PM}}$$

where $n_p\Omega L_{\mathrm{s}}i_{\mathrm{q}}$ and $n_p\Omega L_{\mathrm{s}}i_{\mathrm{d}}$ are cross coupling terms between the two axis and $n_p\Omega\phi_{\mathrm{PM}} = E$ is the back-emf expressed in the d-q reference frame. Finally, the torque expression and the mechanical equation are presented. The first one is obtained by setting an energy balance between the electrical and mechanical part of the system. For isotropic machines, it holds that:

$$m_e = n_p\phi_{\mathrm{PM}}i_{\mathrm{q}} \tag{21.5}$$

where m_e is the electrical torque generated by the PMSM. Instead, the dynamical equations of the mechanical part of the system is obtained from a torque balance as follows:

$$J_{\mathrm{eq}}p\Omega + \beta\Omega = m_e - m_l \tag{21.6}$$

where m_l is the load torque applied in the shaft of the machine, J_{eq} is the equivalent inertia of the PMSM and β is the friction coefficient.

Hint#2: control strategy

The control strategy for PMSM is the vector control for AC brushless machines. As previously mentioned, it is possible to act separately on the torque and on the flux of the machine, thanks to the Park transformation. This peculiarity represents the key of the vector control, which has a similar philosophy with respect to the strategy adopted for DC machines. Indeed, the two transformed windings can be considered separately. One is used to generate torque

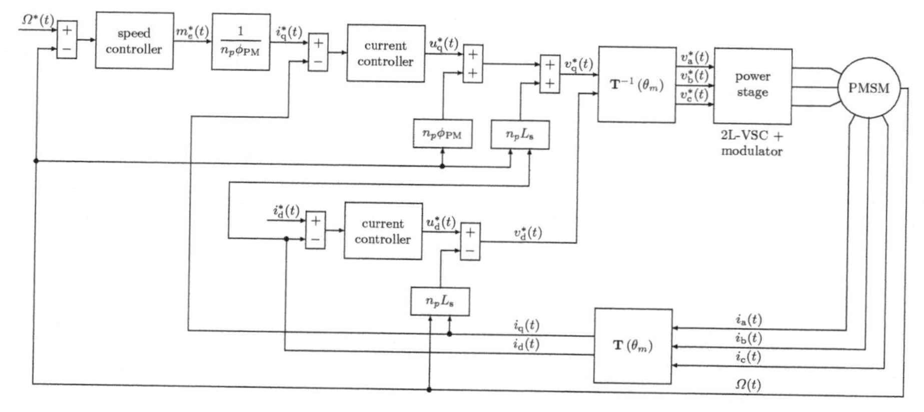

Figure 21.7 Block diagram of a generic PI based vector control for a PMSM.

and, consequently, for regulating the speed of the rotor, whereas the other one is exploited to provide a variation in the magnetic field. For PMSM, this last operation is done by generating currents in the stator windings that partially demagnetize the magnetic field due to the magnets ϕ_{PM}. Carefulness must be paid while demagnetizing real PMSM since too high current may compromise the magnets.

Focusing on the realization of the vector control, the speed regulation is implemented in an external loop and it is cascaded with the current control acting on i_{q}. Indeed, this current is responsible for torque variations (see equation 21.5). The other current control acting on i_{d} can be considered as a loop which acts in parallel to i_{q}. A PI-based scheme of the vector control is reported in Figure 21.7.

For Ω lower than the base speed $i_{\mathrm{d}}^{\mathrm{ref}}$ is set equal to 0. This choice allows the quadrature current i_{q} to reach the maximum sustainable current of the machine i_{max}. Above Ω_{b}, the value of $i_{\mathrm{d}}^{\mathrm{ref}}$ is decreased ($i_{\mathrm{d}}^{\mathrm{ref}} < 0$) until it reaches the value $-\phi_{\mathrm{PM}}/L_{\mathrm{s}}$. Once this limit is reached, the d-axis current is kept constant. On top of that, compensation terms should be added in the control scheme based to the model of the machine previously shown.

Speed Controller

This PI controller is tuned based on the mechanical transfer function of the PMSM. This latter is retrieved starting from equation 21.6, that relates the mechanical speed with the electrical torque, and moving to the Laplace domain. The resulting transfer functions $G_{\Omega}(s)$ reads as:

$$G_{\Omega}(s) = \frac{1}{\beta + J_{eq}s} \tag{21.7}$$

Hence, considering the classical parallel structure for PI controllers reported in Section 5.2, the open loop transfer function is:

$$L_{\Omega}(s) = R_{\Omega}(s)G_{\Omega}(s) = \frac{k_p^{\Omega}s + k_i^{\Omega}}{s}\frac{1}{\beta + J_{eq}s} \tag{21.8}$$

This transfer function shows two poles in zero. Therefore, in order to guarantee asymptotic stability, it is crucial to properly tune k_p^{Ω} and k_i^{Ω}. From the study of $L(s)$, both terms k_p^{Ω} can be directly linked to the bandwidth of the controlled system ω_{Ω} through the following equations:

$$k_p^{\Omega} = J_{eq}\omega_{\Omega} \tag{21.9}$$

$$k_i^{\Omega} = \beta\omega_{\Omega} \tag{21.10}$$

Current Controllers

The tuning procedure of the current controllers is based on the same electrical transfer function of the PMSM. Indeed, the transformed machine in the Laplace domain, it can be noted that the resulting transfer function has the same structure of that one adopted for the same control in a DC machine. As a matter of fact, the cross-coupling terms and the back-emf can be dealt with as if they were disturbances. Namely, for both d and q axes, it holds that the electrical transfer function is:

$$G_i(s) = \frac{1}{R_s + sL_s} \tag{21.11}$$

Therefore, the open loop transfer function reads as:

$$L_i(s) = R_i(s)G_i(s) = \frac{k_p^i s + k_i^i}{s}\frac{1}{R_s + sL_s} \tag{21.12}$$

As for the Speed PI controller, a relationship between the gains of the controller and the bandwidth of the system ω_i can be written as:

$$k_p^i = L_s\omega_i \tag{21.13}$$

$$k_i^i = R_s\omega_i \tag{21.14}$$

Due to the symmetry in the transformed machine, both the current controllers can be tuned with the same parameters. In addition, it must be remembered to select a value at least ten time larger with respect to ω_{Ω} to fulfill the cascade constraint for nested loops. Indeed, the speed loop is cascaded with the current one, as shown in Figure 21.7.

Hint#3: acceleration

The motor can start from 0 speed and accelerate up to the initial speed of the profile shown in Figure 21.4 (a) or an initial speed can be assigned.
Set suitable torque limits and separate the mechanical torque from the electromagnetic one.
Use the block **Saturation Dynamic** from library **Discontinuities** to implement the field weakening logic.

Hint#4: computation of the simulation parameters

Rated parameters of the motor

1. Overall mass of the car and driver:

$$m_{\mathrm{tot}} = m_{\mathrm{T}} + m_{\mathrm{D}} = 1580\,\mathrm{kg}$$

2. Electrical power absorbed by the PMSM:

$$P_{\mathrm{el}} = \frac{P_{\mathrm{r}}}{\eta} = 189.47\,\mathrm{kW}$$

3. Rated stator current (peak value):

$$I_{\mathrm{r_{peak}}} = \frac{2P_{\mathrm{el}}}{3V_{\mathrm{peak}}\cos\varphi} = 546.96\,\mathrm{A}$$

4. Rated motor torque:

$$m_{\mathrm{r}} = \frac{P_{\mathrm{r}}}{\Omega_{\mathrm{r}}} = 450\,\mathrm{N\,m}$$

Mechanical parameters

1. Equivalent inertia of the vehicle seen on the motor shaft:

$$J_{\mathrm{eq}} = m_{\mathrm{tot}}\left(r\rho_{\mathrm{eq}}\right)^2 = 142.20\,\mathrm{kg\,m^2}$$

2. Maximum motor torque:

$$m_{\mathrm{max}} = m_{\mathrm{tot}}a_{\mathrm{max}}r\rho_{\mathrm{eq}} = 632\,\mathrm{N\,m}$$

3. Maximum quadrature-axis current:

$$i_{\mathrm{q_{max}}} = \frac{m_{\mathrm{max}}}{\eta p\phi_{\mathrm{PM}}} = 665.26\,\mathrm{A}$$

4. Base speed of the motor:

$$\Omega_{\text{b}} = \frac{1}{p}\sqrt{\frac{\left(V_{\text{peak}}\sqrt{3/2}\right)^2}{\phi_{\text{PM}}^2 + L_{\text{s}}^2\left(I_{\text{r}_{\text{peak}}}\sqrt{3/2}\right)^2}} = 192.95\,\text{rad/s}$$

5. Motor speed at which the direct-axis current reaches its maximum value:

$$\Omega^* = \frac{1}{p}\sqrt{\frac{\left(V_{\text{peak}}\sqrt{3/2}\right)^2}{-\phi_{\text{PM}}^2 + L_{\text{s}}^2\left(I_{\text{r}_{\text{peak}}}\sqrt{3/2}\right)^2}} = 733.21\,\text{rad/s}$$

6. Minimum direct-axis current:

$$i_{\text{d}_{\text{max}}} = -\frac{\phi_{\text{PM}}}{L_{\text{s}}} = -625\,\text{A}$$

A

Appendix A: Basics of C

In this appendix, some basics of C programming which are useful to understand many operations hidden by the Simulink® code generation are recalled.

A.1 Operations between numbers

A.1.1 Sum and differences

Sum and differences do not pose particular problems as regard the base system representation, provided that the base used for the conversion has a lower value of bits with respect to the actual base implemented in the controller (e.g., if for the conversion it has been considered a 12-bit base, the actual type of the variable in the C code can be chosen equal to 16-bits). In this case, the sum of two variables represented in 12-bit base will not produce an overflow in the result.

A.1.2 Shift operation

Consists in moving a to the left (<<) or to the right (>>) a sequence of bits by a certain number of positions. This type of operation is much faster to be performed for microcontrollers and it can be used to perform multiplication between variables and constants in a faster way with respect to the same operations with floating point structures: this fact holds provided that variables are represented in integer form, with a base system multiple of 2. If a variable i (e.g., a current) whose maximum value is 100 represented by means of `uint16` numeric structure is considered, a 12-bit base is chosen ($2^{12} = 4096$). Thus:

$$\#I_{12} = \frac{i}{100} \cdot 4096$$

Assuming that the actual value of the current is 75, its representation in the 12-bit base is:

$$\#I_{12} = \frac{75}{100} \cdot 4096 = 3072$$

whose corresponding structure in the uint16 format is 0*b*0000110000000000.

DOI: 10.1201/9781003196938-A

If this variable has to be multiplied (or divided) by 2, this operation can be obtained in a simple way by means of a left shit << by one bit (or on the right >> for the division) of the corresponding bit sequence.

Thus, the quantity $(2 \cdot \#I_{12})$ can be simply obtained in the microcontroller as a shift operation $2 \cdot \#I_{12} = \#I_{12} >> 2$. Any multiplication (or division) by a constant quantity can be expressed exploiting the representation in power of 2. In case the variable $\#I_{12}$ has to be multiplied by π, the first step is to express the constant coefficient as a power-two series:

$$\#I_{12} \cdot 3.1416 \cong \#I_{12} \cdot (2^1 + 2^{-1} + 2^{-3} + 2^{-6} + 2^{-11})$$

Then, the shift properties can be exploited, making the calculation of the product very fast. Since the shift operations have lower priority with respect to the sum, parenthesis are required between the terms of the summation in order to avoid errors:

$$\begin{aligned}\#I_{12} \cdot 3.1416 \cong &(\#I_{12} << 1) + (\#I_{12} >> 1) + (\#I_{12} >> 3) + \\ &+ (\#I_{12} >> 6) + (\#I_{12} >> 11)\end{aligned}$$

A.1.3 Multiplication

The multiplication between two variables in the same base can be written as:

$$\#C_{bitbase} = \#(A \cdot B)_{bitbase} = \frac{\#A_{bitbase} \cdot \#B_{bitbase}}{2^{bitbase}}$$

By expressing the constant division with the shift operator, it becomes:

$$\#C_{bitbase} = \#(A \cdot B)_{bitbase} = (\#A_{bitbase} \cdot \#B_{bitbase}) >> bitbase$$

A.1.4 Division

The division of two variables follows the same method introduced for the multiplication and it can be expressed as:

$$\#C_{bitbase} = \#(A/B)_{bitbase} = \frac{\#A_{bitbase}}{\#B_{bitbase}} \cdot \frac{1}{2^{bitbase}}$$

The above expression is equivalent to:

$$\#C_{bitbase} = \#(A/B)_{bitbase} = \frac{\#A_{bitbase} << bitbase}{\#B_{bitbase}}$$

Example: Product operation in C

The base value to be consider has $N_{\text{bit}} = 16$ bit and the processor framework is 16 bit too. Products are computation without any tolerance bandwidth.

The strategy to avoid overflow is to temporary move into 32 bit (`long`) variable through left shifting, making the product computation, and then come back to 16 bit integers. The corresponding C code is as follows:

$$\#x_{16} = (\texttt{int})\Big(\ ((\texttt{long})\#x_{a,12} \cdot \#x_{b,12}) \gg 4\ \Big) \tag{A.1}$$

A.2 Structure of a C program

C-programs for micro-controller applications are generally characterized by a constant structure. Two main types of files contribute to the definition of the code: *header files* (denoted by the extension .h) and *main* program files (whose extension is .c). In header files functions, constants and macros are generally defined, while in the .c file all the aforementioned structures are called and combined together to perform the desired operations.

The inclusion of an header file into the main program is performed by means of the following piece of code placed at the beginning of the .c code:

```
#include filename.h
```

In order to have global visibility for both functions and variables set inside any header file, it is necessary to define them as *extern*.

```
extern int error; // Variable
extern void function1(int,double); // Function
```

Alternatively, it is possible to define constant and macros with the following syntax:

```
#define CONST 1.0341 // Constant

#define SUM(e1, e2) (e1 + e2) // Macro
```

Macros are predefined pieces of code that are are substituted to the corresponding invocation before code execution. In practice, a macro is a fragment of code which has been given a name. Whenever the name is used, it is replaced by the contents of the macro. Macros are useful in order to speed up the coding by avoiding multiple writing of the same pieces of code. Macros can be invoked in the main .c file with a syntax similar to the one adopted for the functions:

```
c = SUM(a,b) // -> c = (a + b)
```

Macros are also useful for simplifying the coding of multiplications between a variable and a constant. Recalling the properties of the shift, as defined in Section 7.3 and previously in this appendix, it is possible to express the multiplication of any variable by π with the following macro:

```
#define TIMES_PI(e) ((e<<1) + (e>>1) + (e>>3) + (e>>6) + (e>>11))
```

A typical structure of a C file is reported here below:

```
//Inclusion of the compiler file, depending on the micro
processor
#include <p30fxxxx.h>
   //Inclusion of the .h file with used-defined structures
#include "localfile.h"
   // Macro definition
   #define MUL2(in)(((int)in<<1))
   // Global variable definition long int error, input, output;
//static long int
   //Main function
   int main () {
   //do something... while(1) {
   //infinite loop... do something } }
   //Interrupt routine (depending on the type of the MCU)
void _ISR _T1Interrupt()
   { //do something... }
```

In this example, it is possible to see the difference between normal variables and *static* ones: while normal variables are deleted each time the code is re-executed (that is, at the end of any complete cycle of the control), *static* ones retain their value in the local memory until a re-assigning operation is performed.

B

Appendix B: Custom Expansion Boards and Hardware Kits

In this Appendix, the main characteristics of the custom expansion boards and assembled kits targeted to the exercises proposed previously are illustrated.

As university initiative, the boards have been developed at the Laboratory of Electrical Drives and Power Electronics, in the Department of Mechanical Engineering of Politecnico di Milano, Italy, in collaboration with Würth Elektronik™ Group and Texas Instruments™ Inc. The aim of this collaboration is to provide to students ready-to-use test benches for the study of power electronics-based applications. In particular, the developed boards and test benches are:

- **extPot3 board**, which provides a direct manipulation of analog signals, conversion chains, and buzzer;
- **extRL(C) board**, which is a configurable load equipped with output filter. It is able to realize *RL*, *RLC*, *R*, *L*, *LC* topologies;
- **RL(C) kit**, which integrates the extRL(C) board with the LaunchPad™ F28069M and the BOOSTXL-DRV8301 BoosterPack boards;
- **DecMot kit**, which allows to test PMDC motors. An encoder sensor is included as well as MCU interface for external power supply. Other kind of motors can be installed and driven by the same hardware as well. Indeed, the housing of this setup is suitable for two generic small motors;
- **B2B-PMDC kit**, which contains two coupled PMDC motors anchored on an aluminum base plate, encoder sensor and MCU interface for external power supply.

It is important to note that, according to the considered application, boards/kits need to be manually configured, e.g., topology settings, motor calibration and cable connections should be edited.

Either custom expansion boards and assembled kits are open to users community. Project files can be shared and/or pre-assembled boards/kit can be directly shipped. In case of interest, please contact:

- Mattia Rossi mattia.rossi@polimi.it
- Nicola Toscani nicola.toscani@polimi.it
- Francesco Castelli Dezza francesco.castellidezza@polimi.it

DOI: 10.1201/9781003196938-B

Figure B.1 extPot3 board. Front and back views.

extPot3 Board

The extPot3 board (see Figure B.1) is a custom-made expansion board which aims to drive three ADC channels of the LaunchPad™ F28069M board. Namely, ADCA3, which is connected to pin 66, ADCB3 which is connected to pin 67 and ADCB5 which is connected to pin 68. These channels are connected to three 20 kΩ linear potentiometers (Pot1, Pot2 and Pot3), which are able to generate analog signals in the range $[0, 3.3]$ V at the ADC inputs. The extPot3 board can be directly mounted on top of the 40-pin plug-in module connector (headers J5-J8) of the LaunchPad™ F28069M board. In addition, the board includes a piezoelectric buzzer. This device can be driven by a generic GPIO or by a ePMW, namely GPIO8 (which is connected to pin 78) and ePWM5B (which is connected to pin 77). A bipolar transistor BJT NPN 25 V 0.5 A is used as buzzer driver to decouple the driving signal from the actuation stage.

The extPot3 board mounts a six pins terminal header arranged to connect hall-effect sensors through pull-up resistors. Starting from the left-hand side of the board (front view of the board, see Figure B.1), the first three pins corresponds to three phases for hall-effect sensor measurements, while the other three to 3.3 V, GND, and 5 V supply, respectively. A SMD green led turns on when the LaunchPad™ F28069M board is supplied with 3.3 V, verifying the correct connection between boards (the 3.3V supply of the extPot3 are taken from the MCU board).

As previously mentioned, the extPot3 board is connected to the LaunchPad™ F28069M board through standard 40-pin plug-in module connector. These pins remains accessible for debugging purpose or further connections. The maximum height of a BoosterPack is 1700 mil. This value becomes 1350 mil if there is no need for accessing header J5. Instead, there is no detailed standard for the inner 20 pins (i.e., header J3 and J4). extPot3 board must be attached with care to the LaunchXL F28069M one to ensure a successful stacking with the various pins and to improve device lifetime as well.

All these pins can be used as GPIOs except VCC, GND, TEST, and RESET. It can be noted that most of the LaunchPad™ F28069M board pins are multiplexed for dual functionality. Thus, attention must be paid to check if any of the ADC pin used by the extPot3 (which is physically connected to one of the linear potetniometers) is reconfigured. Therefore, it is recommended to keep the default pin mapping pins unless it does not restrict the firmware design.

The Bill of Material of the extPot3 board is reported in the following table:

Designator	Description	Manufacturer	Part Number
Pot1, Pot2, Pot3	20 kΩ Potentiometer	Alps Alpine	RK11K1120A5T
J1, J2	Launchpad connectors	SparkFun	PRT-11376
J3, J4, J5	Launchpad connectors	SparkFun	PRT-11376
J6	Spring clamp terminals	Würth Elektronik	691401710006B
ENABLE	Enable header	Würth Elektronik	61300211121
J8	Signal selector	Würth Elektronik	61300311121
R1, R2, R3, R4	2.2 kΩ, SMD 0805	Vishay	CRCW08052K20FKEA
R5	1 kΩ, SMD 0805	Kamaya	RMC1/10K102FTP
B1	Buzzer	RS PRO	617-3081
Q1	Buzzer driver	Infineon	BC818K40E6327HTSA1
D1	Green led	Würth Elektronik	150141GS73100
TP (10x)	Test point	RS PRO	262-2185

Key Features

- Manipulation of analog signals through linear potentiometers;
- Piezoelectric buzzer;
- Pins of LaunchPad™ F28069M board are still accessible;
- Suitable (i.e., without conflicts) for TI BOOSTXL-DRV8301 converter board top mouting.

Test Points

Many test points are available to debug the LaunchXL F28069M board pins:

- TP1: connected to ENC A of QUEP_A;
- TP2: connected to ENC B of QUEP_A;
- TP3: connected to ENC I of QUEP_A;
- TP4: connected to 5 V of QUEP_A;
- TP5: connected to GND of QUEP_A;

- TP6: connected to ENC A of QUEP_B;
- TP7: connected to ENC B of QUEP_B;
- TP8: connected to ENC I of QUEP_B;
- TP9: connected to 5 V of QUEP_B;
- TP10: connected to GND of QUEP_B.

Jumpers

- J8: signal selector from GPIO or ePWM connected to the buzzer;
- ENABLE: to be connected to enable QEP_B functions when using TI BOOSTXL-DRV8323RS converter board.

Connectors

- J6: hall-effect sensor terminal block.

Hint: PCB Design Rules

For custom board, it can be noted that trace width can be estimated through the following steps:

- Area computation:

$$\text{area}\left[\text{mil}^2\right] = \frac{\text{current}\,[\text{A}]}{\left(k \cdot \text{temp.rise}\,[^\circ\text{C}]^b\right)^{\frac{1}{c}}} \tag{B.1}$$

- Width computation:

$$\text{width}\,[\text{mil}] = \frac{\text{area}\left[\text{mil}^2\right]}{1.378\,[\text{mil/oz}] \cdot \text{thickness}\,[\text{oz}]} \tag{B.2}$$

- For IPC-2221 *internal layers*: $k = 0.024$, $b = 0.44$, $c = 0.725$
- For IPC-2221 *external layers*: $k = 0.048$, $b = 0.44$, $c = 0.725$

where k, b, and c are numerical constants resulting from curve fitting to the IPC-2221 curves. For further information on this topic the reader is referenced to `https://www.ipc.org` and `https://www.digikey.it/en/resources/online-conversion-calculators`.

extRL(C) Board

The extRL(C) board (see Figure B.2) is an configurable load which is able to combine different resistors R, inductors L and capacitors C, thus, allowing to operate the system with several equivalent values of load resistance, inductance and capacitance. This board is suitable for both operation in DC or AC. The board includes dedicated current and voltage sensors as well as an additional two-level converter leg with dedicated gate driver. The latter is useful if no external converter board is used (e.g. the TI BOOSTXL-DRV8301 BoosterPack is not present).

The main feature of the extRL(C) load is the possibility to dynamically change parameters values and load topology through three switches mounted at PCB top and one at PCB bottom. Combining the extRL(C) with the TI BOOSTXL-DRV8301 BoosterPack, the user is able to operate the system with both Half or Full-Bridge DC-DC converter, with single-phase loads and performing measurements comparison between the extRL(C) and BOOSTXL-DRV8301 on-board sensors.

The details of the RLC load are reported here in the following:

- The available resistance values $\boldsymbol{R}$ are 6.8 Ω, 3.4 Ω and 0 Ω (no connected resistors). Such value changes are made by acting on switch S1, which can:
  - Put in parallel two 6.8 Ω resistors (decreasing the overall/equivalent load resistance);
  - Keep connected just one of the two 6.8 Ω resistors;
  - Disconnect the passive load (switch S4 `off` is also required).

  The adopted resistances are two Arcol HS50 Series Aluminum Housed Axial Wire Wound Panel Mount Resistor, with nominal value 6.8 Ω±5% and rated power 50 W.

- The available capacitance values $\boldsymbol{C}$ are 100 µF, 10 µF and 0 F (no connected capacitor). Such value change is made by acting on switch S2, which can connect one of the available capacitors at a time or by-pass the capacitor stage (i.e., removing the C from the load). Electrolytic capacitors are used. Their characteristics are reported here below:
  - WCAP-ATET: Aluminum Electrolytic Capacitor, Radial, THT, D8xH11.5mm, rated capacitance 10 µF ± 20%, rated voltage 100 V;
  - WCAP-ATET: Aluminum Electrolytic Capacitor, Radial, THT, D13xH25mm, rated capacitance 100 µF ± 20%, rated voltage 100 V.

  The equivalent series-resistance of the capacitor (ESR) can be computed from the ESR-f characteristic given in the datasheet, e.g. ESR $= R_c = 66\,\text{m}\Omega$

Figure B.2 Different views of the extRL(C) board.

at $f_{sw} = 10\,\text{kHz}$. By disconnecting the capacitor, the extRL(C) can operate as single-phase RL load in AC (besides DC).

- The available inductance values $\boldsymbol{L}$ are 860 µH, 1.72 mH and 3.44 mH. Such value change is made by acting on switch S3, which can:
  - Include just one 860 µH inductor;
  - Connect in series two 860 µH inductors,
  - Connect in series four 860 µH inductors.

  The adopted inductors are four WE-FI Leaded Toroidal Line Choke, with rated inductance 860 µH ± 20%, rated current 3 A and series resistance $R_\ell = 150\,\text{m}\Omega$ at 20 °C.

The extRL(C) board allows to operate with a internal or external converter (i.e., Half/Full-Bridge) through the switch called MODE, placed at PCB top.

In *external mode*, the legs of the TI BOOSTXL-DRV8301 BoosterPack are used as the switching stage, with the extRL(C) board used as configurable filter stage interfacing a passive resistive load. Half- or Full-Bridge converter configurations are possible. The generation of the gating signals is performed through a firmware operating on the LaunchPad™ F28069M board. Note that the diode D1 must not be connected to operate in Full-Bridge configuration.

In *internal mode*, a dedicated switching stage, which consists in a Half-Bridge converter, is used. This is a BTN8962TA by Infineon (aimed to motor drive/automotive applications). Thus, no BOOSTXL-DRV8301 converter is not needed. In particular, the BTN8962TA can be driven by:

- gating signals coming from the LaunchPad™ F28069M board, e.g., ePWM1A and 1B. This option is called PWMext;
- gating signals provided by a timer blox monostable pulse generator integrated circuit LTC6992 by Analog Devices (mounted on the PCB bottom). This option is called PWMint.

The LTC6992 component allows to generate a PWM signal by simply controlling a $[0, 1]$ V analog input (the IC is powered at 3.3 V). To this end, a 10 Ω trimmer is used to vary the trigger values (i.e., the duty cycle) at pin 1.

The auxiliary circuits of the board are powered through the spring clamp connectors EXT and SUPPLY. A fixed step-down regulator module and a linear low drop voltage regulator create different voltage levels (5 V, 3.3 V and 1.65 V) to satisfy the circuit inputs.

The extRL(C) board is equipped withe a ACS711ELCTR-12AB-T current sensor from Allegro. This sensor can operate at 3.3 V or 5 V, with a sensitivity of 110 mV/A for the voltage output value and a bandwidth of 100 kHz. This device consists in a linear hall-effect sensor-based circuit with a copper conduction path located near the surface of the die. The current flows through this copper conduction line that generates a magnetic field which is sensed by the integrated hall-effect sensor-based circuit and converted into a proportional voltage. In addition, the extRL(C) board is also equipped with an isolated voltage sensing circuit. This is a voltage divider plus a rail-to-rail CMOS operational amplifiers OPA350EA operating in a $[0, 3.3]$ V range. The circuit can be used both in DC or AC operations. If the voltage sense is not required, e.g., when the capacitor is disconnected, this stage is by-passed through two test points called BP. Both current and voltage sensing path are provided with *RC* signal filters at the outputs of the related ICs. According to the application, these filters may requires some adjustments by the user, e.g., variations in the R and/or C values. PWMext and measurement feedbacks can be accessed by the lateral header (PCB top) and sent to the LaunchPad™ F28069M board.

Based on the whole parameters limits, the extRL(C) board is designed to operate at 3 A 40 V at rated conditions. Even if this values can be temporary exceeded, the board incorporates a over-current protection (OCP) circuit based on 4 A WE fuse and an over-voltage protection circuit based on zener diodes.

Due to the high current and voltage values, the adopted aluminum housed resistors are separated from the PCB and placed in contact with a heat sink,

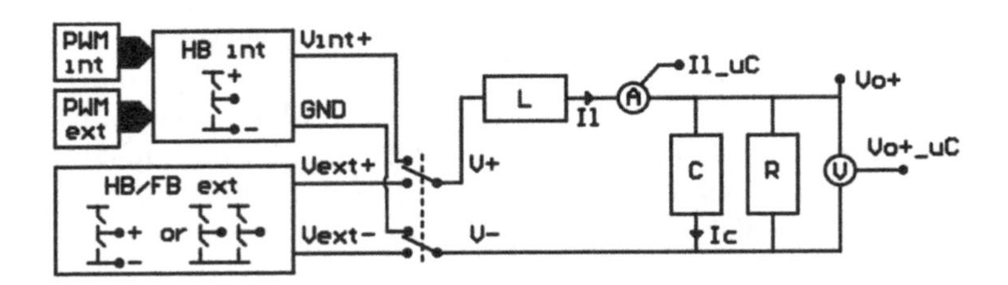

Figure B.3 extRL(C) board features printed on the PCB top surface.

since those are the components heating the most in this load stage. The heat sink ensures a sufficient temperature drop which allows to keep the extRL(C) board on a (student) desk. The (non-scaled) voltage applied across the resistance(s) is accessible from the screws placed on the top of the PCB.

A summary of the extRL(C) features is reported in the scheme printed on the PCB top surface, which is shown in Figure B.3. The Bill of Material of the extRL(C) Board is reported here in the following:

Designator	*Description*	*Manufacturer*	*Part Number*
POT	10 kΩ PWM Trimmer	Bourns	3296W-1-103LF
Q1	High Current PN (HB)	Infineon	BTN8962TA
EXT, SUPPLY	Spring clamp terminals	Würth Elektronik	691401710002B
Vo+, V-	REDCUBE Pressfit	Würth Elektronik	7461096
U1	FDSM Fixed Step Down Regulator Module	Würth Elektronik	173010535
U2	Linear low drop voltage regulator	STmicroelectronic	LD1117S33TR
U3	Pulse Generator	Analog Devices	LTC6992CS6
U4	Current sensor 12AB	Allegro	ACS711ELCTR
U5	rail-to-rail CMOS operational amplifier	Texas Instruments	OPA350EA_250
J1,J2,J3	External connectors	Würth Elektronik	61300211121
PWM	PWM selector	Würth Elektronik	61300311121
PWM_EXT	external PWM header	Würth Elektronik	61300211121
MODE	Mode Switch	C&K	L202011MS02Q
SWITCH L	Impedance Switch	C&K	L103111MS02Q
SWITCH R	Resistance Switch	C&K	L202011MS02Q
SWITCH C	Capacitor Switch	C&K	L103111MS02Q
S1	Switch	C&K	1101M2S3CQE2
L1, L2, L3, L4	Leaded Toroidal Line Choke, 860 µH	Würth Elektronik	7447075
C12	Elec. Capacitor 100 µF	Würth Elektronik	860130878011

C13	Elec. Capacitor 10 µF	Würth Elektronik	860130874006
R18, R19	6.8 Ω Resistance	Arcol	HS50 6R8 J
D1	Diode	Comchip tech.	CDBB5100-HF
D2	Diode	Comchip tech.	CDBB5100-HF
R1	40 kΩ, SMD 0805	Vishay	CRCW080540K2
R2	309 kΩ, SMD 0805	RS PRO	RS-0805-309K-1
R3	3.48 kΩ, SMD 0805	TE Connectivity	CPF0805B3K48E1
R4	976 kΩ, SMD 0805	Panasonic	ERJP06F9763V
R5	182 kΩ, SMD 0805	Vishay	CRCW0805182K
R6, R8, R9	10 kΩ, SMD 0805	TE Connectivity	CRG0805F10K
R7	33 kΩ, SMD 0805	KOA	RK73H2ATTD3302
R11, R12, R13, R14	1 MΩ, SMD 0805	ROHM	MCR10EZPF1004
R15, R16	11 kΩ, SMD 0805	Panasonic	ERJU06F1102V
R17	68 Ω, SMD 0805	Panasonic	ERJU06F68R0V
R20	15 kΩ, SMD 0805	TE Connectivity	CRGH0805F15K
R21, R10[1]	1 kΩ, SMD 0805	Kamaya	RMC1/10K102FTP
C1, C2, C4, C5 C8, C9, C11	100 nF, SMD 0805	Würth Elektronik	885012207128
C3	10 µF, SMD 0805	Würth Elektronik	885012207026
C6[2]	220 nF, SMD 0805	Würth Elektronik	885012207074
C7[1]	47 nF, SMD 0805	Würth Elektronik	885012207126
C10	1 µF, SMD 0805	Würth Elektronik	885012207022
F1	Fuse (3 A)	RS PRO	563-542A
	Fuseholder	Würth Elektronik	696101000002
PWM, Vext+, Vext- Vo+, V-, Vdc, V+, GND 3v3, 5v, GND, bp, bp	Test point	RS PRO	262-2185
External heat sink	Heat Sink	Fischer Elektronik	SK58-50-SA

Key Features

- Different resistance, inductance, capacitance values can be set on-the-fly;
- Choose between internal or external switching stage;
- Choose between internal or external PWM signals generation when the internal Half-Bridge converter is used;
- Over-current and over-voltage protections;
- On-board high-accuracy current sensor;

[1]Varying values depending on the desired filtering action.
[2]To be decided components

- On-board isolated voltage sensing;
- Both DC or AC operation;
- Compact size;
- Nice handling and pin access.

Test Points

- PWM: internal pulse width modulation;
- Vext+: positive pole of external input voltage;
- Vext-: negative pole of external input voltage;
- Vo+: positive pole of output voltage;
- V-: negative pole of converter output voltage (before L);
- Vdc: positive pole of input voltage;
- V+: positive pole of converter output voltage (before L);
- GND: ground;
- 3v3: 3.3 V connected to output pin of linear (low-drop) voltage regulator;
- 5v: 5 V connected to output pin of FDSM fixed step down regulator module;
- GND: ground;
- bp: bypass OPA350EA;
- bp: bypass OPA350EA.

Jumpers

- PWM: selection from internal or external PWM generator circuit.

Connectors

- EXT: Half-Bridge/Full-Bridge external power supply;
- SUPPLY: Half-Bridge internal power supply;
- J2: microcontroller current connector;
- J3: microcontroller voltage connector;
- PWM_EXT: external PWM genertor connector;
- Vo+: positive pole of output voltage;
- V-: negative pole of output voltage.

Figure B.4 Different views of the RL(C) kit and its connections.

RL(C) Hardware Kit

The hardware kit shown in Figure B.4 includes the following devices:

- One LaunchPad™ F28069M board;
- One TI BOOSTXL-DRV8301 BoosterPack;
- One extPot3 board;

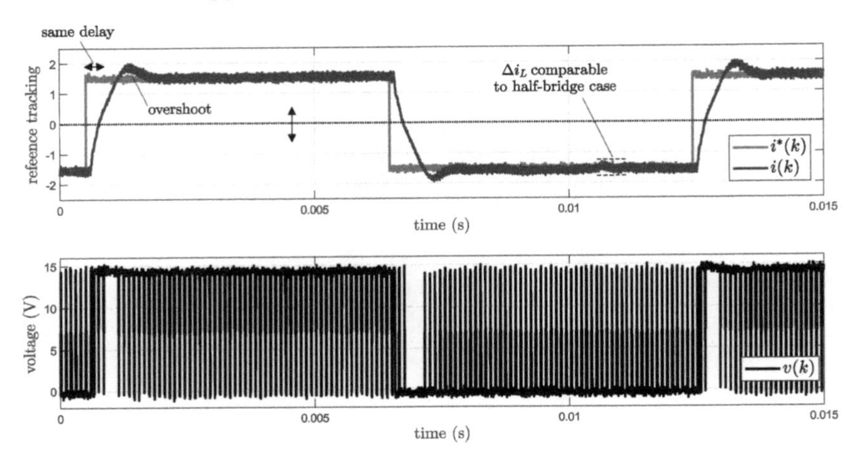

Figure B.5 Closed-loop dynamics measured from the F28069M pins through an oscilloscope (Full-Bridge operating with unipolar voltage switching).

- A mezzanine board to hold the MCU and manage the external power supply;
- One extRL(C) board.

This kit is a complete development test-bench for the energy management of power topologies and passive loads. Digital power management and control capabilities provided by rapid prototyping allows the study and design of adaptable high frequency switching power supplies, e.g. targeting both power processing and EV on-board battery charge applications. This kit is mainly built around the extRL(C) which was previously presented. All the aforementioned features holds. The mezzanine card is a MCU housing based on a custom made design. It can be even separately used as support for the MCU board. The kit is designed to operate with 40 V 3 A input source at rated conditions. Depending on the considered application target, those values can be temporary exceeded. A usage example of this board is reported below.

Example: *RL* Current Control in Full-Bridge Configuration

A Full-Bridge configuration (or H-bridge) refers to use two legs of the adopted TI BOOSTXL-DRV8301 (legs A and B), for which two ePWM modules are required, e.g., ePWM1 and ePWM2. The design of a PI-based current control for an *RL* load (i.e., extRL(C) configured in *RL* mode) is considered in the Simulink® environment. The terminals of the extRL(C) board are connected to the central point of legs A and B (i.e., the boosterpack switches 1_A, 1_B and 2_A, 2_B are operated). Based on to Chapter 17, the actuation of the two legs (i.e., two duty cycles for two modulation stages) can be done by manipulating one modulation signal $d(k)$ only, from which the gating signals of the four switches are generated. Since Chapter 15 already discussed the implementation and hardware settings of such strategy as well as the differences between bipolar and unipolar voltage switching, further details are not reported here. A benefit of a H-bridge configuration with respect to the Half-Bridge one is the chance to regulate even negative currents without changing the electrical connections of the setup. As an example, a bipolar current reference profile may be internally generated in the LaunchPad™ F28069M board in addition to the closed-loop scheme. Figure B.5 denotes the command tracking performances of a Full-Bridge configuration driving an *RL* load realized through the extRL(C) board and operating with unipolar voltage switching. The waveforms are acquired with an oscilloscope.

DecMot Hardware Kit

The hardware kit shown in Figure B.7 includes the following devices:

- One LaunchPad™ F28069M board;
- Two TI BOOSTXL-DRV8301 BoosterPacks;
- One extPot3 board;
- A mezzanine board to hold the MCU and manage the external power supply;
- One PMDC motor;
- One brushless DC (BLDC) motor;
- Encoder LPD3806-600BM-G5-24C;
- Aluminum base plate, motor supports and a joint;

This kit is a complete motor control development test-bench aimed to test single motor. Both PMDC and BLDC motors installed in this configuration can be separately tested with sensored (using an encoder) or sensorless control schemes, targeting automotive and servo-drive applications. The mezzanine card is a MCU housing based on a custom made design. It can be even separately used as support for the MCU board. The kit is designed to work with 24 V 10 A at rated conditions; according to the considered motor, these values can be adjusted in the external power supply.

Encoder Connection

The encoder LPD3806-600BM-G5-24C adopted for the exercise proposed in this book (see Chapter 14) and in the DecMot kit see an encoder interface made by open collector outputs with 20 Ω resistors for protection. This circuit draws around 30 mA. Since it can not source current, the open-collector

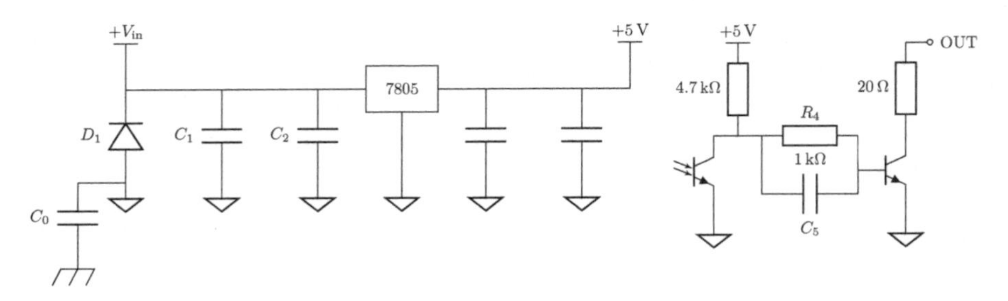

Figure B.6 Internal circuit of the encoder interface, model LPD3806-600BM-G5-24C.

circuit must be connected to positive DC voltage through a pull-up resistor. The eQEP pins of the F28069M LaunchPad™ board have pull-up resistors, thus, the encoder cable can be directly connected to the extPot3 pins related to QEP_A and QEP_B. Th e extPot3 board can also supply the encoder with the available 5 V and GND terminals. Attention must be paid to limit the length of the cable connecting the encoder to the board, since it is preferable to have pull-up resistors in close proximity to the encoder interface to improve noise immunity. Moreover, as shown by the circuit scheme reported in Figure B.6, the encoder interface has a LM7805 5 V linear regulator, which power dissipation has to be kept within reasonable limits. The regulator is in a D-PAK package and it lies on a big ground plane that acts as a heat-sink.[3]

B2B-PMDC Hardware Kit

This hardware kit shown in Figure B.8 includes the following devices:

- One LaunchPad™ F28069M board;
- One or two TI BOOSTXL-DRV8301 BoosterPacks;
- One extPot3 board;
- A mezzanine board to hold the MCU and manage the external power supply;
- Two equal PMDC motors;
- Encoder LPD3806-600BM-G5-24C;
- Aluminum base plate, motor supports and joints.

[3]The encoder case is connected to the chassis and not to circuit ground.

Figure B.7 Different views of the DecMot kit and its connections.

This kit is a complete motor control development test-bench aimed at rapid prototyping PMDC-based sensored (via encoder) or senorless back-to-back control schemes, targeting automotive, industrial and servo-drive applications. One PMDC motor is used as brake to manipulate the load torque. Therefore, back-to-back operations and motor characterization are possible. The mezzanine card is a MCU housing based on a custom made design. It can be even separately used as support for the MCU board. The kit is designed to operate with a 24 V 10 A power supply at rated conditions. Depending on the application target, those values can be temporary exceeded. The previous encoder connection considerations still hold.

Figure B.8 Different views of the B2B-PMDC kit and its connections.

This kit can be modified by substituting a PMDC motor with a BLDC one, resultign in the:

B2B-BLDC Hardware Kit (see Figure B.9)

Therefore, three-phase sensored (via encoder) or senorless control schemes can be tested as well. In this case, the PMDC motor is still used as brake. The rated voltage and current are again 24 V and 10 A, respectively. Such kit is suitable to emulate:

- Electric power-trains; electric fuel, water and oil pumps; engine cooling fans.
- Actuators and valve controls; doors, window lift and seat control.

Figure B.9 Different views of the B2B-BLDC kit and its connections.

Further information can be found at the following webpage:
`https://www.mecc.polimi.it/nc/us/`

Politecnico di Milano
Department of Mechanical Engineering, via La Masa 1, Milan 20156, Italy
Electrical Machines, Drives, and Power Electronics Research Group

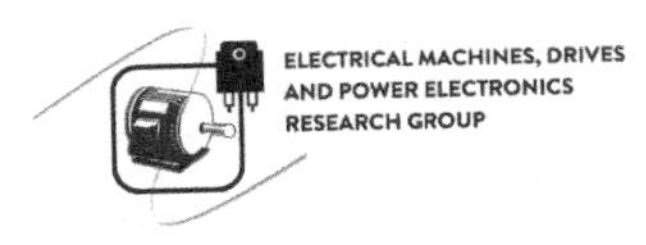

Newer versions of these hardware kits would be distributed by ePEBB^{s} Srl, please contact mattia.rossi@polimi.it and nicola.toscani@polimi.it.

Bibliography

[1] IEEE Standard for Binary Floating-Point Arithmetic. *ANSI/IEEE Std 754-1985*, pages 1–20, 1985.

[2] Paolo Bolzern, Riccardo Scattolini, and Nicola Schiavoni. *Fondamenti di controlli automatici.* McGraw-Hill Education, 4th edition, February 2015.

[3] Bimal K. Bose. *Modern Power Electronics and AC Drives.* Pearson Education, 2001.

[4] Jason Bridgmon and Carolus Andrews. Current Sensing for Inline Motor-Control Applications. Technical Report SBOA172, Texas Instruments, October 2016. Available at `https://www.ti.com/lit/an/sboa172/sboa172.pdf`.

[5] Bill Chou. The Joy of Generating C Code from MATLAB. `https://it.mathworks.com/company/newsletters/articles/the-joy-of-generating-c-code-from-matlab.html`.

[6] Luca Corradini, Dragan Maksimović, Paolo Mattavelli, and Regan Zane. *Digital Control of High-Frequency Switched-Mode Power Converters.* John Wiley & Sons, 22 June 2015.

[7] Texas Instruments. BoosterPack Checker. `https://dev.ti.com/bpchecker/#/`.

[8] Texas Instruments. C2000 Real-Time Control MCUs: Applications. `https://www.ti.com/microcontrollers/c2000-real-time-control-mcus/applications`.

[9] Texas Instruments. Inverter & motor control. `https://www.ti.com/solution/hev-ev-inverter-motor-control?variantid=14291&subsystemid=17097`.

[10] Texas Instruments. Website. `https://www.ti.com`.

[11] Texas Instruments. Meet the TMS320F28069M - LaunchPad Development Kit. Technical Report SPRUI02, Texas Instruments, 2014. Available at `https://www.ti.com/lit/ml/sprui02/sprui02.pdf`.

[12] Texas Instruments. Lift-Off with the TI LaunchPad™ Development Ecosystem. Technical Report SLAT152B, Texas Instruments, 2015.

[13] Texas Instruments. Motor Drive and Control Solutions. Technical Report SLYB165I, Texas Instruments, 2017.

[14] Texas Instruments. DRV8301 Three-Phase Gate Driver With Dual Current Shunt Amplifiers and Buck Regulator. Technical Report SLOS719F, Texas Instruments, August 2011 (revised January 2016). Available at `https://www.ti.com/lit/ds/symlink/drv8301.pdf`.

[15] Texas Instruments. C2000™ Key Technology Guide. Technical Report SPRACN0A, Texas Instruments, August 2019 (revised March 2020). Available at `https://www.ti.com/lit/an/spracn0a/spracn0a.pdf`.

[16] Texas Instruments. User's Guide - LAUNCHXL-F28069M Overview (Rev. B). Technical Report SPRUI11B, Texas Instruments, January 2015 (revised March 2019). Available at `https://www.ti.com/lit/ug/sprui11b/sprui11b.pdf`.

[17] Texas Instruments. TMS320x280x, 2801x, 2804x Serial Communications Interface (SCI). Technical Report SPRUFK7B, Texas Instruments, March 2009 (revised April 2011). Available at `https://www.ti.com/lit/ug/sprufk7b/sprufk7b.pdf`.

[18] Texas Instruments. C2000 MCU JTAG Connectivity Debug. Technical Report SPRACF0, Texas Instruments, May 2018. Available at `https://www.ti.com/lit/an/spracf0/spracf0.pdf`.

[19] Texas Instruments. TMS320x280x, 2801x, 2804x DSP Analog-to-Digital Converter (ADC). Technical Report SPRU716D, Texas Instruments, November 2004 (revised April 2010). Available at `https://www.ti.com/lit/ug/spru716d/spru716d.pdf`.

[20] Texas Instruments. TMS320x280x, 2801x, 2804x Enhanced Quadrature Encoder Pulse (eQEP) Module. Technical Report SPRU790D, Texas Instruments, November 2004 (revised December 2008). Available at `https://www.ti.com/lit/ug/spru790d/spru790d.pdf`.

[21] Texas Instruments. TMS320x280x, 2801x, 2804x Enhanced Pulse Width Modulator (ePWM) Module. Technical Report SPRU791F, Texas Instruments, November 2004 (revised July 2009). Available at `https://www.ti.com/lit/ug/spru791f/spru791f.pdf`.

[22] Texas Instruments. User's Guide - BOOSTXL-DRV8301 Hardware User's Guide. Technical Report SLVU974, Texas Instruments, October 2013. Available at `http://www.ti.com/lit/ug/slvu974/slvu974.pdf`.

[23] Paul Krause, Oleg Wasynczuk, Scott Sudhoff, and Steven Pekarek. *Analysis of Electric Machinery and Drive Systems*. John Wiley & Sons, 3rd edition, 2013.

[24] Werner Leonhard. *Control of Electrical Drives.* Springer, 3rd edition, 2001.

[25] Renato Manigrasso, Ferdinando L. Mapelli, and Marco Mauri. *Azionamenti elettrici vol.1: Generalità e Macchine Rotanti.* Pitagora, 2007.

[26] Renato Manigrasso, Ferdinando L. Mapelli, and Marco Mauri. *Azionamenti elettrici vol.2: Convertitori e Controllo.* Pitagora, 2007.

[27] Ferdinando L. Mapelli, Davide Tarsitano, and Marco Mauri. Plug-In Hybrid Electric Vehicle: Modeling, Prototype Realization, and Inverter Losses Reduction Analysis. *IEEE Transactions on Industrial Electronics*, 57(2):598–607, 2010.

[28] MathWorks. Getting Started—Embedded Coder Support Package for Texas Instruments C2000 Processors. Video available at `https://www.mathworks.com/videos/getting-started-with-embedded-coder-support-package-for-ti-c2000-processors-1573540550102.html`.

[29] MathWorks. Pid tuner documentation. `https://it.mathworks.com/help/control/ref/pidtuner.html`.

[30] MathWorks. Specialized power system documentation. `https://it.mathworks.com/help/physmod/sps/specialized-power-systems.html`.

[31] MathWorks. TI C2000 Support from Embedded Coder. `https://it.mathworks.com/hardware-support/ti-c2000.html`.

[32] MathWorks. White Paper - Hardware-in-the-Loop Testing for Power Electronics Control Design - Featuring Simulink and Speedgoat real-time target machines. Technical report, MathWorks, 2019. Available at `https://it.mathworks.com/campaigns/offers/power-electronics-control-hil-white-paper.html`.

[33] Mathworks. Simulink for Developing Digital Control for Motors, Power Converters, and Battery Systems. Technical report, MathWorks, 2020. Available at `https://www.mathworks.com/content/dam/mathworks/brochure/speed-digital-control-motor-power-converters-systems-simulink-handout.pdf`.

[34] MathWorks. White Paper - Model-Based Design for Embedded Control Systems. Technical report, MathWorks, October 2020. Available at `https://it.mathworks.com/campaigns/offers/model-based-design-embedded-control-systems.confirmation.html?elqsid=1607505922909&potential_use=Education`.

[35] Marco Mauri, Maria Stefania Carmeli, and Mattia Rossi. *Macchine Elettriche. Modelli a regime: teoria ed esercizi.* Società Editrice Esculapio, 2nd edition, 2020.

[36] Robert D. Middlebrook and Slobodan Ćuk. A general unified approach to modelling switching-converter power stages. In *1976 IEEE Power Electronics Specialists Conference*, pages 18–34, 1976.

[37] Ned Mohan, Tore M. Undeland, and William P. Robbins. *Power Electronics: Converters, Applications, and Design.* John Wiley & Sons, 3rd edition, 2002.

[38] Mattia Rossi, Francesco Castelli Dezza, Marco Mauri, and Maria Prandini. Explicit Computation of Indirect Hybrid MPC for Voltage Control in Multilevel DC-DC Converters. In *2018 20th European Conference on Power Electronics and Applications (EPE'18 ECCE Europe)*, pages P.1–P.9, 2018.

[39] Mattia Rossi, Eyke Liegmann, Petros Karamanakos, Francesco Castelli Dezza, and Ralph Kennel. Direct Model Predictive Power Control of a Series-Connected Modular Rectifier. In *2019 IEEE International Symposium on Predictive Control of Electrical Drives and Power Electronics (PRECEDE)*, pages 1–6, 2019.

[40] Mattia Rossi, Luigi Piegari, Francesco Castelli Dezza, Marco Mauri, and Maria S. Carmeli. Voltage Control Comparison for Low-Power DC-DC Converters in EVs: PI and Explicit MPC. In *IECON 2018 - 44th Annual Conference of the IEEE Industrial Electronics Society*, pages 5005–5011, 2018.

[41] Sigurd Skogestad and Ian Postlethwaite. *Multivariate Feedback Control: Analysis and Design.* John Wiley & Sons, 2nd edition, November 2005.

[42] Stephen Umans. *Fitzgerald & Kingsley's Electric Machinery.* McGraw-Hill Education, 7th edition, 16 May 2013.

[43] Peter Vas. *Sensorless Vector and Direct Torque Control.* Oxford University Press, 1998.

[44] J. G. Ziegler and N.B. Nichols. Process lags in automatic control circuits. *Transactions of the A.S.M.E.*, 65(5):433–444, 1943.

Index

Action qualifier, 173–179
actuation variable, 78, 271–273
ADC (analog-to-digital converter)
peripheral, 1, 8, 33, 39, 101, 151–163, 172, 175, 189–195, 257–271
Anti-windup scheme, 77–82
Automatic code generation, 1, 4, 15–25, 42, 101, 109

B2B (back-to-back)
-BLDC hardware kit,
-PMDC hardware kit, 331, 333, 411
setup, 367–376
Baud rate, 103, 106n2, 113–114, 119
Bipolar voltage switching, 246–251, 254–255
Bode diagram, 72, 311
Boot options, 35–36

Carrier signal, 175–178
Cascade control architecture, 336–339
Closed-loop control scheme, 53–66
Closed-loop transfer function, 55, 64
Continuous-time domain, 56–62, 74
Counting modes, 174, 176–178, 183
Current
control, 67–86, 273–298, 336–339, 352, 367–370, 382–383, 401–403, 422
measurement, 36, 40, 206, 208, 257–271
Custom board, 37, 215, 411–428

DAC (digital-to-analog converter)
peripheral, 1, 33, 167, 195–201, 325, 362–365
Data type, 23–24, 87–98, 113, 117, 122, 136–137, 157
DC bus, 36, 39, 233, 279
DC-DC converter, 227, 275, 301–306
DecMot kit, 331–333, 411, 423–425
Discrete-time domain, 62–66, 74, 86, 231
Duty cycle, 5, 36, 144, 168, 228, 233–237, 241–251, 279–286, 303–320, 340–350
Dynamical systems, 54–56

Embedded coder for Texas Instruments C2000 processors, 43
Embedded platform, 11, 86, 97
Enhanced quadrature encoder pulse (eQEP) peripheral, 29, 101, 211–223
Evaluation board, 25
External mode execution, 379, 386–389
extPot3 board, 157–162, 215, 228, 273, 286, 299, 331–334, 411–414
extRL(C) board (RL(C) hardware kit), 273–274, 287, 294, 299–302, 322, 328, 411, 415–423

Firmware
deployment, 22, 107
design, 1–13, 22

- environment, 115–119, 125–131, 135–148, 155–160, 163–164, 182–195, 197–200, 203–205, 217–222, 238–240, 252–255, 261–265, 267, 286–294, 321–328, 355–366, 371–376

Fixed-point data, 87–90, 93–95, 407–409
Floating-point data, 31, 87–93, 95–97, 121–122, 407–409
Full-bridge configuration, 240–255, 275, 340–353, 422–423

GPIO (general purpose input/output), 133–149
- peripheral, 133–135
- DI (digital input), 135–136
- DO (digital output), 136–137

Half-bridge configuration, 233–240, 275, 279–294, 312–328
Hardware kit, 29–40, 411–428

Installation procedure, 43–47

Jumper configuration, 35

LC filter, 310, 328–329
Linear model, 228–232, 306–312, 334–336, 398–401
Locked rotor test, 261–266, 269–271
Low-Side Shunt Current Sensing, 39–40, 257–271

MATLAB, 17–27, 41–47, 60, 90–93
MCU (microcontroller), 1–13, 22, 29–36, 41, 111–112, 208–210, 285, 347, 366, 379–381
Model-based design, 17–20, 70
Motor control, 227–256, 331–376, 391–405
Model reference adaptive system (MRAS) observer, 331, 343, 350–353, 365, 367, 370

Open-loop transfer function, 58–59, 228–231, 275–277, 309–310, 334, 338, 402–403
Optical rotary encoder, 211–215, 331

Parameter identification, 296–297, 354–355, 369–370
Peripherals, 1–4, 101, 113–223
PI (proportional-integral) controller, 21, 51–52, 56, 62, 74, 79–84, 289
- current control 67–86, 273–298, 336–339, 370, 403, 422
- speed control, 336–353, 356–366, 367–376, 402
- tuning, 56–62, 76–79, 84–86
  - pole/zero cancellation 58–59, 69
  - generalized approach 59–60
- voltage control 299–330

PIL (processor-in-the-loop), 25, 379–386, 391
Pin muxing, 32–33
Pin-out table, 32, 39, 133–134
PMDC (permanent magnet DC) motor, 227–255, 240, 260–266, 331–376
PMSM (permanent magnet synchronous motor), 396–405
PWM (pulsed-width modulation), 21, 37, 167–172, 175–181, 228, 236, 241–243, 248, 277–278, 281, 284–286, 291–293, 318, 346
- Logic, 143–149, 158–162, 167–172, 175–181, 193–195, 242, 278, 284
- peripheral, 1, 26, 167–210, 325n1

Power Electronics, 1–3, 9–13, 17, 20–21, 208–210

Rapid prototyping, 17–23
Resistive-inductive-capacitive (RLC) load, 299–312, 328–329

Sample rate, 122–124, 208–210
SCI (serial communication interface), 35, 108, 113–131
  receive, 101, 115–118
  transmit, 101, 118–119
Sensor characterization, 258–260, 266–271
Serial send, 120–121
Serial receive, 121–122
Shift operation, 95, 407
Shunt current sensing, 39–40, 257–258
Simscape (specialized power systems) 17, 20, 27n1, 69–73, 75, 232, 235, 242, 277, 280, 312–313, 339, 342
Simulink 17, 20–26, 43, 69–76, 101, 107–112
Software installation, 41–47
Speed computation, 215–217, 219–222
State-space average modelling method, 306–312
Step-down converter, 233–237, 275, 279–286, 301–306, 312–321
Synchronization, 101, 105n1, 172, 208, 292, 326, 352, 364
  between channels, 162–165, 173, 183–185, 201–205
  between peripherals, 206–210

Testing environment, 109, 119–122, 126–128, 129–131, 142, 148, 161, 263, 267
TI
  BOOSTXL-DRV8301
    BoosterPack, 36–40, 411
  C2000 MCU, 7–9, 24–26, 29, 31, 43–47
  Piccolo F28069M LaunchPad, 8, 29–36
Transfer function, 54–57, 64, 68, 310, 338, 402–403

Unipolar voltage switching, 241–247, 252–254, 422–423

Virtual COM port, 103–105, 113, 115
Voltage
  control, 299–330
  measurement, 323, 326–329
  sensor, 323, 415